Universitext

Universitext

Universitext is a series of textbooks that presents material from a wide variety of mathematical disciplines at master's level and beyond. The books, often well class-tested by their author, may have an informal, personal even experimental approach to their subject matter. Some of the most successful and established books in the series have evolved through several editions, always following the evolution of teaching curricula, to very polished texts.

Thus as research topics trickle down into graduate-level teaching, first textbooks written for new, cutting-edge courses may make their way into *Universitext*.

For further volumes:
http://www.springer.com/series/223

Sergio Benenti

Hamiltonian Structures and Generating Families

Springer

Sergio Benenti
Dipartimento di Matematica
Facoltà di Scienze Matematiche,
Fisiche e Naturali
Università di Torino
10123 Torino
Italy
sergio.benenti@unito.it

ISSN 0172-5939 e-ISSN 2191-6675
ISBN 978-1-4614-1498-8 e-ISBN 978-1-4614-1499-5
DOI 10.1007/978-1-4614-1499-5
Springer New York Dordrecht Heidelberg London

Library of Congress Control Number: 2011936973

Mathematics Subject Classification (2010): 13B10, 37D35, 47A06, 53D20, 53D05, 53D12, 58K99, 70H20, 78A45, 80A05, 00A69, 35Q79, 53Z05, 58Z05, 93C99

Printed on acid-free paper

Springer is part of Springer Science+Business Media (www.springer.com)

Everything has a generating family.

Preface

A *Hamiltonian structure* is a mathematical model of a physical phenomenon in which symplectic geometry plays a basic role. The Hamiltonian formulation of analytical mechanics as well as the Hamiltonian formulation of geometrical optics, place of birth of the Hamilton–Jacobi equation, are well-known examples. Other examples can be added, for instance, the control of static mechanical systems and of the equilibrium states of thermodynamic systems.

A *generating family* is a smooth real function that is able to describe special subsets of a cotangent bundle, here called *Lagrangian sets*. A Lagrangian set may be a *Lagrangian submanifold*. However, as we show, several examples of physically meaningful phenomena are in fact represented by Lagrangian sets that are not submanifolds.

The sense of this dichotomy, *nonsmooth* and *smooth*, becomes clear when we deal with *symplectic relations*, one of the most important tools used in this book. A symplectic relation is defined, at a first stage, as a Lagrangian submanifold of the product of two symplectic manifolds. If these symplectic manifolds are cotangent bundles, then a symplectic relation has (locally or globally) a generating family. Relations can be composed according to a well-defined rule, but the composition of two smooth relations (i.e., submanifolds) may not be smooth; that is, a Lagrangian subset of the product of two cotangent bundles. However, besides the composition of symplectic relations, we have a composition rule of their generating families which yields another smooth generating function. In other words, although the composition of two symplectic relations may produce a nonsmooth object, the composition of their generating families is always smooth. Then, the *symplectic creed* formulated by Alan Weinstein in his article "Symplectic geometry" (1981)

everything is a Lagrangian submanifold,

which means that one should try to express objects in symplectic geometry and mechanics in terms of Lagrangian submanifolds, is here replaced by

everything has a generating family.

In order to make this book self-contained and to clarify the notations, the first two chapters are devoted to those basic notions of calculus on manifolds that are strictly necessary to our purposes.

Chapter 3 is devoted to the notion of symplectic relation within the category of the symplectic manifolds. In Chaps. 4 and 5 we specialize our analysis within the category of the cotangent bundles. Our analysis is based on the notion of a generating family of a Lagrangian set, which is an extension of that of a generating family of a Lagrangian submanifold (or of a symplectic relation). This extension turns out to be necessary in dealing with the composition of symplectic relations.

Indeed, if the composition of two smooth symplectic relations, which are submanifolds of Cartesian products, no longer yields a smooth relation, then we can replace the composition of the relations with the composition of their generating families, which are always smooth objects.

The symplectic formulation of Hamiltonian optics, presented in Chaps. 6 and 7, is based on the fact that, from a geometrical viewpoint, a Hamilton–Jacobi equation is a coisotropic submanifold of a cotangent bundle and that a *geometrical solution* is a Lagrangian set contained in it. The solutions of a Hamilton–Jacobi equation are then described by generating families, and not by an "ordinary" function as in the classical theory.

There are two fundamental symplectic relations associated with a Hamilton–Jacobi equation, the *characteristic relation* and the *characteristic reduction.* The two corresponding generating families are called *Hamilton principal functions* and *complete solutions.*

The characteristic relation is a singular Lagrangian submanifold, thus the Hamilton principal function is necessarily a generating family and not a two-point function as it appears in the classical theory. Furthermore, Cauchy data (or *sources* of systems of rays), mirrors, and lenses are represented by symplectic relations, thus by generating families. Then the Cauchy problem and the action of a lens or of a mirror on a system of rays are translated into the composition of symplectic relations or of generating families.

In Chap. 5 it is shown that the use of generating families cannot be avoided if we want to give a global meaning to the Hamilton characteristic function, from which all solutions of the Hamilton–Jacobi equation can be derived, or if we want to describe very singular optical phenomena.

Symplectic relations and generating families can also play an interesting role in the control theory of static systems, including thermostatic systems. Chapter 8 is devoted to this matter. Our approach is based on the notion of *control relation* and on an extended version of the *virtual work principle* for constrained systems with noncontrolled degrees of freedom (*hidden variables*). Several examples of singular phenomena concerning static and thermostatic systems are illustrated. In particular, it is shown that the *Maxwell rule of equal areas* is a theorem following, through pure mathematical reasoning, from the extended virtual work principle. Thermostatics of simple and composite systems are described here in the four-dimensional state space, with global coordinates (S, V, P, T), entropy, volume, pressure, and absolute temperature, endowed with the natural symplectic structure induced by the first principle of thermodynamics.

Supplementary topics are illustrated in Chap. 9. Chapter 10 is devoted to the calculus of global Hamilton principal functions for the eikonal equations on the two-dimensional sphere $\mathbb{S}_2$ and pseudo-sphere $\mathbb{H}_2$.

***

I am very grateful to the reviewers, and especially to Alan Weinstein, for their criticisms and suggestions, severe, meaningful and useful at the same time.

Special thanks go to Kaitlin Leach, Assistant Editor, who supported me in the many contacts I had with the reviewers and the copy editors.

Sergio Benenti

Torino, June 2011

***

This book is an enhanced version of a first edition in Russian.

Hoc erat in votis
Oratio, Sermones, VI, 1

Contents

Chapter 1
Basic Notions of Calculus on Manifolds

Abstract It is assumed that the reader is acquainted with the notion of a real, finite-dimensional differentiable manifold (smooth manifold). The aim of this chapter is to focus on the basic tools of calculus on manifolds and on the terminology and notation adopted in this book. A particular attention is paid to the concepts of rank of a map, clean and transverse intersection of submanifolds, and derivation of exterior forms.

1.1 Tangent vectors and tangent bundles

Let M be a smooth manifold of dimension m. We denote by:

- T_xM the *tangent space* of M at a point $x \in M$, the linear m-dimensional space of the tangent vectors based at (or applied to) x.
- TM the *tangent bundle* of M, the set of all *tangent vectors* of M.
- $\tau_M : TM \to M$ the *tangent fibration* over M, which maps a tangent vector $v \in TM$ to the point $x \in M$ such that $v \in T_xM$.
- $\mathscr{F}(M) = C^\infty(M, \mathbb{R})$ the ring of all smooth real-valued functions on M.

A *curve* on M is a smooth map $\gamma \colon I \to M$, where $I \subseteq \mathbb{R}$ is an open interval containing 0. We say that the curve is *based* at the point $x = \gamma(0)$.

We refer to two equivalent definitions of *tangent vector*:

1. A tangent vector v at a point $x \in M$ is a *derivation* on $\mathscr{F}(M)$ that is, a map $v \colon \mathscr{F}(M) \to \mathbb{R}$ such that

$$\begin{cases} v(aF + bG) = a\,v(F) + b\,v(G) \quad a, b \in \mathbb{R}, & \text{(linearity)} \\ v(FG) = v(F)\,G(q) + F(q)\,v(G) & \text{(Leibniz rule).} \end{cases}$$

   We use angle brackets to denote the derivative of a function with respect to a vector:

$$v(F) = \langle v, dF \rangle.$$

2. A tangent vector v at a point $x \in M$ is an *equivalence class* $[\gamma]$ of curves. Two curves γ and γ' on M are *equivalent* at a point x if

$$\begin{cases} \gamma(0) = \gamma'(0) = x, \\ D(F \circ \gamma)(0) = D(F \circ \gamma')(0) \ \text{ for all } F \in \mathscr{F}(M). \end{cases}$$

Here, D is the symbol of derivative of a real-valued function on $\mathbb{R}$.

The link between these two definitions is given by

$$v = [\gamma] \iff v(F) = D(F \circ \gamma)(0).$$

If $x \in M$ and (x^i) is a local coordinate system on a domain containing the point x, then the *components* of v with respect to these coordinates are the numbers defined by

$$v^i = \langle v, dx^i \rangle$$

or

$$v^i = D\gamma^i(0).$$

In the first definition, a coordinate x^i is interpreted as a function. In the second definition, equations

$$x^i = \gamma^i(t), \quad t \in I,$$

are the *parametric equations* of the representative curve γ. It follows that

$$\langle v, dF \rangle = v^i \, \partial_i F(x), \quad \partial_i = \frac{\partial}{\partial x^i}.$$

We denote by

$$(x^i, \delta x^i) \ \text{ or } \ (x^i, \dot{x}^i)$$

the coordinates on TM corresponding to coordinates (x^i) on M. They are defined as follows: if $v \in T_x M$ then $x^i(v)$ are the values of the coordinates at the point x and $v^i = \delta x^i(v) = \dot{x}^i$ are the components of the vector in these coordinates.

There is a map $\delta\colon \mathscr{F}(M) \to \mathscr{F}(TM)$, from functions on M to functions on TM, defined by

$$\delta F = \frac{\partial F}{\partial x^i} \, \delta x^i.$$

The function δF is linear on the fibers of TM.

With each curve $\gamma\colon I \to M$ we associate a curve $\dot{\gamma}\colon I \to TM$, called the *tangent lift*, or the *tangent prolongation* of γ, defined by

$$\langle \dot{\gamma}, F \rangle = D(\gamma \circ F).$$

Its local parametric equations are

$$\begin{cases} x^i = \gamma^i(t), \\ \dot{x}^i = D\gamma^i(t). \end{cases}$$

MECHANICAL INTERPRETATION. If a manifold Q represents the n-dimensional *configuration manifold* of a holonomic mechanical system with n degrees of freedom, then a curve $\gamma\colon I \to Q$ represents a *motion* of the system ($t \in I$ is the time). coordinates (q^i) on Q are called *Lagrangian coordinates.* A tangent vector $v \in T_qQ$ represents a *virtual displacement* or a *virtual velocity* of the system at the configuration $q \in Q$. The curve $\dot{\gamma}(t)$ represents the velocity of the system at each instant t.

1.2 The tangent functor

Any smooth map $\varphi\colon M \to N$ can be extended to its *tangent map*

$$T\varphi\colon TM \to TN$$

between the corresponding tangent bundles. The tangent map $T\varphi$ is defined by the equation

$$\langle T\varphi(v), dF\rangle = \langle v, d(F \circ \varphi)\rangle, \quad \text{for all } F \in \mathscr{F}(N),$$

where a vector $v \in T_xM$, $x \in M$, is interpreted as a derivation, or by the equation

$$T\varphi([\gamma]) = [\varphi \circ \gamma],$$

where a vector is interpreted as a class of curves on M. Because $v \in T_xM$ implies $T\varphi(v) \in T_yN$, $y = \varphi(x)$, we have

$$\tau_N \circ T\varphi = \varphi \circ \tau_M,$$

where τ_M and τ_N denote the tangent fibrations. This means that the diagram

$$\begin{array}{ccc} TM & \xrightarrow{\;T\varphi\;} & TN \\ {\scriptstyle\tau_M}\big\downarrow & & \big\downarrow{\scriptstyle\tau_N} \\ M & \xrightarrow{\;\varphi\;} & N \end{array}$$

is commutative. The functorial rules

$$T(\mathrm{id}_M) = \mathrm{id}_{TM}, \quad T(\varphi \circ \psi) = T\varphi \circ T\psi$$

hold. Then the operator T which associates with any manifold M its tangent bundle TM and with any map φ between manifolds its tangent map $T\varphi$ is a covariant functor, called the *tangent functor*.

If in local coordinates φ is represented by equations $y^a = \varphi^a(x^i)$, then $T\varphi$ is represented by equations

$$y^a = \varphi^a(x^i), \quad \dot{y}^a = \frac{\partial \varphi^a}{\partial x^i}\, \dot{x}^i. \tag{1.1}$$

1.3 Rank of a map and special maps

We denote by $T_x\varphi\colon T_xM \to T_{\varphi(x)}N$ the restriction of $T\varphi\colon TM \to TN$ to the tangent space T_xM. It is a linear map.

Definition 1.1. The *rank* of a map $\varphi\colon M_m \to N_n$ at a point $x \in M$ is the dimension of the image of the linear map $T_x\varphi$,

$$\mathrm{rank}_x\varphi = \dim(\text{image of } T_x\varphi). \qquad \heartsuit$$

If $y^a = \varphi^a(x^i)$ is a coordinate representation of φ, then $T\varphi$ is represented by Eqs. (1.1), so that

$$\mathrm{rank}_x\varphi = \text{rank of the matrix} \left[\frac{\partial \varphi^a}{\partial x^i}\right]_{m\times n} \quad \text{at } x.$$

By means of the notion of *rank* we can distinguish some *special maps*:

Definition 1.2. A smooth map $\varphi\colon M_m \to N_n$, is
an *immersion* if $T_x\varphi$ is injective for all $x \in M$,
a *submersion* if $T_x\varphi$ is surjective for all $x \in M$,
a *subimmersion* if $T_x\varphi$ has constant rank. $\heartsuit$

Remark 1.1. Notice that:

- φ is an immersion $\iff$ $m \le n$ and $\mathrm{rank}_x\varphi = m$ for all $x \in M$,
- φ is a submersion $\iff$ $m \ge n$ and $\mathrm{rank}_x\varphi = n$ for all $x \in M$,
- immersions and submersions are subimmersions.

Then, for any local representation $y^\alpha = \varphi^\alpha(x^i)$ of φ,

$$\begin{aligned}
\varphi \text{ is an immersion} &\iff \operatorname{rank}\left[\frac{\partial \varphi^\alpha}{\partial x^i}\right] = m \leq n,\\
\varphi \text{ is a submersion} &\iff \operatorname{rank}\left[\frac{\partial \varphi^\alpha}{\partial x^i}\right] = n \leq m,\\
\varphi \text{ is a subimmersion} &\iff \operatorname{rank}\left[\frac{\partial \varphi^\alpha}{\partial x^i}\right] = \text{constant}.
\end{aligned}$$

This classification does not involve topology. ◇

There are other special maps whose definition involves topology:

Definition 1.3. An *embedding* is an immersion that is also a homeomorphism onto the image $\varphi(M) \subseteq N$, in the induced topology. A *diffeomorphism* is an embedding $\varphi\colon M \to N$ admitting the inverse $\varphi^{-1}\colon N \to M$ which is also an embedding.[1] Two manifolds M and N are *diffeomorphic* if there exists a diffeomorphism $\varphi\colon M \to N$. A *transformation* of a manifold M is a diffeomorphism $\varphi\colon M \to M$. ♡

Definition 1.4. A *fibration* is a surjective map $\varphi\colon M \to N$ such that for each $q \in N$ there exist a neighborhood U and a manifold F such that the set $\varphi^{-1}(U) \subseteq M$ is diffeomorphic to the product $U \times F$ in such a way that the restriction of φ to $\varphi^{-1}(U)$ coincides with the canonical projection of $U \times F$ over U. ♡

This is illustrated by the commutative diagram

$$\begin{array}{ccc}
M \supset \varphi^{-1}(U) & \longrightarrow & U \times F \\
\Big\downarrow \varphi & & \Big\downarrow \mathrm{pr}_U \\
N \supset U & \xrightarrow{\ \mathrm{id}_U\ } & U
\end{array}$$

A fibration is a surjective submersion. The manifolds M and N are respectively called the *fiber bundle* and the *base manifold* of the fibration. If for all $q \in N$ the corresponding manifolds F are diffeomorphic (this happens for instance when the manifold N is connected), then F is called the *fiber-type* of the fibration. A fibration $\varphi\colon M \to N$ is *trivial* if the commutative diagram above holds for $U = N$. This means that, up to a diffeomorphism, $M = N \times F$.

A map $S\colon N \to M$ such that $\varphi \circ S = \mathrm{id}_N$ (i.e., $S(p) \in \varphi^{-1}(p)$) is a *section* of the fibration.

The tangent fibration $\tau_M\colon TM \to M$ is an example of fibration. When the tangent fibration is trivial (i.e., $TM = M \times \mathbb{R}^m$) then the manifold M is said to be *parallelizable.*

[1] In this case, $m = n$.

1.4 The rank theorem

A basic tool for the analysis of maps is the so-called *rank theorem*:

Theorem 1.1. *Let* $\varphi\colon M_m \to N_n$ *be a smooth map and* x_0 *be a point of* M. (i) *If* $\operatorname{rank}_{x_0}\varphi = r$, *then there exist coordinates* (x^i) *around* x_0 *and coordinates* (y^a) *around* $y_0 = \varphi(x_0)$ *such that* φ *is represented by equations*

$$\begin{cases} y^1 = x^1, \\ y^2 = x^2, \\ \dots \\ y^r = x^r, \end{cases} \qquad \begin{cases} y^{r+1} = \varphi^{r+1}(x^i), \\ y^{r+2} = \varphi^{r+2}(x^i), \\ \dots \\ y^n \quad = \varphi^n(x^i). \end{cases} \tag{1.2}$$

(ii) *If* $\operatorname{rank}_x\varphi = r$ *in a neighborhood of* x_0, *then the representation* (1.2) *of* φ *can be reduced to the form*

$$\begin{cases} y^1 = x^1, \\ y^2 = x^2, \\ \dots \\ y^r = x^r, \end{cases} \qquad \begin{cases} y^{r+1} = 0, \\ y^{r+2} = 0, \\ \dots \\ y^n \quad = 0. \end{cases} \tag{1.3}$$

Proof. (i) Let $y^\alpha = \varphi^\alpha(\bar{x}^i)$ any representation of φ around a point $x_0 \in M$. If $\operatorname{rank}_{x_0}\varphi = r$ then we can always assume that

$$\det\left[\frac{\partial\varphi^a}{\partial\bar{x}^b}\right]_{x_0} \neq 0, \quad \text{for } a, b = 1, \dots, r.$$

Consequently, in a neighborhood of x_0 we can define a new coordinate system (x^i),

$$\begin{cases} x^1 = \varphi^1(\bar{x}^i), \\ x^2 = \varphi^2(\bar{x}^i), \\ \dots \\ x^r = \varphi^r(\bar{x}^i), \end{cases} \qquad \begin{cases} x^{r+1} = \bar{x}^{r+1}, \\ x^{r+2} = \bar{x}^{r+2}, \\ \dots \\ x^m \quad = \bar{x}^m. \end{cases}$$

In these coordinates the representation of φ assumes the form (1.2).

(ii) For a representation of the form (1.2) the matrix

$$\left[\frac{\partial\varphi^\alpha}{\partial\bar{x}^i}\right]$$

is composed of four submatrices,

$$\begin{bmatrix} \mathbf{1}_r & \mathbf{0} \\ \dfrac{\partial\varphi^a}{\partial x^b} & \dfrac{\partial\varphi^a}{\partial x^c} \end{bmatrix},$$

where $a = r+1, \ldots, n$, $b = 1, \ldots, r$, and $c = r+1, \ldots, m$. Assume that $\operatorname{rank}_x \varphi = r$ in a neighborhood of x_0. We observe that if an element of the submatrix

$$\left[\frac{\partial \varphi^a}{\partial x^c}\right], \quad a = r+1, \ldots, n, \quad c = r+1, \ldots, m,$$

does not vanish, then the rank of the whole matrix would be greater than r. Hence, all the elements of this submatrix must vanish:

$$\frac{\partial \varphi^a}{\partial x^c} = 0, \quad a = r+1, \ldots, n, \quad c = r+1, \ldots, m.$$

This means that in a neighborhood of x_0 the functions $\varphi^{r+1}, \ldots, \varphi^n$ depend on the coordinates $x^1, x^2, \ldots, x^r$ only. At this point the representation (1.2) assumes the form

$$\begin{cases} y^1 = x^1, \\ y^2 = x^2, \\ \ldots \\ y^r = x^r, \end{cases} \qquad \begin{cases} y^{r+1} = \varphi^{r+1}(x^1, x^2, \ldots, x^r), \\ y^{r+2} = \varphi^{r+2}(x^1, x^2, \ldots, x^r), \\ \ldots \\ y^n \;\;\;= \varphi^n(x^1, x^2, \ldots, x^r). \end{cases} \tag{1.4}$$

Taking into account these two sets of equations, we can perform the following transformation of the coordinates y^α:

$$\begin{cases} \bar{y}^1 = y^1, \\ \bar{y}^2 = y^2, \\ \ldots \\ \bar{y}^r = y^r, \end{cases} \qquad \begin{cases} \bar{y}^{r+1} = y^{r+1} - \varphi^{r+1}(y^1, y^2, \ldots, y^r), \\ \bar{y}^{r+2} = y^{r+2} - \varphi^{r+2}(y^1, y^2, \ldots, y^r), \\ \ldots \\ \bar{y}^n \;\;\;= y^n - \varphi^n(y^1, y^2, \ldots, y^r). \end{cases}$$

In the coordinates $(x^i, \bar{y}^\alpha)$ we get a representation of the kind (1.3). □

1.5 Submanifolds

There are several notions of submanifold. For the purposes of this book it is not necessary to go into details of a fine analysis. It is sufficient to base our approach on the following definition.

Definition 1.5. A *submanifold* S of a smooth manifold M_m is a subset $S \subset M$ having this property: around each point $x_0 \in S$ there exists a coordinate system (x^i) on M, with domain U containing x_0, such that the intersection $S \cap U$ is described by the $m - s$ equations

$$x^{s+1} = 0, \quad x^{s+2} = 0, \quad \ldots, \quad x^m = 0. \tag{1.5}$$

These coordinates are said to be *adapted* to the submanifold. The pair (U, x^i) is called an *adapted chart.* ♡

Remark 1.2. Any adapted chart (U, x^i) generates a chart on S:

$$(S \cap U, x^1, \ldots, x^s).$$

In this way, adapted charts generate an atlas of S, and S becomes a manifold of dimension s and of *codimension* $m - s$. ♢

Remark 1.3. An open subset $U \subseteq M$ is a submanifold of dimension m. ♢

Definition 1.6. A vector $v \in TM$ is *tangent to a submanifold* $S \subset V$ if $v(f) = 0$ for all functions $f \in \mathscr{F}(M)$ constant on S, or equivalently, if it can be represented by a curve $\gamma\colon I \to M$ lying on S: $\gamma(I) \subset S$. ♡

Remark 1.4. The *tangent of a submanifold* $S \subset Q$, which is also called the *tangent prolongation* of S, is the set $TS \subset TQ$, made of all vectors tangent to S. This is a submanifold of dimension $2(n-k)$, where k is the codimension of S. ♢

Submanifolds arise and are described in various ways. Here we look at the most common cases.

Theorem 1.2. *Let $\varphi\colon M_m \to N_n$ be a subimmersion of rank r.*[2] *Then for any $x_0 \in M$ there exists a neighborhood U such that the image $\varphi(U)$ is a submanifold of N of dimension equal to r.*

Proof. Since φ has constant rank r, the assumptions of item (ii) of the rank theorem 1.1 are fulfilled and a local coordinate representation of the kind (1.3) holds for φ,

$$\begin{cases} y^1 = x^1, \\ y^2 = x^2, \\ \ldots \\ y^r = x^r, \end{cases} \qquad \begin{cases} y^{r+1} = 0, \\ y^{r+2} = 0, \\ \ldots \\ y^n \quad = 0. \end{cases}$$

The second set of these equations, compared with Eq. (1.5) of Definition 1.5, shows that (y^i) are adapted coordinates of $\varphi(U) \subseteq N$ and that $\dim \varphi(U) = r$. □

This theorem shows that the image of a subimmersion is a piecewise submanifold. Self-intersections or other strange phenomena may occur.

Theorem 1.3. *Let $\varphi\colon M_m \to N_n$ be a smooth map and y_0 a fixed point of N. If φ has constant rank r in a neighborhood of the set $S = \varphi^{-1}(y_0)$ the S is a submanifold of codimension r and moreover,*

$$T_x S = \ker(T_x\varphi) \quad \text{for all } x \in X. \tag{1.6}$$

[2] Recall that immersions and submersions are special cases of subimmersions; Remark 1.1.

Proof. The restriction of the map φ to the neighborhood of S is a subimmersion. We can apply item (ii) of Theorem 1.1 and consider coordinates in which the representation (1.3) of φ holds:

$$\begin{cases} y^1 = x^1, \\ y^2 = x^2, \\ \dots \\ y^r = x^r, \end{cases} \qquad \begin{cases} y^{r+1} = 0, \\ y^{r+2} = 0, \\ \dots \\ y^n \quad = 0. \end{cases}$$

Let $c^\alpha = y^\alpha(y_0)$ be the coordinates of the point y_0, $\alpha = 1, \dots, n$. It follows that

$$c^{r+1} = 0, \ \ c^{r+2} = 0, \ \ \dots \ \ c^n = 0.$$

Hence, the system of coordinates $(\bar{x}^i)$ on M defined by

$$\begin{cases} \bar{x}^1 = x^1 - c^1, \\ \bar{x}^2 = x^2 - c^2, \\ \dots \\ \bar{x}^r = x^r - c^r, \end{cases} \qquad \begin{cases} \bar{x}^{r+1} = x^{r+1}, \\ \bar{x}^{r+2} = x^{r+2}, \\ \dots \\ \bar{x}^m \quad = x^m, \end{cases}$$

are adapted to S; see Definition 1.5 and Remark 1.2. In order to prove Eq. (1.6) we observe that

$$v \in \ker T_x\varphi \iff v(f \circ \varphi) = 0 \ \text{ for all } f \in \mathscr{F}(N)$$

and that

$$v \in T_xS \iff v(g) = 0 \ \text{ for all } g \in \mathscr{F}(M) \text{ constant on } S.$$

Any composition of the type $f \circ \varphi$ is constant on S, therefore the second condition implies the first; that is $T_xS \subseteq \ker T_x\varphi$. Conversely, if $f = f(y^\alpha)$ is the representation of f in the coordinates considered above, then $f(x^1, \dots, x^r, 0, \dots, 0)$ is the representation of $f \circ \varphi$. It follows that the condition $v(f \circ \varphi) = 0$ for all $f \in \mathscr{F}(N)$ is equivalent to equation

$$v^1 \frac{\partial g}{\partial x^1} + v^2 \frac{\partial g}{\partial x^2} + \dots + v^r \frac{\partial g}{\partial x^r} = 0$$

for all functions $g(x^1, \dots, x^r)$. This means that $v^1 = v^2 = \dots = v^r = 0$; that is $v \in TS$. Hence, $\ker T_x\varphi \subseteq T_xS$. □

The following theorem can be proved in a similar way.

Theorem 1.4. *If $\varphi\colon M_m \to N_n$ is a subimmersion and S_0 is a submanifold of N then $S = \varphi^{-1}(S_0)$ is a submanifold of M.*

1.5.1 Submanifolds defined by equations

Definition 1.7. Let $F^a \in \mathscr{F}(M)$, $a = 1, \cdot, r$, be a set of r functions. They are called independent at a point $x \in M$ if their differentials $d_x F^a$ are linear independent. ♡

A system of equations like

$$F^a(x^i) = 0: \qquad \begin{cases} F^1(x^i) = 0, \\ F^2(x^i) = 0, \\ \dots \\ F^n(x^i) = 0, \end{cases} \tag{1.7}$$

where (x^i) are local coordinates on a manifold M_m with domain U, define a subset $S \subset M$. Here, $F^a(x^i)$ are local representatives of real-valued function F^a on M.

Definition 1.8. Equations (1.7) are called *independent* if the differentials $d_x F^a$ are linearly independent at each point $x \in S$. ♡

Remark 1.5. The functions F^a are independent if and only if the matrix

$$\left[\frac{\partial F^a}{\partial x^i} \right]_{n \times m} \tag{1.8}$$

has a maximal rank r in the domain of the coordinates. Note that if the functions are independent at a point x, then they are independent in a neighborhood of x. ♢

Question: when is S a submanifold of M according to Definition 1.5? A first answer is the following.

Theorem 1.5. *If the matrix (1.8) has a constant rank r in the domain of the coordinates, then $S \cap U$ is a submanifold of dimension r.*

Proof. Consider the map $\varphi \colon M \to \mathbb{R}^n \colon (x^i) \mapsto S^\alpha(x^i)$ and apply Theorem 1.3 for $y_0 = 0$. □

Example 1.1. Take $M = \mathbb{R}^3$, $(x^i) = (x, y, z)$ and equations

$$F^1 \doteq z = 0, \quad F^2 \doteq z - x^2 y^2 = 0.$$

In this case, $U = \mathbb{R}^3$ and S is the union of the x-axis and the y-axis. S is not a submanifold. Because $F^1(x, y, z) = z$ and $F^2(x, y, z) = z - x^2 y^2$, we have

$$\left[\frac{\partial F^a}{\partial q^i} \right] = \begin{bmatrix} 0 & 0 & 1 \\ -2xy^2 & -2yx^2 & 1 \end{bmatrix}$$

and

$$\operatorname{rank}\left[\frac{\partial F^a}{\partial q^i}\right]_S = \operatorname{rank}\begin{bmatrix} 0\,0\,1 \\ 0\,0\,1 \end{bmatrix} = 1.$$

The rank is constant on S but not in a neighborhood of S. This example shows that if the rank of the matrix (1.6) is not the same on S and in a neighborhood of S, then S may not be a submanifold. ♢

Definition 1.9. When $r = n$, that is when the rank of the matrix is maximal (equal to number of equations) then the equations are called *independent.* ♡

Remark 1.6. If $S \subset M$ is described by Eqs. (1.6) then TS is described by the equations

$$F^a(x^i) = 0, \quad \frac{\partial F^a}{\partial x^i}\,\dot{x}^i = 0. \quad \diamondsuit \tag{1.9}$$

1.5.2 Clean intersection

This concept plays an important role in dealing with the reduction of submanifolds, Sect. 2.3.1 and Sect. 3.7.

Definition 1.10. Two submanifolds S_1 and S_2 of a manifold M have a *clean intersection* if:[3]

- $S_1 \cap S_2$ is a submanifold
- $T(S_1 \cap S_2) = TS_1 \cap TS_2$; that is,

$$T_x(S_1 \cap S_2) = T_xS_1 \cap T_xS_2 \text{ for all } x \in S_1 \cap S_2. \quad \heartsuit \tag{1.10}$$

Remark 1.7. (i) The inclusion

$$T(S_1 \cap S_2) \subseteq TS_1 \cap TS_2$$

is always valid for two submanifolds such that their intersection is a submanifold. Indeed, if $v = [\gamma] \in T(S_1 \cap S_2)$ is a vector represented by a curve γ on $S_1 \cap S_2$, then γ is a curve on S_1 and on S_2 simultaneously, so that $v \in TS_1$ and $v \in TS_2$. (ii) $S_1 \subset S_2$ and $S_1 \cap S_2 = \emptyset$ are two cases of clean intersection. ♢

Theorem 1.6. *Let $F_1 \colon M \to \mathbb{R}^{n_1}$ and $F_2 \colon M \to \mathbb{R}^{n_2}$ be of constant rank r_1 and r_2 on a neighborhood of*

$$S_1 = F_1^{-1}(0) \;\; and \;\; S_2 = F_2^{-1}(0),$$

respectively. If the map

[3] (Bott 1954), (Weinstein 1973, 1977).

$$(F_1, F_2): M \to \mathbb{R}^{n_1} \times R^{n_2}$$

defined by

$$(F_1, F_2)(x) = (F_1(x), F_2(x))$$

has a constant rank in a neighborhood of $S_1 \cap S_2$*, then* S_1 *and* S_2 *have a clean intersection.*

Proof. Because $S_1 \cap S_2 = (F_1, F_1)^{-1}(0)$, $S_1 \cap S_2$ is a submanifold due to Theorem 1.3. By an elementary property of linear algebra,

$$\ker T_x F_1 \cap \ker T_x F_2 = \ker T_x (F_1, F_2) \text{ for all } x \in M. \tag{1.11}$$

Then we have:

$$\begin{aligned} T_x(S_1 \cap S_2) &= \ker T_x(F_1, F_2), && \text{Theorem 1.3,} \\ &= \ker T_x F_1 \cap \ker T_x F_2, && \text{Formula (1.11),} \\ &= T_x S_1 \cap T_x S_2, && \text{Theorem 1.3.} \end{aligned}$$

□

The preceding theorem can be reinterpreted as follows.

Theorem 1.7. *If* S_1 *and* S_2 *are two submanifolds of* M *defined by equations*

$$\begin{cases} F_1^\alpha(x^i) = 0, & \alpha = 1, \ldots, n_1, \\ F_2^a(x^i) = 0, & a = 1, \ldots, n_2, \end{cases} \tag{1.12}$$

such that in neighborhoods of S_1 *and* S_2 *the matrices*

$$\left[\frac{\partial F_1^\alpha}{\partial x^i}\right], \quad \left[\frac{\partial F_2^a}{\partial x^i}\right] \tag{1.13}$$

and

$$\left[\frac{\partial F_1^\alpha}{\partial x^i} \;\middle|\; \frac{\partial F_2^\alpha}{\partial x^i}\right] \tag{1.14}$$

have constant rank, then the intersection of S_1 *and* S_2 *is clean.*

Example 1.2. Take $M = \mathbb{R}^3$, $(x^i) = (x, y, z)$, and equations

$$F_1 = z = 0, \quad F_2 = z - x^2 y^2 = 0.$$

These are just the equations of example 1.1 but now interpreted as equations of two submanifolds S_1 and S_2. The two matrices (1.13) are now

$$[0, 0, 1] \text{ and } [-2x\, y^2\ ,\ -2y\, x^2\ ,\ 1].$$

They have constant rank everywhere. The matrix (1.14) is

$$\begin{bmatrix} 0 & 0 & 1 \\ -2xy^2 & -2yx^2 & 1 \end{bmatrix}.$$

On $S_1 \cap S_2$ it becomes

$$\begin{bmatrix} 0 & 0 & 1 \\ 0 & 0 & 1 \end{bmatrix}$$

and its rank is 1. But outside $S_1 \cap S_2$ its rank may be greater than 1. In other words, its rank is constant on $S_1 \cap S_2$ but not in a neighborhood of $S_1 \cap S_2$. This example shows the relevance of the assumption that the matrix (1.14) also must have a constant rank. ♢

1.5.3 Transverse intersection

Definition 1.11. Two submanifolds S_1 and S_2 of a manifold M have a *transverse intersection* if for each point $x \in S_1 \cap S_2$ the tangent spaces $T_x S_1$ and $T_x S_2$ together span the space $T_x M$:

$$T_x S_1 + T_x S_2 = T_x M \quad \text{for all } x \in S_1 \cap S_2. \tag{1.15}$$

♡

Theorem 1.8. *A transverse intersection implies a clean intersection.*

Proof. Let S_1 and S_2 be defined by independent equations (1.12),

$$\begin{cases} F_1^\alpha(x^i) = 0, & \alpha = 1, \ldots, n_1, \\ F_2^a(x^i) = 0, & a = 1, \ldots, n_2. \end{cases}$$

The matrices (1.13) have maximal rank, n_1 and n_2, respectively. The tangent spaces $T_x S_1$ and $T_x S_2$ are defined by equations

$$\frac{\partial F_1^\alpha}{\partial x^i}\, \dot{x}^i = 0 \quad \text{and} \quad \frac{\partial F_2^a}{\partial x^i}\, \dot{x}^i = 0,$$

respectively. Then the differential forms

$$\phi_1^\alpha = \frac{\partial F_1^\alpha}{\partial x^i}\, dx^i \quad \text{and} \quad \phi_2^a = \frac{\partial F_2^a}{\partial x^i}\, dx^i$$

annihilate the vectors of $T_x S_1$ and $T_x S_2$, respectively, and any linear combination $\lambda_\alpha\, \phi_1^\alpha + \lambda_a\, \phi_2^a$ annihilates any vector belonging to the intersection $T_x S_1 \cap T_x S_2$. But, due to (1.15), such a form must be the zero-form:

$$\lambda_\alpha\, \frac{\partial F_1^\alpha}{\partial x^i} + \lambda_a\, \frac{\partial F_2^a}{\partial x^i} = 0.$$

This is a system of m equations, linear in the coefficients λ, which must admit the trivial solution $\lambda_\alpha = \lambda_a = 0$ only. This occurs if and only if the matrix

$$\left[\frac{\partial F_1^\alpha}{\partial x^i} \;\middle|\; \frac{\partial F_2^\alpha}{\partial x^i} \right]$$

has maximal rank. Then we can apply Theorem 1.7. □

1.6 Vector fields

A *vector field* on a manifold M is a section of the tangent bundle TM; that is a smooth map $X\colon M \to TM$ which assigns to each point $x \in M$ a vector $X(x) \in T_xM$ at that point. Such a section is locally described by equations

$$\dot{x}^i = X^i(x).$$

The functions X^i are the *components* of the vector field X in the coordinates (x^i).

There is an equivalent definition: a vector field is a *derivation* on $\mathscr{F}(M)$; that is a map $X\colon \mathscr{F}(M) \to \mathscr{F}(M)$ such that

$$\begin{cases} X(aF + bG) = a\,X(F) + b\,X(G), \quad a, b \in \mathbb{R} & \text{(linearity)}, \\ X(FG) = X(F)\,G + F\,X(G) & \text{(Leibniz rule)}. \end{cases}$$

We use the notation

$$X(F) = \langle X, dF \rangle.$$

This function is called the *derivative* of F with respect to X. The link between these two definitions is given by equation

$$\langle X, dF \rangle(x) = \langle X(x), dF \rangle.$$

The components of a vector field X are the derivatives of the coordinates,

$$X^i = \langle X, dx^i \rangle,$$

so that

$$\langle X, dF \rangle = X^i\, \partial_i F.$$

We denote by $\mathscr{X}(M)$ the set of the smooth vector fields on M. It is a module over the ring $\mathscr{F}(M)$ and an infinite-dimensional vector space over $\mathbb{R}$, the sum and the product by a function being defined by

$$(X + Y)(x) = X(x) + Y(x), \quad (fX)(x) = f(x)X(x).$$

1.7 Integral curves and flows

Let X be a vector field on a manifold M. An *integral curve* of X is a curve on M, $\gamma\colon I \to M$, such that $\dot\gamma(t) = X(\gamma(t))$ for all $t \in I$ (i.e., $\dot\gamma = X \circ \gamma$). The integral curves of X are locally represented by the solutions of a first-order differential system in normal form,

$$\frac{dx^i}{dt} = X^i(x).$$

Hence, a vector field can be interpreted as a *dynamical system*. We say that an integral curve is *based* at a point x if $\gamma(0) = x$. For smooth vector fields the Cauchy theorem asserts that for each point x there exists a unique *maximal integral curve* $\gamma_x\colon I_x \to M$ based on x, such that any other integral curve based at x is defined on an interval $I \subseteq I_x$. When $I_x = \mathbb{R}$ for all x, then the field is said to be *complete*.

A *flow* on a manifold M is a smooth map

$$\varphi\colon \mathbb{R} \times M \to M\colon (t, x) \mapsto \varphi(t, x)$$

such that for all $t, s \in \mathbb{R}$ the map

$$\varphi_t\colon M \to M\colon x \mapsto \varphi(t, x)$$

is a transformation of M and

$$\varphi_t \circ \varphi_s = \varphi_{t+s}.$$

It follows that

$$\begin{cases} \varphi_0 &= \mathrm{id}_M, \\ \varphi_t \circ \varphi_s &= \varphi_s \circ \varphi_t, \\ \varphi_{-t} &= (\varphi)^{-1}. \end{cases}$$

The set of all φ_t, $t \in \mathbb{R}$, is said to be a *one-parameter group of transformations*.

A complete vector field X generates a flow $\varphi^X\colon \mathbb{R} \times M \to M$ defined by

$$\varphi^X(t, x) = \gamma_x(t).$$

Conversely, a flow φ generates a complete vector field X by setting

$$X(x) = \dot\gamma_x(0), \tag{1.16}$$

where $\gamma_x\colon \mathbb{R} \to M$ is the curve defined by

$$\gamma_x(t) = \varphi(t, x).$$

These curves are the maximal integral curves of X. A noncomplete vector field generates *local flows*, defined on open subsets of $\mathbb{R} \times M$. If

$$x^i = \varphi^i(t, x_0^h) \tag{1.17}$$

is a local representation of a flow φ in local coordinates (x^i) then, according to Eq. (1.16), the components of the associated vector field at the point x_0 are given by

$$X^i(x_0) = D\varphi^i(0, x_0),$$

where D represents the derivative with respect to the variable t.

If φ_t^X is the one-parameter group of transformations generated by a complete vector field X, then

$$T\varphi_t^X : TM \to TM$$

is a one-parameter group of transformations of TM, generating a vector field on TM which we denote by $\dot{X}$. The vector field $\dot{X}$ is projectable onto X. This means that the following diagram is commutative,

$$
\begin{array}{ccc}
TTM & \xrightarrow{\;T\tau_M\;} & TM \\[4pt]
\dot{X} \uparrow & & \uparrow X \\[4pt]
TM & \xrightarrow{\;\tau_M\;} & M
\end{array}
$$

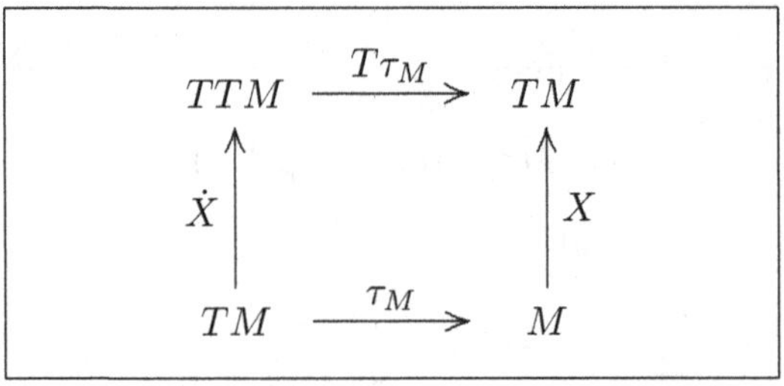

that is

$$T\tau_M \circ \dot{X} = X \circ \tau_M.$$

The components of $\dot{X}$ in coordinates $(x^i, \dot{x}^i)$ of TM are $(X^i, \dot{X}^i)$ where X^i are the components of X and

$$\dot{X}^i(x_0^h, \dot{x}_0^j) = \dot{\varphi}_j^i(0, x_0^h)\, \dot{x}_0^j,$$

where

$$\varphi_j^i(t, x_0^h) = \frac{\partial \varphi^i}{\partial x_0^j},$$

being $\varphi^i(t, x_0^h)$ the local representative of φ_t^X; see Eq. (1.17).

1.8 First integrals

A *first integral* or *integral function of a vector field* X on M is a function $F \in \mathscr{F}(M)$ such that

$$\langle X, dF \rangle = 0.$$

The first integrals can be locally determined by integrating the first-order linear partial differential equation

$$X^i \partial_i F = 0.$$

There is an equivalent definition: a first integral is a function F that takes a constant value along any integral curve:

$$D(F \circ \gamma_x) = 0.$$

Indeed, the local expression of this condition is

$$\frac{d}{dt} F(\gamma^i(t)) = \partial_i F \, \dot{\gamma}^i(t) = \partial_i F \, X^i(t) = 0.$$

A vector field may not have global first integrals. However,

Theorem 1.9. *In a neighborhood of a nonsingular point $x \in M$ $(X(x) \neq 0)$ there exist $n - 1$ independent first integrals.*

This follows from the next theorem.

Theorem 1.10. *In a neighborhood of a non-singular point $x \in M$ $(X(x) \neq 0)$ there exists a coordinate system (x^i) such that $X = \partial/\partial x^1$.*

These coordinates are said to be *adapted* to X.

1.9 Lie bracket

The *Lie-bracket* $[X, Y]$ of two vector fields is the vector field defined by

$$[X, Y]F = X(YF) - Y(XF).$$

In local coordinates,

$$[X, Y]^i = X^k \, \partial_k Y^i - Y^k \, \partial X^i.$$

This operation satisfies the following properties.

$$\begin{cases} [X,Y] = -[Y,X] & \text{(anticommutativity)}, \\ [aX+bY,Z] = a[X,Z]+b[Y,Z], \quad a,b \in \mathbb{R} & \text{($\mathbb{R}$-linearity)}, \\ [X,[Y,Z]] + [Y,[Z,X]] + [Z,[X,Y]] = 0 & \text{(cyclic or Jacobi identity)}. \end{cases}$$

Thus, the space of the vector fields $\mathscr{X}(M)$ endowed with the Lie bracket is a Lie-algebra. We say that two vector fields *commute* if $[X,Y]=0$. Indeed, the following theorem can be proved.

Theorem 1.11. *The flows of two (complete) vector fields commute, that is*

$$\varphi_t^X \circ \varphi_s^Y = \varphi_s^Y \circ \varphi_t^X,$$

for all $t,s \in \mathbb{R}$, *if and only if* $[X,Y]=0$.

A vector field is *tangent to a submanifold* $S \subset M$ when all its values $X(x)$ are vectors tangent to S. This holds if and only if any integral curve intersecting S lies on S. The following can be proved.

Theorem 1.12. *If two vector fields X and Y are tangent to a submanifold S, then $[X,Y]$ is also tangent to S.*

1.10 One-forms

A *one-form* on a manifold M is a map $\theta\colon TM \to \mathbb{R}$ linear on each tangent space T_xM. We use the notation

$$\theta(v) = \langle v, \theta \rangle.$$

An equivalent definition is the following. A one-form is a linear map from vector fields to functions, $\theta\colon \mathscr{X}(M) \to \mathscr{F}(M)$. We use the notation

$$\theta(X) = \langle X, \theta \rangle.$$

The link between these two definitions is

$$\langle X, \theta \rangle(x) = \langle X(x), \theta \rangle.$$

The linearity implies that

$$\langle X, \theta \rangle = X^i\,\theta_i,$$

where θ_i are functions called the *components* of θ (with respect to the coordinates (x^i)). It follows that

$$\theta_i = \langle \partial_i, \theta \rangle.$$

We can define the sum of two one-forms and the product of a one-form with a function (or a number) in an obvious way.

A special case of one-form is the *differential of a function* dF. It is defined by

$$\langle X, dF \rangle = XF$$

and its components are

$$(dF)_i = \partial_i F.$$

It follows that in a coordinate system any one-form can be represented by a linear combination of the differentials dq^i,

$$\theta = \theta_i \, dx^i.$$

Thus, a one-form is also called a *linear differential form*. We call an *elementary one-form* a one-form of the kind $F\,dG$, where F and G are smooth functions on M.

1.11 Exterior forms

Let $\times_M^p TM$ be the subset of the Cartesian power $(TM)^p$ made of ordered sets of p tangent vectors applied to a same point. It is a manifold of dimension $(p+1)\dim M$. An *exterior form of order* p, briefly a p*-form*, on a manifold M is a multilinear skew-symmetric smooth map from this space to $\mathbb{R}$,

$$\omega\colon \times_M^p TM \to \mathbb{R}\colon (v_1, \ldots, v_p) \mapsto \omega(v_1, \ldots, v_p).$$

The value $\omega(v_1, \ldots, v_p)$ changes in sign by interchanging any two arguments. It follows that for linearly dependent vectors $\omega(v_1, \ldots, v_p) = 0$. Thus, any p-form for $p > n$ vanishes identically.

A *zero-form* ($p = 0$) is a function $F\colon M \to \mathbb{R}$. For $p = 1$ we get the definition of one-form.

An equivalent definition is the following. A p-form is a multilinear and skew-symmetric smooth map from the Cartesian power $(\mathscr{X}(M))^p$ of the space of vector fields to $\mathscr{F}(M)$,

$$\omega\colon (\mathscr{X}(M))^p \to \mathscr{F}(M)\colon (X_1, \ldots, X_p) \mapsto \omega(X_1, \ldots, X_p).$$

The sum of two p-forms and the multiplication of a p-form with a function or a real number are defined in an obvious way. We denote by $\Phi^p(M)$ the linear space of all p-forms. It is a module on the ring $\mathscr{F}(M)$. In particular, $\Phi^0(M) = \mathscr{F}(M)$ and $\Phi^p(M) = 0$ for $p > n$. We set $\Phi^p(M) = 0$ for $p < 0$ and denote by $\Phi(M)$ the direct sum of all these spaces,

$$\Phi(M) = \bigoplus_{p=-\infty}^{+\infty} \Phi^p(M).$$

An *exterior* or *differential form* is an element of this space.

1.12 Exterior algebra

The *exterior product* $\varphi \wedge \psi$ of a p-form φ times a q-form ψ is the $p+q$-form defined by

$$\varphi \wedge \psi = \frac{(p+q)!}{p!\,q!} \mathsf{A}(\varphi \otimes \psi),$$

A being the *antisymmetrization operator*. On any p-linear form $\eta\colon \times_M^p TM \to \mathbb{R}$ it is defined by

$$\mathsf{A}\eta = \frac{1}{p!} \sum_{S \in G_p} \varepsilon_S \eta \circ S,$$

where G_p is the permutation group of order p and $\varepsilon_S = \pm 1$ is the signature of the permutation S. For a 0-form (function), $\mathsf{A}f = f$. For $p < 0$ or $q < 0$, $\varphi \wedge \psi = 0$. If one of the two forms is a function, then

$$\varphi \wedge \psi = \varphi\, \psi.$$

By a linear extension of the exterior product to the direct sum $\Phi(M)$ we get the *exterior algebra*. It is a commutative and associative graded algebra,

$$\begin{cases} \Phi^p(M) \wedge \Phi^q(M) \subset \Phi^{p+q}(M), \\ \varphi \wedge \psi = (-1)^{pq} \psi \wedge \varphi, \\ (\varphi \wedge \psi) \wedge \varphi = \varphi \wedge (\psi \wedge \varphi). \end{cases}$$

For two one-forms the exterior product is anticommutative, $\varphi \wedge \psi = -\psi \wedge \varphi$.

An *elementary p-form* is a p-form of the kind

$$\omega = F\, dG_1 \wedge \cdots \wedge dG_p,$$

where $F, G_1, \ldots, G_p$ are functions. Then the exterior product of two elementary exterior forms is obtained by applying the associative rule and the commutation rules $F \wedge dG = dG \wedge F$ and $dF \wedge dG = -dG \wedge dF$. Any p-form can be locally expressed as a sum of elementary p-forms. Indeed, in any coordinate system we have the representation

$$\omega = \frac{1}{p!}\, \omega_{i_1 \cdots i_p}\, dx^{i_1} \wedge \cdots \wedge dx^{i_p},$$

where $\omega_{i_1 \cdots i_p} = \omega(\partial_{i_1}, \ldots, \partial_{i_p})$ are the components of ω.

1.13 Pullback of forms

Let $\alpha\colon M_1 \to M_2$ be a smooth map. For each $p \in \mathbb{Z}$ we define a linear map

$$\alpha^*\colon \Phi^p(M_2) \to \Phi^p(M_1)$$

by setting

$$\begin{cases} \alpha^*\omega(v_1, \ldots, v_p) = \omega(T\alpha(v_1), \ldots, T\alpha(v_p)), & p > 0, \\ \alpha^*\omega = \omega \circ \alpha, & p = 0, \\ \alpha^*\omega = 0, & p < 0. \end{cases}$$

By a linear extension we get a linear map

$$\alpha^*\colon \Phi(M_2) \to \Phi(M_1),$$

called a *pullback*, with the properties:

$$\begin{cases} \alpha^*(\omega \wedge \psi) = \alpha^*\omega \wedge \alpha^*\psi, \\ \mathrm{id}_M^* = \mathrm{id}_{\Phi(M)}, \\ (\beta \circ \alpha)^* = \alpha^* \circ \beta^*. \end{cases}$$

The last two properties show that the operator

$$^*\colon \quad \begin{cases} M \mapsto \Phi(M), \\ \alpha \mapsto \alpha^* \end{cases}$$

is a covariant functor from the category of the smooth manifolds into the category of the graded algebras, the *exterior functor*.

If $\iota\colon S \to M$ is the canonical injection of a submanifold $S \subset M$, then the pullback $\iota^*\omega$ of a form on M is the *restriction* of ω to S and it is also denoted by $\omega|S$. In fact, it is the restriction of $\omega\colon \times_M^p TM \to \mathbb{R}$ to the submanifold $\times_S^p TS$.

If in local coordinates the map α is represented by equations $y_2^a = \alpha^a(x_1^i)$, then the pullback of a form is obtained by replacing these functions in the local coordinate representation. It follows that the pullback is locally represented by equations

$$(\alpha^*\omega)_{i_1 \cdots i_p} = \omega_{a_1 \cdots a_p} \frac{\partial \alpha^{a_1}}{\partial x_1^{i_1}} \cdots \frac{\partial \alpha^{a_p}}{\partial x_1^{i_p}}.$$

1.14 Derivations

Definition 1.12. A *derivation of degree* $r \in \mathbb{Z}$ on the exterior algebra $\Phi(M)$ is a map $D\colon \Phi(M) \to \Phi(M)$ satisfying the following rules.

$$\begin{cases} D\Phi^p(M) \subset \Phi^{p+r}(M), & p \in \mathbb{Z}, \\ D(a\varphi + b\psi) = aD\varphi + bD\psi, & a, b \in \mathbb{R}, \\ D(\varphi \wedge \psi) = D\varphi \wedge \psi + (-1)^{pr}\, \varphi \wedge D\psi, & \varphi \in \Phi^p(M). \end{cases} \tag{1.18}$$

♡

Hence, D maps a p-form to a $(p+r)$-form, it is $\mathbb{R}$-linear and satisfies a *graded Leibniz rule.* From the linearity and the Leibniz rule it follows that $Da = 0$ for any number $a \in \mathbb{R}$ interpreted as a constant 0-form.

The general theory of derivations is due to (Frölicher and Nijenhuis 1956) and it is based on the following theorems.

Theorem 1.13. *Let D be a derivation. If $\varphi, \psi \in \Phi(M)$ are two exterior forms such that $\varphi|U = \psi|U$ in an open subset $U \subset M$, then $D\varphi|U = D\psi|U$ (locality of a derivation).*

Theorem 1.14. *Any derivation is uniquely determined by its action on $\Phi^0(M)$ and $\Phi^1(M)$* (i.e., on functions and one-forms).

In other words: any map $D\colon \Phi^0(M) \oplus \Phi^1(M) \to \Phi(M)$ satisfying the rules (1.18) is extended in a unique way to a derivation of degree r on $\Phi(M)$. Note that if $r = -2$, then D has an image in $\Phi^{-2} \oplus \Phi^{-1} = 0 \oplus 0$, so that its extension is necessarily the zero-map. Thus,

Theorem 1.15. *Any derivation of degree $r < -1$ is trivial: $D = 0$.*

Definition 1.13. The *commutator* of two derivations D_1 and D_2, of degree r_1 and r_2, respectively, is the derivation of degree $r_1 + r_2$ defined by

$$[D_1, D_2] = D_1 D_2 - (-1)^{r_1 r_2} D_2 D_1. \qquad ♡ \tag{1.19}$$

Indeed, the composition $D_1 D_2 = D_1 \circ D_2$ is linear but it does not satisfy the graded Leibniz rule, which is instead satisfied by the operator defined in (1.19).

There are three special important derivations: the differential, the interior product, and the Lie derivative.

1.15 The differential

Definition 1.14. The *differential* is the derivation d of degree 1 whose action on functions and one-forms is defined by

$$\langle X, df \rangle = Xf$$

and

$$d\theta(X, Y) = \langle X, d\langle Y, \theta \rangle \rangle - \langle Y, d\langle X, \theta \rangle \rangle - \langle [X, Y], \theta \rangle,$$

respectively. ♡

As a consequence, it can be proved that

$$d(\varphi \wedge \psi) = d\varphi \wedge \psi + (-1)^p \varphi \wedge d\psi, \quad \varphi \in \Phi^p(M),$$

and

$$d^2 = 0.$$

For an elementary p-form $\omega = F\, dG_1 \wedge \cdots \wedge dG_p$,

$$d\omega = dF \wedge dG_1 \wedge \cdots \wedge dG_p.$$

The pullback α^* associated with a map α commutes with the differential,

$$d\alpha^* \omega = \alpha^* d\omega.$$

In particular, the differential commutes with the restriction of forms to submanifolds,

$$d(\omega | S) = (d\omega) | S.$$

A p-form ω is said to be *closed* if $d\omega = 0$, and*exact* if there exists a $(p-1)$-form ϕ, called a *potential form*, such that $\omega = d\phi$. *An exact form is closed,* because $d^2 = 0$. Conversely, it can be proved that *a closed form is locally exact* (*Poincaré–Volterra lemma*).

A derivation D is called a i_**-type* derivation if it is trivial on functions: $Df = 0$. It is called a d_**-type* derivation if it commutes with the differential:

$$Dd = (-1)^r dD.$$

Theorem 1.16. (i) *Any derivation can be decomposed in a unique way as a sum of a derivation of type i_* and a derivation of type d_*.* (ii) *Any derivation of type d_* is uniquely determined by its action on functions.*

We have two fundamental derivations of type i_* and d_* associated with a vector field: the interior product and the Lie derivative.

1.16 Interior product

Definition 1.15. The *interior product* (or the *Cartan product*) with respect to a vector field X is the derivation i_X of degree -1 and type i_* defined by the following action on functions and one-forms,

$$i_X f = 0, \;\; i_X \theta = \langle X, \theta \rangle. \qquad \heartsuit$$

It has the following properties,

$$i_X(\varphi \wedge \psi) = i_X\varphi \wedge \psi + (-1)^p \, \varphi \wedge i_X\psi,$$

where p is the degree of φ;

$$i_X f\varphi = f\, i_X\varphi, \;\; i_Y i_X \omega = \omega(X, Y)$$

for a two-form. A similar formula holds for any p-form.

$$i_X i_Y = -i_Y i_X, \qquad i_X^2 = 0.$$

In local coordinates,

$$i_X\, dx^i = X^i, \;\; (i_X\omega)_{i_2 \cdots i_p} = X^{i_1}\omega_{i_1 i_2 \cdots i_p}.$$

1.17 Lie derivative

Definition 1.16. The *Lie derivative* with respect to a vector field X is the derivation of type d_* and degree 0 defined by

$$dd_X = d_X d \;\text{ and }\; d_X f = i_X df. \qquad \heartsuit$$

For the Lie derivative there are other two (equivalent) definitions,

$$d_X = [i_X, d] = i_X d + d i_X, \;\; d_X\omega = \lim_{t \to 0} \frac{1}{t} \left(\varphi_t^* \omega - \omega\right).$$

The first equation is known as the *Cartan formula* (the Lie derivative is the commutator of the Cartan product and the differential). The Lie derivative has the following properties,

$$\begin{aligned} d_X(\varphi \wedge \psi) &= d_X\varphi \wedge \psi + \varphi \wedge d_X\psi, \\ [d_X, d_Y] &= d_X d_Y - d_Y d_X = d_{[X,Y]}, \\ [d_X, i_Y] &= d_X i_Y - i_Y d_X = i_{[X,Y]}, \\ d\theta(X, Y) &= d_X i_Y\theta - d_Y i_X\theta - i_{[X,Y]}\theta. \end{aligned}$$

For a two-form ω,

$$d\omega(X, Y, Z) = i_{[X,Y]} i_Z\omega - d_X i_Y i_Z\omega + c.p.,$$

where *c.p.* means the sum of the similar terms obtained by all cyclic permutations of the vector fields.

A form $\omega \in \Phi(M)$ is said to be *invariant* with respect to a transformation $\varphi\colon M \to M$ if $\varphi^*\omega = \omega$. The following can be proved.

Theorem 1.17. *A form ω is invariant with respect to the group φ_t generated by a (complete) vector field X if and only if $d_X\omega = 0$.*

Chapter 2
Relations

Abstract We examine the notion of "relation", which is central to this book, at various progressive levels: relations on sets (Sect. 2.1), linear relations on vector spaces (Sect. 2.2), smooth relations and reductions (Sect. 2.3), linear symplectic relations (Sect. 3.1), symplectic relations on symplectic manifolds (Chap. 3), and symplectic relations on cotangent bundles (Chap. 4).

2.1 Relations on sets

A *relation* R between two sets A and B is a subset of their Cartesian product:

$$R \subseteq B \times A.$$

The sets A and B are the *domain* and the *codomain* of the relation, respectively. For a relation we use the notation[1]

$$R \colon B \leftarrow A, \quad \text{or} \quad A \overset{R}{\leftarrow} B.$$

The composition of two relations $R \colon B \leftarrow A$ and $S \colon C \leftarrow B$ is the relation

$$S \circ R \colon C \leftarrow A$$

defined by

$$(c, a) \in S \circ R \quad \Longleftrightarrow \quad \begin{cases} \text{There exists } b \in B \text{ such that} \\ (b, a) \in R \text{ and } (c, b) \in S. \end{cases}$$

[1] This is not the most common convention, but the use of the backward arrow (i.e., from right to left) turns out to be more convenient in dealing with the composition of relations.

The composition of relations is associative:

$$(S \circ R) \circ Q = S \circ (R \circ Q).$$

Then *"sets" and "relations" are objects and morphisms of a category.*

2.1.1 The transposition functor

With a relation $R \subseteq B \times A$ we associate the *transpose relation* or *inverse relation*

$$R^\top \subseteq A \times B,$$

made of the same pairs of R, but in reverse order. The *contravariant transposition rule*

$$(S \circ R)^\top = R^\top \circ S^\top$$

holds. Hence, if we put by definition $A^\top = A$ for all sets, the transposition operator $^\top$ is a contravariant functor in the category of set relations.

A relation $R \subseteq B \times A$ is *symmetric* if $A = B$ and $R^\top = R$.

A map $\rho\colon A \to B$ can be interpreted as a relation by its graph, $R = \mathrm{graph}(\rho) \subset B \times A$:

$$(b, a) \in R \iff b = \rho(a).$$

Hence, a relation $R \subseteq B \times A$ is a map if and only if

$$\begin{cases} R^\top \circ B = A, \\ (b, a) \in R,\ (b', a) \in R \implies b' = b. \end{cases}$$

The *diagonal* of a product $A \times A$ is denoted by Δ_A,

$$\Delta_A = \{(a, a') \in A \times A \mid a = a'\}.$$

It behaves as the *identity relation* over the set A; if $R : B \leftarrow A$ then $R \circ \Delta_A = R$ and $\Delta_B \circ R = R$.

Remark 2.1. In the category of relations it is convenient to interpret a subset $S \subseteq A$ as a relation $S \subseteq A \times \{0\}$ where $\{0\}$ is a *singleton*, an arbitrary set made of a single element. If $R\colon B \leftarrow A$ then $R \circ S$ is the *image of the subset* S by the relation R. In particular $R \circ A \subseteq B$ is the *image* of the relation R and $R^\top \circ B \subseteq A$ is the *inverse image* of R. ♢

2.2 Linear relations

Definition 2.1. A *linear relation* $R\colon B \leftarrow A$ is a linear subspace of the direct sum $B \oplus A$ of two vector spaces A and B. ♡

The direct sum is the Cartesian product endowed with the natural structure of a vector space. This definition is suggested by the fact that *a map $f\colon A \to B$ is linear if and only if its graph R is a linear subspace of $B \oplus A$*. It can be shown that *the composition of two linear relations is a linear relation.* Vector spaces and linear relations form a category (Benenti and Tulczyjew 1979).

2.3 Smooth relations and reductions

Definition 2.2. A *smooth relation* is a submanifold $R \subseteq M_2 \times M_1$ of the product of two smooth manifolds M_1 and M_2. ♡

The composition of two smooth relations may not be a smooth relation; that is $S \circ R$ may not be a submanifold. The graph of a smooth map is a special case of smooth relation.

With a smooth relation $R\colon M_2 \leftarrow M_1$ we associate the *tangent relation*

$$TR \subset TM_2 \times TM_1 \simeq T(M_2 \times M_1).$$

This is always a smooth relation. If R is the graph of a map $\rho\colon M_1 \to M_2$, then TR is the graph of the tangent map $T\rho\colon TM_1 \to TM_2$.

Definition 2.3. A *reduction* is a smooth relation $R\colon M_0 \leftarrow M$ which is the graph of a surjective submersion $\rho\colon C \to M_0$ from a submanifold $C \subseteq M$ onto M_0. The transpose $R^\top$ of a reduction R is called *coreduction*. A *fiber of a reduction* $R\colon M_0 \leftarrow M$ is the inverse image of a point of M_0: $R^\top\{p\}$, $p \in M_0$. ♡

Remark 2.2. A fiber of a reduction is a submanifold (Theorem 1.3). If all fibers are connected and $S_0 \subseteq M_0$ is a submanifold, then $S = R^\top \circ S_0 \subseteq M$ is a submanifold. ◇

Remark 2.3. (i) A surjective submersion is a reduction (case $C = M$). (ii) The transpose of the injection of a sumanifold $C \subseteq M$ is a reduction (case $C = M_0$). (iii) A reduction is always the compositions of reductions of type (ii) and (i). ◇

Remark 2.4. The composition of two reductions is a reduction. Reductions are morphisms of a category (Benenti 1983). ◇

2.3.1 Reduction of submanifolds

Let $R\colon M_0 \leftarrow M$ be a reduction, $\rho\colon C \to M_0$ the associated submersion, and $\Lambda \subset M$ a submanifold. We call $\Lambda_0 = R \circ \Lambda$ the *reduced set*. In general, Λ_0 is not a submanifold of M_0. This depends on the way Λ intersects C and the fibers of R.

In order to give an answer to this question, let us consider, for each point $x \in \Lambda \cap C$, the subspace $V_x \subset T_x C$ of the vectors tangent to the fiber of the submersion ρ passing through x (*vertical vectors*). Furthermore, let us denote by $\rho'\colon \Lambda \cap C \to M_0$ the restriction of ρ to the intersection $\Lambda \cap C$.

Theorem 2.1. *Assume that* (i) *Λ and C have a clean intersection,*[2] *and that* (ii) *$\dim(V_x \cap T_x\Lambda)$ does not depend on the point $x \in \Lambda \cap C$. Then ρ' is a subimmersion and for each point $x_0 \in N \cap C$ there exists a neighborhood U of x_0 in $N \cap C$ such that $R \circ U$ is a submanifold of M_0.*

Proof. Let $L\colon A \to B$ be a linear map. We define

$$\begin{cases} \operatorname{rank}(L) \doteq \dim(\text{image of } L), \\ \ker(L) \quad \doteq L^{-1}(0). \end{cases}$$

We know that $\dim\ker(L) + \dim(\text{image of } L) = \dim A$. Then,

$$\operatorname{rank}(L) = \dim A - \dim(\ker(L)).$$

Apply this formula to $L = T_x\rho'\colon T_x(\Lambda \cap C) \to T_{\rho'(x)}M_0$:

$$\operatorname{rank}(T_x\rho') = \dim(T_x(\Lambda \cap C)) - \dim(\ker(T_x\rho')).$$

Observe that $\ker(T_x\rho') = V_x \cap T_x(\Lambda \cap C)$. Then,

$$\operatorname{rank}(T_x\rho') = \dim(T_x(\Lambda \cap C)) - \dim(V_x \cap T_x(\Lambda \cap C))$$

Due to the clean intersection,

$$V_x \cap T_x(N \cap C) = V_x \cap T_x N \cap T_x C = V_x \cap T_x\Lambda,$$

inasmuch as $V_x \subset T_x C$. Then,

$$\operatorname{rank}(T_x\rho') = \dim(T_x(\Lambda \cap C)) - \dim(V_x \cap T_x\Lambda). \tag{2.1}$$

Hence, due to assumption (ii), $\operatorname{rank}(T_x\rho') = \text{constant}$. This proves that ρ' is a subimmersion. Then apply Theorem 1.2. □

---

[2] Recall Definition 1.10 of a clean intersection: $\Lambda \cap C$ is a submanifold and $T_x(\Lambda \cap C) = T_x\Lambda \cap T_x C$ for all $x \in \Lambda \cap C$.

Chapter 3
Symplectic Relations on Symplectic Manifolds

Abstract In this chapter we examine the notion of "relation" in the presence of a symplectic structure. To continue with the study of the relations between symplectic manifolds, we begin with the simplest but fundamental case of linear relations between symplectic vector spaces.

3.1 Linear symplectic relations

3.1.1 Symplectic vector spaces

A *symplectic vector space* is a pair (E, ω) where E is a real even-dimensional vector space and $\omega\colon E \times E \to \mathbb{R}$ is a real-valued bilinear, nondegenerate, and skew-symmetric form,

$$\begin{cases} \omega(u,v) = -\,\omega(v,u), \\ \omega(u,v) = 0 \ \text{ for all } \ v \in E \quad \Longrightarrow \quad u = 0. \end{cases}$$

A linear map $\varphi\colon E_1 \to E_2$ between two symplectic spaces (E_1, ω_1) and (E_2, ω_2) is *symplectic* if

$$\omega_2\left(\varphi(\boldsymbol{u}), \varphi(\boldsymbol{v})\right) = \omega_1(\boldsymbol{u}, \boldsymbol{v})$$

for all pairs of vectors $\boldsymbol{u}$, $\boldsymbol{v} \in E_1$. It can be proved that *symplectic spaces and linear symplectic maps form a category.*

3.1.2 The symplectic dual functor

Let (E, ω) be a symplectic vector space. The dual space E^* of E is the space of *covectors* (i.e., of the linear maps $f\colon E \to \mathbb{R}$). The map $\flat\colon E \to E^*\colon u \mapsto u^\flat$ defined by[1]

$$\langle v, u^\flat \rangle = \omega(u, v)$$

is a linear isomorphism, because ω is nondegenerate. Then we can define on E^* the *dual symplectic form*

$$\omega^*(u^\flat, v^\flat) = \omega(u, v).$$

The *dual symplectic space* of (E, ω) is the pair (E^*, ω^*).

If $\omega_{\mathrm{AB}} = \omega(e_{\mathrm{A}}, e_{\mathrm{B}}) \in \mathbb{R}$ are the components of ω in any basis (e_{A}) of E, then the elements of the inverse matrix of $[\omega_{AB}]$, defined by $\omega^{\mathrm{AB}}\omega_{\mathrm{CB}} = \delta^{\mathrm{A}}_{\mathrm{C}}$ are the components of ω^*, and the map $\flat$ corresponds to the operation of lowering the indices, $(v^\flat)_{\mathrm{A}} = v^{\mathrm{B}}\omega_{\mathrm{BA}}$.

The *dual map* of a linear map $\alpha\colon E \to F$ is the linear map $\alpha^*\colon F^* \to E^*$ defined by

$$\langle u, \alpha^*(f) \rangle = \langle \alpha(u), f \rangle.$$

It can be proved that *the operator* * *defined by*

$$\begin{cases} *\colon (E, \omega) \mapsto (E^*, \omega^*), \\ *\colon f \mapsto F^*, \end{cases}$$

is a contravariant functor on the category of symplectic spaces.

3.1.3 The symplectic polar operator

With each subspace A of a vector space E we associate the *dual subspace* $A^\circ \subseteq E^*$ or *polar subspace* defined by

$$A^\circ = \{\alpha \in E^* \ \text{ such that } \ \langle u, \alpha \rangle = 0 \ \text{ for all } \ u \in A\}.$$

The *polar operator* $\circ\colon A \mapsto A^\circ$ satisfies the following fundamental rules:

$$\begin{cases} \dim(A) + \dim(A^\circ) = \dim(E), \\ A^\circ \subset B^\circ \iff B \subset A, \\ (A + B)^\circ = A^\circ \cap B^\circ, \\ A^\circ + B^\circ = (A \cap B)^\circ, \\ A^{\circ\circ} = \iota(A), \end{cases} \tag{3.1}$$

[1] We use the *pairing* $\langle u, f \rangle$ for the evaluation between vectors and covectors.

where $\iota\colon E \to E^{**}$ is the natural isomorphism defined by $\langle \alpha, \iota(v)\rangle = \langle v, \alpha\rangle$, $\alpha \in E^*$, $v \in E$.

If E is endowed with a symplectic structure, then with a subspace A we associate the *symplectic polar subspace*[2] $A^\S \subseteq E$ defined by

$$A^\S = \{v \in E \ \text{ such that } \ \omega(u,v) = 0 \ \text{ for all } \ u \in A\}.$$

The *symplectic polar operator* $\S\colon A \mapsto A^\S$ satisfies rules that are formally similar to (3.1):

$$\begin{cases} \dim(A) + \dim(A^\S) = \dim(E), \\ A^\S \subset B^\S \quad \Longleftrightarrow \quad B \subset A, \\ (A+B)^\S = A^\S \cap B^\S, \\ A^\S + B^\S = (A \cap B)^\S, \\ A^{\S\S} = A. \end{cases} \tag{3.2}$$

The correspondence between (3.1) and (3.2) follows from equation $\flat(A^\S) = A^\circ$. Notice that $A^{\S\circ} = A^{\circ\S}$.

3.1.4 Special subspaces

Definition 3.1. A subspace A of a symplectic space (E, ω) is called

$$\begin{cases} \textit{isotropic} & \text{if } A \subseteq A^\S, \\ \textit{coisotropic} & \text{if } A^\S \subseteq A, \\ \textit{Lagrangian} & \text{if } A^\S = A, \\ \textit{symplectic} & \text{if } A^\S \cap A = 0, \end{cases} \qquad \heartsuit$$

By means of the formulae above we can prove the following properties.[3]

1. If A is isotropic (coisotropic, Lagrangian) then $A^\S$ is coisotropic (isotropic, Lagrangian).
2. If A is isotropic (coisotropic, Lagrangian) then $\dim A \leq \frac{1}{2} \dim E$ ($\geq$, $=$, respectively).
3. A subspace A is isotropic if and only if $\omega(u,v) = 0$ for all $u, v \in A$.
4. A subspace of dimension 1 (codimension 1) is isotropic (coisotropic).
5. A subspace $A \subseteq E$ is coisotropic (isotropic, Lagrangian) if and only if its polar $A^\circ \subseteq E^*$ is isotropic (coisotropic, Lagrangian) in the symplectic dual space (E^*, Ω).

---

[2] Many authors use the term *symplectic orthogonal subspace* and the notation $A^\perp$. This notation may conflict with that used for orthogonal subspaces in Euclidean spaces.

[3] This list is rather long, but very important for the applications that follow.

6. A subspace $A \subseteq E$ contained in (containing) a Lagrangian subspace L is isotropic (coisotropic).

7. A Lagrangian subspace L of a coisotropic subspace C contains the symplectic polar $C^{\S}$.

3.1.5 Linear symplectic relations

Let (A, α) and (B, β) be symplectic vector spaces. On the direct sum $B \oplus A$ a bilinear skew-symmetric form $\beta \ominus \alpha$ is defined by

$$(\beta \ominus \alpha)\left((b, a), (b', a')\right) = \beta(b, b') - \alpha(a, a').$$

This form is non-singular, thus it is a symplectic form. With a linear relation $R \subseteq B \times A$ we associate its *symplectic dual relation* $R^{\S} \subseteq B \times A$, where § is the symplectic polar operator with respect to $\beta \ominus \alpha$, namely,

$$(b, a) \in R^{\S} \iff \beta(b, b') - \alpha(a, a') = 0 \text{ for all } (b', a') \in R. \tag{3.3}$$

According to the terminology used for special subspaces, a linear relation $R \subseteq B \times A$ between symplectic vector spaces is called *Lagrangian* if $R^{\S} = R$, *isotropic* if $R \subseteq R^{\S}$, or *coisotropic* if $R^{\S} \subseteq R$. A Lagrangian linear relation is also called *symplectic* or *canonical.* Note that if R is a linear symplectic relation, then $R^{\top}$ is symplectic.

A basic property of linear relations between symplectic vector spaces is the following *symplectic functorial rule.*

Theorem 3.1. *If $S \circ R$ is the composition of two linear relations between symplectic vector spaces, then*

$$\boxed{(S \circ R)^{\S} = S^{\S} \circ R^{\S}} \tag{3.4}$$

A proof of this fundamental formula is given in Sect. 9.3.[4] By using this formula we can easily prove the following two fundamental statements.

Theorem 3.2. *The composition of two linear symplectic relations is a linear symplectic relation.*[5]

Proof. If $R^{\S} = R$ and $S^{\S} = S$, then $(S \circ R)^{\S} = S^{\S} \circ R^{\S} = S \circ R$. □

Theorem 3.3. *The image $R \circ K$ of an isotropic (coisotropic, Lagrangian) subspace K by a linear symplectic relation R is an isotropic (coisotropic, Lagrangian) subspace.*

---

[4] It is taken from (Benenti 1988) and it is rather cumbersome. Finding a simpler proof is desirable.

[5] Symplectic spaces and symplectic linear relations form a category (Benenti and Tulczyjew 1981).

Proof. Let $R\colon B \leftarrow A$ be a symplectic linear relation and $K \subset A$ an isotropic subspace: $K \subseteq K^{\S}$. Then $R \circ K \subseteq R \circ K^{\S}$, and due to (3.4), $(R \circ K)^{\S} = R \circ K^{\S} \supseteq R \circ K$. This shows that the image $R \circ K$ is isotropic. There is a similar proof for K coisotropic. □

Definition 3.2. A *linear symplectic reduction* is a Lagrangian subspace $R \subseteq B \times A$ that is the graph of a linear surjective map from a subspace $K \subseteq A$ onto B. ♡

Theorem 3.4. *If $R\colon B \leftarrow A$ is a linear symplectic reduction then $K = R^{\top} \circ B$ is a coisotropic subspace, and $R^{\top} \circ \{0\} = K^{\S}$.*

Proof. By the functorial rule (3.4) we have $K^{\S} = (R^{\top} \circ B)^{\S} = R^{\S})^{\top} \circ B = R^{\top} \circ B^{\S} = R^{\top} \circ \{0\} \subset K$. □

3.2 Symplectic manifolds

A *symplectic manifold* is a pair (M, ω) consisting of an even-dimensional manifold M endowed with a *symplectic form* ω (i.e., a nondegenerate closed two-form).

In coordinates (x^{A}), $\mathrm{A} = 1, \ldots, m$, $m = \dim M$, any two-form on M admits the representation

$$\omega = \tfrac{1}{2}\, \omega_{\mathrm{AB}}\, dx^{\mathrm{A}} \wedge dx^{\mathrm{B}},$$

with

$$dx^{\mathrm{A}} \wedge dx^{\mathrm{B}} = dx^{\mathrm{A}} \otimes dx^{\mathrm{B}} - dx^{\mathrm{B}} \otimes dx^{\mathrm{A}}, \quad \omega_{\mathrm{AB}} = \omega(\partial_{\mathrm{A}}, \partial_{\mathrm{B}}).$$

Here, ∂_{A} denotes the partial derivative $\partial/\partial x^{\mathrm{A}}$ interpreted as a vector field. The components $\omega_{AB}(\underline{x})$ form a skew-symmetric $m \times m$ matrix, $[\omega_{\mathrm{AB}}]$, $\omega_{\mathrm{AB}} = -\,\omega_{\mathrm{BA}}$. A two-form is *nondegenerate* if $\omega(u, v) = 0$ for all vectors v implies $u = 0$. This is equivalent to

$$\det[\omega_{\mathrm{AB}}] \neq 0.$$

This shows that the dimension of a symplectic manifold is even. A two-form is *closed* if $d\omega = 0$. This is equivalent to $\partial_{\mathrm{A}} \omega_{\mathrm{BC}}\, dx^{\mathrm{A}} \wedge dx^{\mathrm{B}} \wedge dx^{\mathrm{C}} = 0$; that is to

$$\partial_{\{\mathrm{A}} \omega_{\mathrm{BC}\}} = 0,$$

where $\{\cdots\}$ denotes the sum over the cyclic permutations of the indices.

A *symplectic map* between two symplectic manifolds (M_1, ω_1) and (M_2, ω_2) is a smooth map $\varphi\colon M_1 \to M_2$ that "preserves" the symplectic forms, that is such that

$$\varphi^* \omega_2 = \omega_1.$$

It can be shown that *symplectic manifolds and symplectic maps are objects and morphisms of a category.* Isomorphisms and automorphisms in this category are called *symplectomorphisms* and *canonical transformations.*

It is well-known that a symplectic form ω gives rise to two basic operations:

• An $\mathbb{R}$-linear map from the space $\mathscr{F}(M)$ of smooth real-valued functions on M to the space $\mathscr{X}(M)$ of smooth vector fields on M,

$$\mathscr{F}(M) \to \mathscr{X}(M) \colon H \mapsto X_H,$$

defined by equation

$$\boxed{i_{X_H}\omega = -\,dH} \tag{3.5}$$

where i_* is the interior product (Sect. 1.16). The vector field X_H is called the *Hamiltonian vector field* generated by the *Hamiltonian* H.

• A binary internal operation $\{F, G\}$, called the *Poisson bracket*, on the space $\mathscr{F}(M)$ is defined by equation

$$\boxed{\{F, G\} = \omega(X_F, X_G)} \tag{3.6}$$

equivalent to equation

$$\boxed{\{F, G\} = i_{X_G} i_{X_F}\omega = X_F G = \langle X_F, dG\rangle} \tag{3.7}$$

The Poisson bracket satisfies the following properties.

$$\begin{aligned}
&\{F, G\} = -\,\{G, F\} \\
&\{a\,F + b\,G, H\} = a\,\{F, H\} + b\,\{G, H\} && a, b \in \mathbb{R}, \\
&\{F, \{G, H\}\} + \{G, \{H, F\}\} + \{H, \{F, G\}\} = 0 && \text{(Jacobi identity)}, \\
&\{F, GH\} = \{F, G\}\,H + \{F, H\}\,G && \text{(Leibniz rule)}, \\
&\{F, G\} = 0, \ \text{ for all } F \quad \Longrightarrow \quad dG = 0 && \text{(regularity)}.
\end{aligned} \tag{3.8}$$

The first three properties show that the space $\mathscr{F}(M)$ endowed with the Poisson bracket is a Lie-algebra. Because of the Leibniz rule (and the $\mathbb{R}$-linearity) the Posisson bracket is a *biderivation* (on functions).

In components with respect to any coordinate system (x^{A}), Eq. (3.5) reads

$$X^{\mathrm{A}}\omega_{\mathrm{AB}} = -\,\partial_{\mathrm{B}} H.$$

If we introduce the inverse matrix $[\omega^{\mathrm{AB}}]$ of the matrix components $[\omega_{\mathrm{AB}}]$, defined by

$$\omega^{\mathrm{AB}}\,\omega_{\mathrm{CB}} = \delta^{\mathrm{A}}_{\mathrm{C}},$$

then we get the explicit definition of the components of X_H,

$$X^A = -\omega^{\mathrm{AB}}\,\partial_{\mathrm{B}}H = \partial_{\mathrm{B}}H\,\omega^{\mathrm{BA}},$$

and of the Poisson bracket,

$$\boxed{\{F, G\} = \omega^{\mathrm{AB}}\,\partial_{\mathrm{A}}F\,\partial_{\mathrm{B}}G} \tag{3.9}$$

Remark 3.1. It can be shown that the Jacobi identity is equivalent to $d\omega = 0$. ◇

Two functions are said to be *in involution* if $\{F, G\} = 0$. Due to the regularity property, if a function is in involution with all other functions, then it is constant on the connected components of M.

A manifold endowed with a bracket on functions satisfying conditions (3.8), except the regularity, is called a *Poisson manifold* (for further information and references see, for instance, (Weinstein 1998), (Libermann and Marle 1987), and (Vaisman 1994).

A remarkable property, relating the Poisson bracket of functions to the Lie bracket of Hamiltonian vector fields, is expressed by the formula

$$\boxed{[X_F, X_G] = X_{\{F,G\}}} \tag{3.10}$$

which shows the following.

Theorem 3.5. *The Lie bracket $[X_F, X_G]$ of two Hamiltonian vector fields is the Hamiltonian vector field generated by the Poisson bracket $\{F, G\}$.*

This means that the map $H \mapsto X_H$ is a Lie-algebra homomorphism. An equivalent form of (3.10) is

$$\boxed{i_{[X_F, X_G]}\omega = -\,d\{F, G\}} \tag{3.11}$$

3.3 Special submanifolds

For each point $p \in M$ of a symplectic manifold (M, ω), the tangent space T_pM is a symplectic vector space: the symplectic two-form is given by the restriction ω_p of ω to the vectors of T_pM. Then the notion of "special subspaces" given in Sect. 3.1.4 can be extended to that of "special submanifolds".

Let K be a submanifold of a symplectic manifold (M_{2n}, ω). For each point $p \in K$ we define

$$T_p^{\S}K = \{v \in T_pM \text{ such that } p \in K \text{ and } \omega(u, v) = 0 \text{ for all } u \in T_pK\}.$$

We consider the set TK of all vectors tangent to K and the set

$$T^{\S}K = \bigcup_{p\in K} T_p^{\S}K.$$

Both TK and $T^{\S}K$ are submanifolds of TM. If $\dim K = k$, then

$$\dim(TK) = 2k, \ \ \dim(T^{\S}K) = k + (2n - k) = 2n.$$

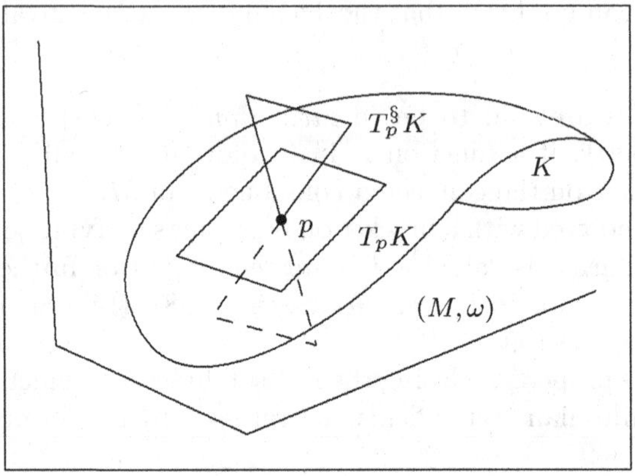

Fig. 3.1 Submanifold K of a symplectic manifold (M, ω)

Definition 3.3. A submanifold K is

$$\begin{cases} \textit{isotropic} & \text{if} \ \ TK \subseteq T^{\S}K, \\ \textit{coisotropic} & \text{if} \ \ T^{\S}K \subseteq TK, \\ \textit{Lagrangian} & \text{if} \ \ T^{\S}K = TK. \end{cases} \qquad \heartsuit$$

In these three cases we have respectively,

$$\dim K \leq n, \ \ \dim K \geq n, \ \ \dim K = n.$$

Note that a Lagrangian submanifold is simultaneously coisotropic and isotropic, and that it is an isotropic submanifold of maximal dimension as well as a coisotropic submanifold of minimal dimension.

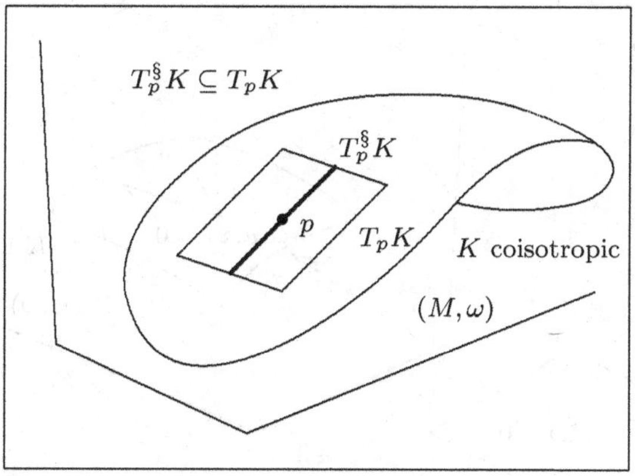

Fig. 3.2 Coisotropic submanifold

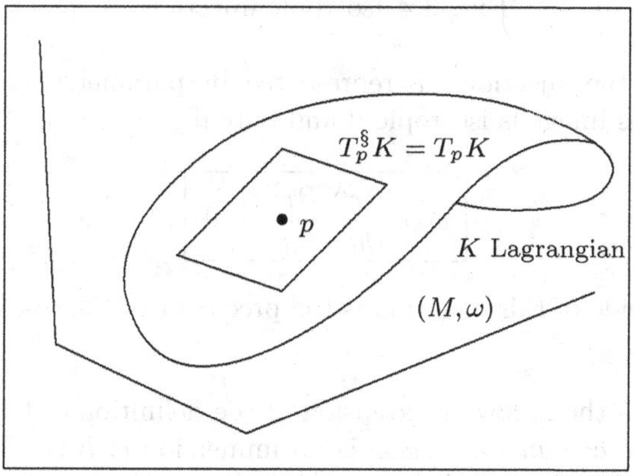

Fig. 3.3 Lagrangian submanifold

As shown by the following theorems, the isotropy of a submanifold is characterized by means of the symplectic form, and the coisotropy is characterized by means of the Poisson bracket.

Theorem 3.6. *A submanifold K is isotropic*
(i) *if and only if $\omega|K = 0$; that is $\omega(u, v) = 0$ for all $u, v \in TK$,*

or

(ii) *if and only if $\iota^*\omega = 0$, where $\iota\colon I \to M$ is any injection (injective immersion) of a manifold I into M with image $\iota(I) = K$.*

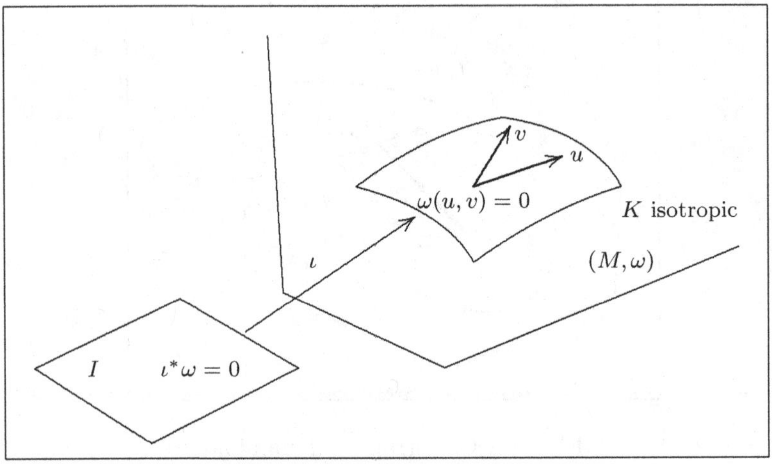

Fig. 3.4 Isotropic injection

Remark 3.2. If the injection ι is represented by parametric equations $x^{\mathrm{A}} = x^{\mathrm{A}}(\kappa^{\alpha})$, then its image is isotropic if and only if

$$\boxed{\omega_{\mathrm{AB}} \frac{\partial x^{\mathrm{A}}}{\partial \kappa^{\alpha}} \frac{\partial x^{\mathrm{B}}}{\partial \kappa^{\beta}} = 0} \tag{3.12}$$

The left-hand side of this equation is the precursor of the so-called *Lagrange bracket.* ◇

This suggests the following extension of the definition of Lagrangian submanifold: a *Lagrangian immersion* is an immersion $\iota\colon K \to M$ into a symplectic manifold such that $\dim K = \frac{1}{2} \dim M$ and $\iota^*\omega = 0$. An *immersed Lagrangian submanifold* is the image $\Lambda = \iota(K)$ of a Lagrangian immersion. If the immersion is an embedding, then we have a Lagrangian submanifold in the ordinary sense.[6]

Theorem 3.7. *A submanifold K is coisotropic if and only if*

$$\{F, G\}|K = 0$$

for all functions $F, G \in \mathscr{F}(M)$ whose restrictions $F|K$ and $G|K$ to K are constant.

Proof. Take any point $x \in M$. The space T_xK is coisotropic in the tangent symplectic space (T_xM, ω_x) if and only if the polar space $T_x^{\circ}K$ is isotropic

[6] For a detailed discussion and further references on Lagrangian immersions and Lagrangian embeddings, as well as for special submanifolds of symplectic manifolds, see, for example e.g. (Marmo et al. 1990).

in the dual vector space (T_x^*M, Ω_x),[7] see Sect. 3.1.4. Observe that T_x^*M is spanned by the differentials d_xF, at the point x, of the functions on M and, in particular, that $T_x^\circ M$ is spanned by the differentials d_xF of the functions constant on K, being $\langle v, d_xF\rangle = 0$ for all $v \in T_xK$. Hence, $T_x^\circ M$ is isotropic in (T_x^*M, Ω_x) if and only if $\Omega_x(d_xF, d_xG) = 0$ for all pairs (F, G) of such functions. But, by definition of the Poisson bracket, the last equation is equivalent to $\{F, G\}_x = 0$. □

As corollaries of this theorem we the following theorems.

Theorem 3.8. *If K is defined by equations $K^a = 0$, then it is coisotropic if and only if $\{K^a, K^b\}|K = 0$.*

Theorem 3.9. *A submanifold of dimension* 1 *is isotropic. A submanifold of codimension* 1 *is coisotropic.*

Theorem 3.10. *On a symplectic manifold of dimension* 2n, *the maximal number of independent functions in involution indexIdependent!functions in involution is n (i.e., if $n + k$ functions are in involution then they are necessarily dependent).*

3.4 Characteristic foliation of a coisotropic submanifold

If $C_m \subseteq M_{2n}$ is a coisotropic submanifold, then $T^\S C$ is a subbundle of TC, whose fibers have dimension $r = 2n - m$, equal to the codimension of C. In other words, $T^\S C$ is a *regular distribution* on C of rank $r = 2n - m$, which we call the *characteristic distribution* of C and denote by Γ_C. Note that it is an isotropic distribution.

A *characteristic vector field* of C is a vector field X on M, tangent to C and such that its image is contained in the characteristic distribution,

$$X(p) \in T_p^\S C, \quad \text{for all } p \in C.$$

This is equivalent to equation $\omega(X, Z) = 0$ or $\langle Z, i_X\omega\rangle = 0$ for all vectors Z tangent to C.

The characteristic vector fields form a linear subspace of $\mathscr{X}(M)$. As we show below–Theorem 3.12 and Remark 3.4–this subspace is also a Lie subalgebra.

Theorem 3.11. (i) *A Hamiltonian vector field X_H is a characteristic vector of a coisotropic submanifold C if and only if H is constant on C.* (ii) *A Hamiltonian vector field X_H is tangent to a coisotropic submanifold C if and only if H is constant on the characteristics of C.*

[7] In accordance with a notation adopted below, by T_x^*M we actually mean the space $(T_xM)^*$ dual of T_xM.

Proof. By applying the equality $\omega(X_H, v) = \langle v, i_{X_H}\omega\rangle = -\langle v, dH\rangle$ for all $v \in TC$ and for all $v \in T^{\S}C$ we get items (i) and (ii), respectively. □

Now we can state two fundamental geometrical properties of the coisotropic submanifolds.

Theorem 3.12. *The characteristic distribution of a coisotropic submanifold is completely integrable.*

This means that for each point $p \in C$ there exists an *integral manifold* of Γ_C; that is a submanifold of dimension $r = \mathrm{codim}(C)$ containing p and tangent to Γ_C. The integral manifolds of Γ_C are called *characteristics* of C. A maximal connected integral manifold is called a *maximal characteristic* of C. Thus, any coisotropic submanifold admits a *characteristic foliation* made of maximal characteristics.

Proof. Let the submanifold C be (locally) described by $r = 2n - m$ independent equations $C^a = 0$. Because of the coisotropy, the functions C^a are in involution (at least on C), $\{C^a, C^b\}|C = 0$, Theorem 3.8. It follows that the Hamiltonian vector fields X_a generated by the functions C^a commute, $[X_a, X_b]|C = 0$, formula (3.10). The differentials dC^a are pointwise linearly independent, thus these vector fields are pointwise independent characteristic vector fields of C, Theorem 3.11. Thus, they span the characteristic distribution. The corresponding (local) flows φ^a_t commute. Thus, starting from any fixed point $x_0 \in C$, the set of points $x \in C$ such that $x = \phi^1_{t_1} \circ \cdots \circ \varphi^r_{t_r}(x_0)$ defines a submanifold of dimension r which is tangent at each point to the vector fields X_a. Hence, this submanifold is an integral manifold of the characteristic distribution of C. □

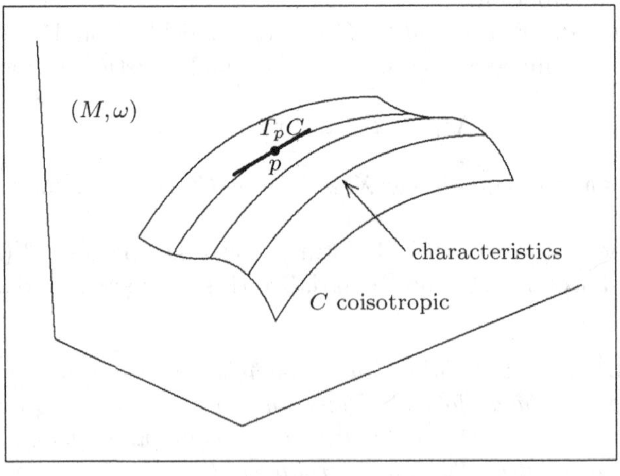

Fig. 3.5 Characteristic foliation of a coisotropic submanifold

Theorem 3.13. *A Lagrangian submanifold Λ contained in a coisotropic submanifold C is the union of characteristics of C.*

This property is known as the *absorption principle* (Vinogradov and Kuperschmidt 1977).

Proof. It is a consequence of a property of linear symplectic algebra: if $L \subseteq K \subseteq E$, where E is a symplectic space, K is a coisotropic subspace and L is Lagrangian, then $K^{\S} \subseteq L^{\S} = L$. □

Remark 3.3. (i) The dimension of the characteristics is equal to the codimension of C. The characteristics are isotropic submanifolds.

(ii) Any submanifold of codimension 1 is coisotropic; hence, its characteristics are (one-dimensional) curves.

(iii) If C is Lagrangian then the maximal characteristics coincide with the connected components of C. ◇

Remark 3.4. In Sect. 9.1 we apply Theorem 3.12 for proving the Frobenius theorem concerning the complete integrability of regular distributions. In reverse, we can use the Frobenius theorem to prove Theorem 3.12. We show that the characteristic vector fields form a Lie-subalgebra of $\mathscr{X}(M)$. (i) The Lie bracket of two vector fields tangent to a submanifold C is tangent to C (this is a general property of the Lie bracket of vector fields). (ii) The intrinsic definition of the differential of a two-form is expressed by the following formula.

$$\begin{aligned} d\omega(X,Y,Z) &= d_X\,(\omega(Y,Z)) + d_Y\,(\omega(Z,X)) + d_Z\,(\omega(X,Y)) \\ &\quad - \omega([X,Y],Z) - \omega([Y,Z],X) - \omega([Z,X],Y). \end{aligned}$$

If ω is closed, X and Y are characteristic vectors and Z is tangent to C, then we get $0 = \omega([X,Y],Z)$ for each vector Z tangent to C. This shows that the Lie bracket of two characteristic vector fields is a characteristic vector field. ◇

3.5 Symplectic relations

Let (M_1, ω_1) and (M_2, ω_2) be two symplectic manifolds. The product $M_2 \times M_1$ is then endowed with four symplectic forms,

$$\pm\, \mathrm{pr}_2^* \omega_2 \pm \mathrm{pr}_1^* \omega_1,$$

where $\mathrm{pr}_1\colon M_2 \times M_1 \to M_1$ and $\mathrm{pr}_2\colon M_2 \times M_1 \to M_2$ are the canonical projections. However, one of them turns out to be of particular interest.

Definition 3.4. A *symplectic relation* (also called *canonical relation*)

$$R\colon (M_2, \omega_2) \leftarrow (M_1, \omega_1)$$

from a symplectic manifold (M_1, ω_1) to a symplectic manifold (M_2, ω_2) is a Lagrangian submanifold of the symplectic manifold $(M_2 \times M_1, \omega_2 \ominus \omega_1)$, where[8]

$$\omega_2 \ominus \omega_1 = \mathrm{pr}_2^* \omega_2 - \mathrm{pr}_1^* \omega_1. \tag{3.13}$$

The symplectic manifold $(M_2 \times M_1, \omega_2 \ominus \omega_1)$ is also denoted by $(M_2, \omega_2) \times (M_1, -\omega_1)$. ♡

This definition is suggested by the following property.

Theorem 3.14. *A diffeomorphism $\varphi\colon M_1 \to M_2$ between two symplectic manifolds is a symplectomorphism; that is it preserves the symplectic forms,*

$$\varphi^* \omega_2 = \omega_1,$$

if and only if its graph $R \subset M_2 \times M_1$ is a Lagrangian submanifold of $(M_2 \times M_1, \omega_2 \ominus \omega_1)$.[9]

Proof. The graph R is the image set of the injective map

$$\iota = (\varphi, \mathrm{id}_{M_1})\colon M_1 \to M_2 \times M_1 : a \mapsto (\varphi(a), a)\,.$$

Since $\mathrm{pr}_1 \circ \iota = \mathrm{id}_{M_1}$ and $\mathrm{pr}_2 \circ \iota = \varphi$, it follows that

$$\begin{aligned}\iota^*(\omega_2 \ominus \omega_1) &= \iota^*(\mathrm{pr}_2^* \omega_2 - \mathrm{pr}_1^* \omega_1) = \iota^* \mathrm{pr}_2^* \omega_2 - \iota^* \mathrm{pr}_1^* \omega_1 \\ &= (\mathrm{pr}_2 \circ \iota)^* \omega_2 - (\mathrm{pr}_1 \circ \iota)^* \omega_1 = \varphi^* \omega_2 - \omega_1.\end{aligned}$$

Thus, R is isotropic if and only if $\varphi^* \omega_2 = \omega_1$. For a diffeomorphism, $\dim R = \frac{1}{2}(\dim M_2 + \dim M_1)$. □

Remark 3.5. A Lagrangian submanifold Λ of a symplectic manifold (M, ω) can be considered as a symplectic relation, because it is a Lagrangian submanifold of the symplectic manifold $M \times \{0\}$. ◊

In general, symplectic relations do not compose nicely, inasmuch as this may occur for the smooth relations. However, it can be shown[10] that

Theorem 3.15. *Let $R\colon (M_2, \omega_2) \leftarrow (M_1, \omega_1)$ and $S\colon (M_3, \omega_3) \leftarrow (M_2, \omega_2)$ be symplectic relations. Assume that:*

(i) $S \circ R$ is a submanifold,

(ii) $T_{(p_3, p_1)}(S \circ R) = T_{(p_3, p_2)} S \circ T_{(p_2, p_1)} R$ for all (p_3, p_2, p_1) such that $(p_3, p_2) \in S$ and $(p_2, p_1) \in R$, .

Then $S \circ R$ is a symplectic relation.

---

[8] Later on we also use the symplectic form $\omega_2 \oplus \omega_1 = \mathrm{pr}_2^* \omega_2 + \mathrm{pr}_1^* \omega_1$.

[9] (Sniatycki and Tulczyjew 1972), (Tulczyjew 1974, 1977b).

[10] (Sniatycki and Tulczyjew 1972).

3.6 Symplectic reductions

We have already introduced the notion of "reduction", at the level of smooth relations, Definition 2.3, and at the level of linear symplectic reductions, Definition 3.2. Now we extend this notion to the level of smooth symplectic relations.

Definition 3.5. A *symplectic reduction* is a symplectic relation that is a reduction according to Definition 2.3. ♡

Reductions and symplectic reductions are morphisms of categories (Benenti 1983a, b, c). The notion of symplectic reduction plays a fundamental role in the global symplectic formulation of the Cauchy problem for a first-order partial differential equation (Sect. 3.9) and of the Jacobi theorem (Sect. 6.6).

Theorem 3.16. *Let* $R\colon M_0 \leftarrow M$ *be a symplectic reduction. Then:*

(i) *The inverse image* $R^\top \circ N \subseteq M$ *of a coisotropic (isotropic, Lagrangian) submanifold* $N \subseteq M_0$ *is a coisotropic (isotropic, Lagrangian) submanifold.*

(ii) $C = R^\top \circ M_0$ *is a coisotropic submanifold.*

(iii) *A fiber* $R^\top \circ \{p_0\}$, $p_0 \in M_0$, *is an integral manifold of the characteristic distribution of* C.

Proof. The inverse image of a submanifold by a submersion is a submanifold. A reduction R is the graph of a (surjective) submersion $\rho\colon C \to M_0$. Thus, the inverse image $R^\top \circ N$ of any submanifold $N \subseteq M_0$ (in particular of a point) is a submanifold of $C = R^\top \circ M_0$. Let $(p_0, p) \in R$. Then $T_{(p_0,p)}R \subset T_{p_0}M_0 \times T_pM$ is a linear symplectic relation; that is a Lagrangian subspace,

$$(T_{(p_0,p)}R)^\S = T_{(p_0,p)}R.$$

Because of the definition of submersion, this linear relation is the graph of a surjective linear map. Thus, $T_{(p_0,p)}R$ is a linear symplectic reduction, with inverse image T_pC.

$$(T_{(p_0,p)}R)^\top \circ T_{p_0}M_0 = T_pC.$$

Due to Theorem 3.3, items (i) and (iii) are proved (note that M_0 is coisotropic). Moreover,

$$(T_{(p_0,p)}R)^\top \circ \{0\} = T_p^\S C, \quad 0 \in T_{p_0}M_0.$$

Let us consider the fiber $I_{p_0} = R^\top \circ \{p_0\}$. We have

$$\begin{aligned} T_pI_{p_0} &= \{v \in T_pC \text{ such that } T\rho(v) = 0\} \\ &= \{v \in T_pC \text{ such that } T_{(p,p_0)}R \circ \{v\} = 0\} \\ &= (T_{(p_0,p)}R)^\top \circ \{0\} = T_p^\S C. \end{aligned}$$

This shows that the tangent space of a fiber at a point p coincides with the tangent space of the characteristic containing that point. This proves item (iii). □

3.7 Reduction of Lagrangian submanifolds

The operation considered in Theorem 3.16 is called *coreduction of a submanifold.* A coreduction preserves the submanifold structure and the symplectic type. On the contrary, the operation of *reduction of a submanifold* is more delicate and requires a closer analysis.

Here we limit ourselves to the most interesting case: the symplectic reduction of a Lagrangian submanifold. We have at our disposal Theorem 2.1, that we have to adapt to the present case: (i) the vertical spaces $V_x \subset T_x C$ tangent to the fibers of the submersion ρ now coincide with $T_x^\S C$ and (ii) $T_x \Lambda = T_x^\S \Lambda$, because Λ is Lagrangian. Then formula (2.1), $\operatorname{rank}(T_x \rho') = \dim(T_x(\Lambda \cap C)) - \dim(V_x \cap T_x \Lambda)$, where ρ' is the restriction of ρ to $\Lambda \cap C$, gives

$$\begin{aligned}\operatorname{rank}(T_x \rho') &= \dim(T_x(\Lambda \cap C)) - \dim(T_x^\S C \cap T_x \Lambda)\\ &= \dim(T_x(\Lambda \cap C)) - \dim(T_x^\S C \cap T_x^\S \Lambda) = \cdots\end{aligned}$$

because $T_x \Lambda = T_x^\S \Lambda$. Now, recall formulae (3.2):

$$\begin{aligned}&= \dim(T_x(\Lambda \cap C)) - \dim(T_x C + T_x \Lambda)^\S\\ &= \dim(T_x(\Lambda \cap C)) - \operatorname{codim}(T_x C + T_x \Lambda)\\ &= \dim(T_x(\Lambda \cap C)) + \dim(T_x C + T_x \Lambda) - \dim M = \cdots .\end{aligned}$$

Due to the clean intersection and the Grassmann formula,

$$\begin{aligned}\cdots &= \dim(T_x \Lambda \cap T_x C) + \dim(T_x C + T_x \Lambda) - \dim M\\ &= \dim T_x \Lambda + \dim T_x C - \dim M\\ &= \dim \Lambda + \dim C - \dim M.\\ &= n + \dim C - 2n = n - c,\end{aligned}$$

where $n = \dim(\Lambda) = \frac{1}{2} \dim M$ and $c = \operatorname{codim} C$. The result of this calculation,

$$\operatorname{rank}(T_x \rho') = n - c = \text{constant}, \tag{3.14}$$

not only shows that ρ' is a subimmersion but also highlights two facts:

(1) In the present case, item (ii) in Theorem 2.1 is redundant.

(2) Since the dimension of M_0 is $2(n-c)$, the dimension of the image of $T_x\rho'$, is just $\frac{1}{2}\dim M_0$, which is the dimension of any Lagrangian submanifold of M_0. Hence, due to the property expressed in Theorem 3.3 at the level of linear symplectic relations, we conclude the following.

Theorem 3.17. *Let* $R\colon M_0 \leftarrow M$ *be a symplectic reduction and* $\Lambda \subset M$ *a Lagrangian submanifold. If* Λ *and* $C = R^\top \circ M_0$ *have a clean intersection then for each* $x_0 \in \Lambda \cap C$ *there exists a neighborhood* U *of* x_0 *in* $\Lambda \cap C$ *such that* $R \circ U$ *is a Lagrangian submanifold of* M_0.[11]

But we can say something more if we consider the special case of transverse intersection of Λ and C, Definition 1.11:

$$T_x\Lambda + T_xC = T_xM \ \text{ for all } x \in \Lambda \cap C. \tag{3.15}$$

Remark 3.6. Since $T_x^\S\Lambda = T_x\Lambda$, this condition is equivalent to

$$T_x\Lambda \cap T_x^\S C = 0. \quad \diamondsuit \tag{3.16}$$

Recall that a transverse intersection is clean, Theorem 1.8, $T_x(\Lambda \cap C) = T_x\Lambda \cap T_xC$. Then,

$$T_x^\S(\Lambda \cap C) = T_x^\S\Lambda + T_x^\S C = T_x\Lambda + T_x^\S C,$$

and, due to (3.16),

$$\dim(T_x^\S(\Lambda \cap C)) = \dim(T_x\Lambda) + \dim(T_x^\S C).$$

This shows that

$$\text{codim }(\Lambda \cap C) = n + \text{codim } C = n + c,$$

that is, due also to (3.14),

$$\dim(\Lambda \cap C) = n - c = \text{rank}\,(T_x\rho').$$

This means that the dimension $\rho'\colon \Lambda \cap C \to M_0$ is an immersion, and from Theorem 3.17 we get the following.

Theorem 3.18. *If* Λ *and* $C = R^\top \circ M_0$ *have a transverse intersection then* $\Lambda_0 = R \circ \Lambda$ *is an immersed Lagrangian submanifold of* M_0.

If the intersection of Λ and $C = R^\top \circ M_0$ is not clean, then it may happen that $\Lambda_0 = R \circ \Lambda$ is not a submanifold. As we show, mainly in Chaps. 5 and 6, this circumstance occurs in fact in many practical applications. We are thus led to introduce the notion of *Lagrangian set*.

---

[11] (Weinstein 1977).

Definition 3.6. A *Lagrangian set* is the symplectic reduction $\Lambda_0 = R \circ \Lambda$ of a smooth Lagrangian submanifold Λ.[12] ♡

3.8 Symplectic relations generated by a coisotropic submanifold

Let $C_m \subseteq M_{2n}$ be a coisotropic submanifold. Let us denote by M_C the *reduced set* of M by C (i.e., the set of the maximal connected characteristics of C). A coisotropic submanifold $C \subseteq M$ generates two relations:[13]

- the *characteristic relation* $D_C : M \leftarrow M$,
- the *characteristic reduction* $R_C : M_C \leftarrow M$.

D_C is made of pairs of points (p, p') belonging to a same characteristic of C,

$$(p, p') \in D_C \iff \text{there exists } \gamma \in M_C \text{ such that } p, p' \in \gamma.$$

R_C is defined by

$$(\gamma, p) \in R_C \iff p \in \gamma.$$

It follows that

$$\boxed{D_C = R_C^{\top} \circ R_C} \tag{3.17}$$

According to its definition, D_C is a relation from C to C, because it involves points of C only. However, it is convenient to consider D_C as a relation in M. Indeed, by using local coordinates adapted to the characteristics (whose existence is due to the local Frobenius theorem) it can be proved that it is (locally) a $2n$-dimensional submanifold of $M \times M$ and moreover, that it is a Lagrangian submanifold with respect to the symplectic form $\omega \ominus \omega$.[14]

In general, the reduced set M_C is not a smooth manifold, so that R_C is not a smooth relation. However,

Theorem 3.19. *If the reduced set M_C has a differentiable structure such that the canonical projection*

$$\rho \colon C \to M_C : p \mapsto \gamma \ \textit{ such that } \ p \in \gamma$$

is a surjective submersion, then:

(i) *There is a unique reduced symplectic form ω_C such that* [15]

---

[12] One can extend this definition by considering the composition $\Lambda_0 = R \circ \Lambda$, where R is a symplectic relation. Alan Weinstein pointed out (private communication) the possibility of other more general definitions. However, the definition adopted here is sufficient for the purposes of this book.

[13] (Tulczyjew 1975).

[14] See, for instance, (Benenti and Tulczyjew 1980).

[15] (Lichnerowicz 1975) (Weinstein 1977).

$$\omega|C = \rho^*(\omega_C). \tag{3.18}$$

(ii) *With respect to this symplectic form $R_C \subset M_C \times M$ is a symplectic relation.*
(iii) $D_C = R_C^\top \circ R_C$ *is a symplectic relation.*

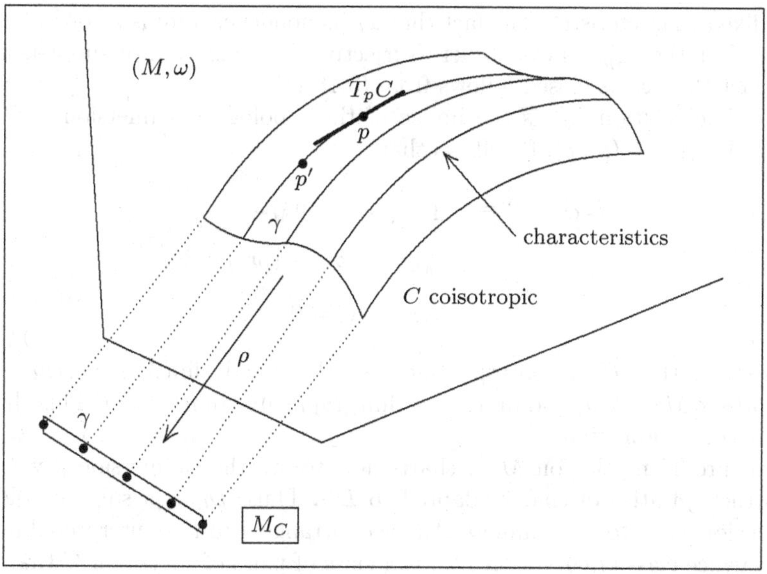

Fig. 3.6 The reduced set M_C

We call (M_C, ω_C) the *reduced symplectic manifold* and R_C the *symplectic reduction* associated with C. Note that

$$\dim M_C = \dim C - \operatorname{codim} C = 2 \dim C - \dim M = 2\,(m-n).$$

Proof. (i) The reduced symplectic form is defined by equation

$$\begin{aligned} \omega_C(v, v') = \omega(w, w'),\ \text{for all } (w, w') \in TC \times_C TC \\ \text{such that } T\rho(w) = v,\ T\rho(w') = w. \end{aligned} \tag{3.19}$$

By the definition of pullback this is equivalent to (3.18). Definition (3.19) does not depend on the choice of the vectors (w, w'). Consider a point $p \in C$ and $w, w', \bar{w}, \bar{w}' \in T_pC$ with $T\rho(w) = v = T\rho(\bar{w})$, $T\rho(w') = v' = T\rho(\bar{w}')$. Then the vectors $w - \bar{w}$ and $w' - \bar{w}'$ project onto the zero-vector; this means that they are tangent to the characteristic at p. Hence, $w-\bar{w},\ w'-\bar{w}' \in T_p^{\S}C$. This subspace is isotropic, thus $\omega(w - \bar{w}, w' - \bar{w}') = 0$ and $\omega(w, w') = \omega(\bar{w}, \bar{w}')$. This proves the independence of Definition (3.19) from the choice of the vectors (w, w') at a fixed point p. Consider the (local) flow $\phi_t\colon C \to C$ of a

Hamiltonian characteristic vector field (generated by a Hamiltonian constant on C). For all admissible $t \in \mathbb{R}$ we have $p_t = \phi_t(p) \in C$ and $\rho \circ \phi_t = \rho$. Then the vectors $w_t = T\phi_t(w) \in T_{p_t}C$ and $w'_t = T\phi_t(w') \in T_{p_t}C$ still project onto the vectors (v, v'). Since ϕ_t is symplectic, $\omega(w, w') = \omega(w_t, w'_t)$. Moreover, we observe that any two points p and p' on the same characteristic can be joined by a finite number of integral curves of characteristic Hamiltonian vector fields. This proves the independence of (3.19) from the choice of the point p on a fixed characteristic. The fact that ω_C is nondegenerate is a consequence of the fact that $T_p\rho$ is everywhere surjective (by definition of submersion). The fact that ω_C is closed follows from (3.18).

(ii) The relation R_C is the image of the topological immersion $\iota\colon C \to M_C \times M\colon p \mapsto (\rho(p), p)$. It follows that

$$
\begin{aligned}
\iota^*(\omega_C \ominus \omega) &= \iota^*(\mathrm{pr}^*_{M_C}\omega_C - \mathrm{pr}^*_M\omega)\\
&= (\mathrm{pr}_{M_C} \circ \iota)^*\omega_C - (\mathrm{pr}_M \circ \iota)^*\omega\\
&= \rho^*\omega_C - \omega|C = 0.
\end{aligned}
$$

This shows that R_C is isotropic. On the other hand, $\dim M_C = 2\,(m-n)$, $\dim(M_C \times M) = 2m$, and $\dim R_C = \dim \mathrm{graph}(\rho) = \dim C = m$. This shows that R_C is Lagrangian.

(iii) From an atlas on M of charts adapted to the submersion ρ we can construct an atlas of charts adapted to D_C. Thus, D_C is a submanifold of dimension equal to $2n = \dim M$. Any vector tangent to D_C, interpreted as an equivalence class of curves on D_C, is a class of pairs of curves on C that are pointwise projected by ρ on a same curve of M_C. This implies that a vector w tangent to D_C at a point (p, p') is a pair of vectors $(v, v') \in T_{(p,p')}C \subset T_{(p,p')}M$ whose images by $T\rho$ coincide, $T\rho(v) = T\rho(v')$. The projections of $w = (v, v')$ onto the first and second factor M are $v \in T_pC \subset T_pM$ and $v' \in T_{p'}C \subset T_{p'}M$, respectively, therefore it follows that for two vectors w and $\bar{w}$ at the same point (p, p'),

$$
\omega \ominus \omega(w, \bar{w}) = \omega(v, \bar{v}) - \omega(v', \bar{v}').
$$

By definition of ω_C, this last difference is equal to

$$
\omega_C(T\rho(v), T\rho(\bar{v})) - \omega_C(T\rho(v') - T\rho(\bar{v}')).
$$

But these two terms are equal, because $T\rho(v') = T\rho(v)$ and $T\rho(\bar{v}') = T\rho(\bar{v})$. Thus, $\omega \ominus \omega(w, \bar{w}) = 0$ and D_C is isotropic. □

The *reduced Poisson bracket* associated with the reduced symplectic form can be directly defined as follows.

Theorem 3.20. *Under the same assumptions as in Theorem 3.19, a Poisson bracket $\{f, g\}_C$ on the reduced manifold M_C is defined by setting*

$$\rho^*\{f,g\}_C = \{F,G\}|C, \tag{3.20}$$

where F and G are any two local extensions to M of the functions $\rho^* f$ and $\rho^* g$, and $\{F,G\}|C$ is the restriction to C of the Poisson bracket $\{F,G\}$ computed on M.

This theorem is based on a lemma.

Lemma 3.1. *Let C be a coisotropic submanifold of a symplectic manifold M.* (i) *If F and G are two functions on M constant on the characteristics of C, then $\{F,G\}$ is constant on the characteristics.* (ii) *Let (F,F') and (G,G') be two pairs of functions on M coinciding on C: $F|C = F'|C$, $G|C = G'|C$. Then their Poisson brackets also coincide on C: $\{F,G\}|C = \{F',G'\}|C$.*

Note that, thanks to this lemma, Eq. (3.20) makes sense: (i) since $\{F,G\}$ is constant on the characteristics, it is reducible to a function on M_C; (ii) $\{f,g\}_C$ does not depend on the extensions F and G.

Proof. If F and G on M are constant on the characteristics of C, then the corresponding Hamiltonian vector fields X_F and X_G are tangent to C – Theorem 3.11, item (ii). Thus, also $[X_F, X_G]$ is tangent to C. But $[X_F, X_G]$ is the Hamiltonian vector field generated by $\{F,G\}$: Theorem 3.5, formula (3.10). Hence, by applying Theorem 3.11, item (ii), once more, $\{F,G\}$ is constant on the characteristics. The proof of item (ii) is similar. □

Proof. PROOF OF THE THEOREM. After the lemma, it is straightforward to prove that the bracket defined by (3.20) fulfills all the characteristic properties of a Poisson bracket. In particular, the regularity condition follows from the fact that, if a function F is constant on the characteristics and if $\{F,G\}_C = 0$ for all functions G constant on the characteristics, then $i_{X_G} dF|C = 0$. But the Hamiltonian fields of the kind X_G generate at each point of C the tangent spaces of C. It follows that $dF|C = 0$ and for the reduced function $df = 0$. □

Remark 3.7. Canonical symplectic structure of the complex projective spaces.[16] The quotient set of $\mathbb{C}^{n+1}$ with respect to the equivalence relation

$$(z_\alpha) \equiv (z'_\alpha) \iff z_\alpha = \lambda\, z'_\alpha, \;\; \lambda \in \mathbb{C} - 0, \;\; \alpha = 0, 1, \ldots, n,$$

is denoted by $\mathbb{C}P^n$. For each $\kappa = 0, 1, \ldots, n$ we define

$$U_\kappa = \{(z_\alpha) \in \mathbb{C}^{n+1};\; z_\kappa \neq 0\}$$

and

$$\phi_\kappa \colon U_\kappa \to \mathbb{C}^n \colon (z_i/z_\kappa), \;\; i \neq \kappa.$$

Each pair (U_κ, ϕ_κ) defines a chart on $\mathbb{C}P^n$ of dimension $2n$. These $n+1$ charts define an analytic atlas on $\mathbb{C}P^n$. On $M = \mathbb{C}^{n+1}$ we consider the canonical symplectic form

---

[16] (Weinstein 1977), (Benenti 1988), (Libermann and Marle 1987).

$$\omega = dy_\alpha \wedge dx^\alpha = -\frac{i}{2}\, dz_\alpha \wedge d\bar{z}_\alpha,$$

where $z_\alpha = x^\alpha + i\, y_\alpha$ and $\bar{z}_\alpha = x^\alpha - i\, y_\alpha$. In M we consider the unitary sphere $C = \mathbb{S}_{2n+1}$, defined by equation

$$\sum_\alpha z_\alpha \bar{z}_\alpha = \sum_\alpha [(x^\alpha)^2 + (y_\alpha)^2] = 1.$$

Being of codimension 1, C is coisotropic. Furthermore, a surjective submersion $\rho\colon C \to \mathbb{C}P^n$ which maps each (z_α) to its equivalence class. It can be proved that

- The characteristics of C are the fibers of ρ.
- The characteristics of C are orbits of the action of $\mathbb{R}$ on M defined by $(t, z_\alpha) \mapsto e^{it}\, z_\alpha$.
- The characteristics of C are maximal circles.

Then, the manifold $\mathbb{C}P^n$ coincides with the reduced manifold M_C and consequently, it is endowed with a reduced symplectic form ω_C such that $\rho^*\omega_C = \omega|C$. If we take for instance the coordinates (ξ^i, η_j) associated with the chart (U_0, ϕ_0), which are defined by the equations

$$\xi^i + i\,\eta_i = \frac{z_i}{z_0},$$

then, taking into account that these equations give the local representation of the submersion ρ, we can find the local representation of the symplectic form ω_C:

$$\omega_C = \gamma^2 \left[d\eta_i \wedge d\xi^i - \gamma^2 \left(\xi^h\, d\xi^h + \eta_h\, d\eta_h\right) \wedge \left(\eta_k\, d\xi^k - \xi^k\, d\eta_k\right)\right] \tag{3.21}$$

where

$$\gamma^{-2} = 1 - \sum_i [(\xi^i)^2 + (\eta_i)^2]. \qquad \diamondsuit \tag{3.22}$$

Remark 3.8. For $n = 1$ we have $C = \mathbb{S}_3$, $\mathbb{C}P^1 = \mathbb{S}_2$, and the surjective submersion $\rho\colon \mathbb{S}_2 \to \mathbb{S}_2$ is the *Hopf fibration.* Eqs. (3.21) and (3.22) now give

$$\omega_C = \frac{1}{(1 + \xi^2 + \eta^2)^2}\, d\eta \wedge d\xi.$$

This shows an interesting fact: ω_C is the area-element of the unitary sphere $\mathbb{S}_2$, because (ξ, η) are stereographic coordinates. These coordinates are not canonical. A canonical coordinate system (a, b), in which $\omega_C = db \wedge da$, is defined by

$$a = \frac{\xi^2 + \eta^2}{1 + \xi^2 + \eta^2}, \quad b = \tfrac{1}{2} \arcsin \frac{\eta}{\sqrt{\xi^2 + \eta^2}}. \qquad \diamondsuit$$

Remark 3.9. The mechanical interest of the complex projective spaces is also due to the fact that the symplectic manifold (M_C, ω_C) describes the isoenergetic Hamiltonian system, with energy $\frac{1}{2}$, of the harmonic oscillator of dimension $n+1$. ◇

Remark 3.10. **The orbit-manifold of the Kepler motions.** The germ of the notion of reduced symplectic manifold dates back to Lagrange, when he studied the *variations of the arbitrary constants* (Souriau 1970) for the planetary motions. In the case of the n-dimensional Kepler motions the phase space is the cotangent bundle $M = T^*Q$ of the configuration manifold $Q = \mathbb{R}^n - 0$. Then, $M = (\mathbb{R}^n - 0) \times \mathbb{R}^n$. Let $\boldsymbol{x}$ be the vector of the Cartesian coordinates of Q and $\boldsymbol{y}$ the vector of the Cartesian coordinates on the fibers $\mathbb{R}^n$. Let us consider on M the natural symplectic form $\omega = dy_i \wedge dx^i$ and the Hamiltonian

$$H(\boldsymbol{x}, \boldsymbol{y}) = \tfrac{1}{2}\,\boldsymbol{y}^2 - \frac{1}{|\boldsymbol{x}|}.$$

For each value of the energy constant E, equation $H = E$ defines a coisotropic submanifold C_E of M. For each dimension n, we have three types of such submanifolds, determined by the three conditions $E > 0$ (hyperbolic case), $E = 0$ (parabolic case), and $E < 0$ (elliptic case).

It is not easy to find the reduced manifold of C_E, that is, the manifold of the orbits.[17] However, Pham Mau Quan (1980) was able to find a trick, valid for any dimension n, that simplifies the matter considerably. He observed that, for $E = -1/2$, the submanifold $C_{-1/2}$, defined by equation

$$|\boldsymbol{x}|\,(1 + \boldsymbol{y}^2) = 2,$$

can be also described by equation

$$K(\boldsymbol{x}, \boldsymbol{y}) = \frac{1}{2}\,(1 + \boldsymbol{y}^2)\,|\boldsymbol{x}|,$$

and that, for $n = 2$, $K(\boldsymbol{x}, \boldsymbol{y})$ is the Hamiltonian of the geodesics on the sphere $\mathbb{S}_2 \subset \mathbb{R}^3$, in the coordinates $\boldsymbol{y}$ corresponding to the stereographic projection with pole $(1, 0, 0)$. Hence, the orbits are in a one-to-one correspondence with the *oriented* maximal circles of the sphere. In turn, these circles are in a one-to-one correspondence with the unit vectors issued from the center of the sphere and thus, with the points of the sphere itself. Consequently, it has been proved that the orbit manifold with negative energy (hence, of all the bounded orbits) of the two-dimensional Kepler problem is isomorphic to the sphere $\mathbb{S}_2$. Inasmuch as $\mathbb{S}_2$ has a natural symplectic form given by its Euclidean area-element, we are induced to think of this form as the reduced symplectic form (up to a factor). It can be proved that it is so. ◇

---

[17] This problem is related to that of the *regularization of the Kepler problem*. See (Moser 1970), (Pham Mau Quan 1980, 1983a, 1983b), (Souriau 1983), and (Cordani 1986, 2003).

3.9 Symplectic background of the Cauchy problem

The following theorem is an important application of the results of the preceding section.

Theorem 3.21. *Let $\Lambda_I \subset M$ be a Lagrangian submanifold having a clean intersection with a coisotropic submanifold $C \subseteq M$. Then there exists a unique connected immersed Lagrangian submanifold Λ contained in C and containing $C \cap \Lambda_I$. This Lagrangian submanifold is defined by the composition formula*

$$\Lambda = D_C \circ \Lambda_I,$$

thus, it is the union of the maximal characteristics of C intersecting Λ_I.

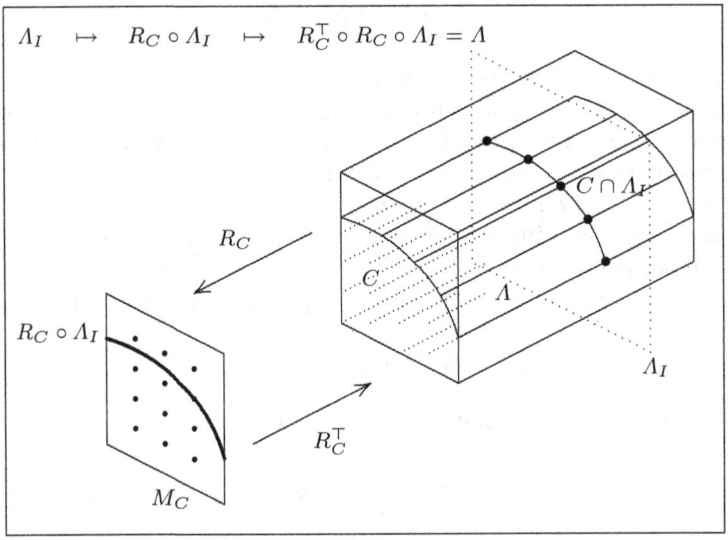

Fig. 3.7 The Cauchy problem

Proof. Due to Theorem 3.17, $R_C \circ \Lambda_I$ is an immersed Lagrangian submanifold of M_C and, due to Theorem 3.16, $\Lambda = R_C^\top \circ R_C \circ \Lambda_I$ is a Lagrangian submanifold of M. The uniqueness follows from the absorption principle. □

Remark 3.11. As we show, when applied to cotangent bundles, this statement can be interpreted as a *symplectic background of the Cauchy problem*: C represents a first-order PDE, Λ_I (or $C \cap \Lambda_I$) the *initial* or *boundary conditions*, and Λ the corresponding solution. However, the case in which the intersection $C \cap \Lambda_I$ is not clean occurs in many interesting applications of this theory. In this case, $R_C \circ \Lambda_I$ and $\Lambda = D_C \circ \Lambda_I$ may not be Lagrangian submanifolds. ♢

Remark 3.12. A remarkable application of the above considerations concerns the composition of symplectic relations.[18] Let $R_1 \subset (M_1 \times M_0, \omega_1 \ominus \omega_0)$ and $R_2 \subset (M_2 \times M_1, \omega_2 \ominus \omega_1)$ be smooth symplectic relations; then the composed relation can be interpreted as a reduced set,

$$R_2 \circ R_1 = R_C \circ (R_2 \times R_1),$$

where $R_2 \times R_2 \subset M = M_2 \times M_1 \times M_1 \times M_0$ is interpreted as a Lagrangian submanifold with respect to the symplectic form

$$\omega = \omega_2 \ominus \omega_1 \oplus \omega_1 \ominus \omega_0,$$

and R_C is the reduction relation generated by the coisotropic submanifold

$$C = M_2 \times \Delta_{M_1} \times M_0,$$

where $\Delta_{M_1} \subset M_1 \times M_1$ is the diagonal. In this case the reduced symplectic manifold M_C is just $(M_2 \times M_0, \omega_2 \ominus \omega_0)$. It follows that if C and $R_2 \times R_1$ have a clean (or transverse) intersection, then the composite relation $R_2 \circ R_1$ is a smooth (possibly immersed) symplectic relation. ◇

3.10 Isomorphism of symplectic reductions

Theorem 3.22. *Assume that a symplectic reduction $R \subset M_0 \times M$ has connected fibers. Then the symplectic reduction R_C associated with the coisotropic submanifold $C = R^T \circ M_0$ is isomorphic to R; that is, the composition*

$$R_C \circ R^{\top} : M_C \leftarrow M_0$$

is the graph of a symplectomorphism $\varphi \colon M_0 \to M_C$ from (M_0, ω_0) to the reduced symplectic manifold (M_C, ω_C).

Proof. It is clear that $\Phi = R_C \circ R^{\top} \subseteq M_C \times M_0$ is a one-to-one smooth relation. In order to prove that it is the graph of a symplectomorphism we have to prove that Φ is a Lagrangian submanifold. For this we consider the relation $R_C \times R$ from $(M, \omega) \times (M, -\omega)$ to $(M_C, \omega_C) \times (M_0, -\omega_0)$. We observe that it is a symplectic reduction with inverse image $C \times C$ and that Φ is the image by R_C of the characteristic relation $D_C \subset C \times C$. Because D_C is a Lagrangian submanifold having a clean intersection with $C \times C$ (inasmuch as it is contained in $C \times C$), it follows that Φ is a Lagrangian submanifold; recall Remark 1.7 and Theorem 3.17. □

Let us repeat the process.

[18] (Weinstein 1977).

1. Consider a symplectic reduction from M to M_0.
2. Take the inverse image of M_0 by R. This is a coisotropic submanifold C of M.
3. Consider the symplectic reduction R_C associated with C and the reduced symplectic manifold M_C.
4. The two symplectic manifolds M and M_C are symplectomorphic.

Chapter 4
Symplectic Relations on Cotangent Bundles

Abstract Any cotangent bundle is endowed with a canonical structure of exact symplectic manifold. Then it becomes "natural" to apply what we have designed for the symplectic manifolds in general in the previous chapter to the special case of cotangent bundles. The material associated with this reduction becomes very rich. Among the various terms that arise spontaneously, the main one is the "generating function" of a Lagrangian submanifold, which is extended to the more general notion of "generating family". This notion is in fact the *fulcrum* around which the entire analysis of the following chapters is built up.

4.1 Cotangent bundles

A *tangent covector* on a manifold Q is a linear map

$$f\colon T_qQ \to \mathbb{R}\colon v \mapsto \langle v, f\rangle.$$

We denote by

$$T_q^*Q$$

the *cotangent space* at the point q (i.e., the dual space of the tangent space T_qQ). The *cotangent bundle* T^*Q of a manifold Q_n is the set of all covectors (i.e., the union of all cotangent spaces). It is a $2n$-dimensional manifold. We denote by

$$\pi_Q\colon T^*Q \to Q$$

the *cotangent fibration* of Q, which maps a covector $f \in T^*Q$ to the point $q \in Q$ where it is applied. We denote by

$$(q^i, p_i)$$

the *canonical coordinates* on T^*Q corresponding to coordinates $q = (q^i)$ $(i = 1, \ldots, n)$ on Q. They are defined as follows. If f is a covector at a point q in the domain of the coordinates, then $q^i(f)$ are the coordinates of q and $p_i(f)$ are the components of the covector in these coordinates, such that for all $v \in T_qQ$,

$$\langle v, f \rangle = \dot{q}^i\, p_i = p_i\, \delta q^i.$$

There are two mechanical interpretations of a cotangent bundle.

(1) If Q is a configuration manifold, then a covector $f \in T^*Q$ represents a *force* and the evaluation $\langle v, f \rangle$ the *virtual work* of the force corresponding to the virtual displacement v (or the *virtual power* if v is interpreted as a *virtual velocity*).

(2) If Q is a configuration manifold, then a covector $p \in T^*Q$ represents an *impulse* and the cotangent bundle T^*Q the *phase space* of the mechanical system.

It is useful to consider the vectors tangent to a cotangent bundle: they form the manifold $T(T^*Q) = TT^*Q$. Natural canonical coordinates (q^i, p_i) on T^*Q generate coordinates

$$(q^i, p_i, \dot{q}^i, \dot{p}_i) = (q^i, p_i, \delta q^i, \delta p_i)$$

on TT^*Q, where $(\dot{q}^i, \dot{p}_i) = (\delta q^i, \delta p_i)$ are the components of the tangent vectors $w \in TT^*Q$.

A vector w tangent to T^*Q is called *vertical* if is tangent to a fiber. A vector is vertical if and only if $T\pi_Q(w) = 0$. The vertical vectors are characterized by equations $\delta q^i = 0$.

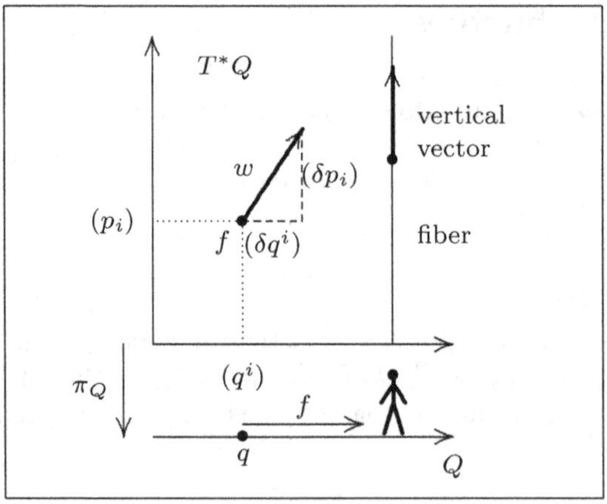

Fig. 4.1 The structure of a cotangent bundle

4.2 The canonical symplectic structure of a cotangent bundle

Definition 4.1. Let Q be a manifold. The *Liouville one-form* or *fundamental form* θ_Q on T^*Q is defined by one of the following three equations:

$$\boxed{\langle w, \theta_Q \rangle = \langle T\pi_Q(w), \tau_{T^*Q}(w) \rangle} \tag{4.1}$$

for all tangent vectors $w \in T(T^*Q)$,

$$\boxed{\sigma^* \theta_Q = \sigma} \tag{4.2}$$

for all one-form σ on Q, and

$$\boxed{\theta_Q = p_i \, dq^i} \tag{4.3}$$

for any choice of the coordinates (q^i) on Q. ♡

Each one of these definitions needs an explanation.

DEFINITION (4.1) – With each vector w tangent to a cotangent bundle T^*Q we can associate "in a natural way" a real number. Let $w \in T_p(T^*Q)$. The point p where w is attached is an element of T^*Q; that is, a covector. As a covector, p is attached at a point $q \in Q$. Let u be the image of the vector w by the tangent map $T\pi_Q \colon TT^*Q \to TQ$ of the cotangent fibration $\pi_Q \colon T^*Q \to Q$. It is a vector $u \in T_qQ$ attached at the same point q of p. Thus, the evaluation $\langle u, p \rangle$ makes sense. This is just the number associated with w. If we consider the map defined in this way,

$$\theta_Q \colon TT^*Q \to \mathbb{R} \colon w \mapsto \langle u, p \rangle, \tag{4.4}$$

we can see that it is linear over the fibers of $T(T^*Q)$, so that it can be interpreted as a one-form over T^*Q, and instead of (4.4) we can write

$$\langle w, \theta_Q \rangle = \langle u, p \rangle. \tag{4.5}$$

By the process illustrated above we have

$$u = T\pi_Q(w), \quad p = \tau_{T^*Q}(w),$$

where

$$\tau_{T^*Q} \colon T(T^*Q) \to T^*Q$$

is the tangent fibration over T^*Q. It follows from (4.5) that the *intrinsic definition* of the Liouville form is just (4.1).

DEFINITION (4.2) – The one-form $\sigma \in \Phi^1(Q)$ is interpreted at the left-hand side as a section $\sigma\colon Q \to T^*Q$: Eq. (4.2) means that the pullback by the map σ of the Liouville form θ_Q, which is a one-form on T^*Q, is a one-form on Q coinciding with σ itself.

DEFINITION (4.3) – In this definition, (q^i, p_i) are the *natural canonical coordinates* corresponding to coordinates (q^i) of Q. Let us consider another system $(\bar{q}^i, \bar{p}_i)$ of natural canonical coordinates. From the transformation law

$$\bar{p}_j = \frac{\partial q^i}{\partial \bar{q}^j}\, p_i,$$

it follows that

$$p_i\, dq^i = p_i \left(\frac{\partial q^i}{\partial \bar{q}^j}\, d\bar{q}^j \right) = \bar{p}_j\, d\bar{q}^j.$$

This shows that Definition (4.3) does not depend on choice of the coordinates.

Now, let us look at the equivalence of the three definitions.

• DEFINITION (4.1) $\iff$ DEFINITION (4.2). This is shown by the following reversible calculation:

$$\begin{aligned}
\langle u, \sigma^*\theta_Q\rangle &= \langle T\sigma(u), \theta_Q\rangle && \text{definition of pullback,}\\
&= \langle T\pi_Q(T\sigma(u)), \tau_{T^*Q}(T\sigma(u))\rangle && \text{equation (4.1),}\\
&= \langle T(\pi_{Q\circ\sigma})(u), \sigma(\tau_Q(u))\rangle && \text{because } \tau_{T^*Q}\circ T\sigma = \sigma\circ\tau_Q,\\
&= \langle u, \sigma(\tau_Q(u))\rangle && \text{because } \pi_Q\circ\sigma = \mathrm{id}_Q,\\
&= \langle u, \sigma\rangle.
\end{aligned}$$

• DEFINITION (4.1) $\iff$ DEFINITION (4.3). Any one-form on T^*Q is of the type $\theta_Q = \theta_i\, dq^i + \theta^i\, dp_i$. Hence, in components, Eq. (4.5) is equivalent to $w^i\theta_i + w_i\theta^i = p_i u^i$. However, $w^i = u^i$ due to the condition $u = T\pi_Q(w)$, and we get $w^i(\theta_i - p_i) + w_i\theta^i = 0$. This holds for all values of the components (w^i, w_i). Hence $\theta^i = 0$ and $\theta_i = p_i$.

Definition 4.2. The differential of the fundamental one-form is the *canonical symplectic form* on T^*Q,

$$\boxed{\omega_Q = d\theta_Q} \tag{4.6}$$

In coordinates,

$$\boxed{\omega_Q = dp_i \wedge dq^i} \tag{4.7}$$

♡

Note that ω_Q is closed, because it is exact. Moreover, it is nonsingular, as can be seen easily from (4.7).

The corresponding expression of the *canonical Poisson bracket* is

$$\boxed{\{F, G\} = \frac{\partial F}{\partial p_i}\frac{\partial G}{\partial q^i} - \frac{\partial F}{\partial q^i}\frac{\partial G}{\partial p_i}} \tag{4.8}$$

and the first-order equations corresponding to a Hamiltonian vector field X_H are the *Hamilton equations*

$$\boxed{\dot{q}^i = \frac{\partial H}{\partial p_i}, \quad \dot{p}_i = -\frac{\partial H}{\partial q^i}} \tag{4.9}$$

There is an important link between closed one-forms and Lagrangian submanifolds.

Theorem 4.1. *The image $\sigma(Q) \subset T^*Q$ of a one-form (interpreted as a section) is a Lagrangian submanifold if and only if σ is closed, $d\sigma = 0$.*

We can give two proofs of this theorem. The first is related to the definition in coordinates (4.7).

Proof. The image $\Lambda = \sigma(Q)$ is a submanifold of dimension $n = \frac{1}{2}\dim T^*Q$. If we consider the canonical symplectic form restricted to Λ, then we get

$$\omega_Q|\Lambda = d\sigma_i \wedge dq^i = \partial_j \sigma_i \, dq^j \wedge dq^i.$$

It follows that $\omega_Q|\Lambda = 0$ (isotropy condition) if and only if $\partial_j\sigma_i = \partial_i\sigma_j$; that is, $d\sigma = 0$. □

The second proof is related to Definition (4.2) of the Liouville form.

Proof. The differential commutes with the pullback, thus we have $\sigma^* d\theta_Q = d\sigma^*\theta_Q = d\sigma$. The section σ is an embedding of Q into T^*Q, therefore its image is an isotropic submanifold if and only if $d\sigma = 0$. Moreover, the dimension of this image is one half of the dimension of T^*Q. □

Remark 4.1. Let $G\colon Q \to \mathbb{R}$ be a smooth function. Its differential dG is an exact, thus closed, one-form. Then its image

$$\Lambda = dG(Q),$$

which is locally described by the n equations

$$p_i = \frac{\partial G}{\partial q^i},$$

is a Lagrangian submanifold. This is the case when the closed one-form σ is exact: $\sigma = dG$. Then G is said to be a *global generating function* of Λ. Of course any other function $G +$ constant is a generating function. ◇

4.3 Basic observables and canonical Poisson bracket on a cotangent bundle

There is a one-to-one correspondence between the vector fields X on a manifold Q and the first-degree homogeneous functions P_X on T^*Q defined by

$$P_X \colon T^*Q \to \mathbb{R} \colon p \mapsto \langle X, p \rangle = X^i(q)\, p_i.$$

Moreover, any function $f \colon Q \to \mathbb{R}$ can be interpreted as a function $f \colon T^*Q \to \mathbb{R}$ constant on the fibers (we use the same symbol for simplicity). We call these functions the *basic observables*. On the basic observables we define an internal operation $\{\cdot\,,\cdot\}$ by setting

$$\begin{cases} \{f, g\} & = 0, \\ \{P_X, f\} & = Xf, \\ \{P_X, P_Y\} & = P_{[X,Y]}. \end{cases} \tag{4.10}$$

We observe that these rules are fulfilled by the Poisson bracket associated with the canonical symplectic form on T^*Q. Conversely, assuming the rules (4.10) as fundamental, we can extend the operation $\{\cdot\,,\cdot\}$ in a unique way to a Poisson bracket on functions over T^*Q. The resulting Poisson bracket coincides with the canonical one. Hence, Eqs. (4.10) characterize the canonical Poisson bracket on a cotangent bundle, and provide a direct definition that avoids the use of the canonical symplectic form.

4.4 One-forms as sections of cotangent bundles

A *one-form* (or *linear differential form*) on a manifold Q can be interpreted in three equivalent ways.

(i) As a map

$$\sigma \colon TQ \to \mathbb{R} \colon v \mapsto \langle v, \sigma \rangle, \tag{4.11}$$

which is linear when restricted to each tangent space T_qQ, $q \in Q$.

(ii) As a *section* of the cotangent bundle, that is, as a map

$$\sigma \colon Q \to T^*Q \colon q \mapsto \sigma(q), \tag{4.12}$$

such that $\sigma(q)$ is a covector in T^*_qQ (in this interpretation, we can say that a one-form is a *field of covectors*.

(iii) As an object locally expressed as a linear combination of the differentials of coordinates

$$\sigma = \sigma_i(q)\, dq^i, \tag{4.13}$$

where σ_i are the *components* of σ. The link between (4.11) and (4.13) is given by $\langle v, \sigma \rangle = \dot{q}^i \sigma_i$. The link between (4.12) and (4.13) is given by $p_i = \sigma_i(q)$.

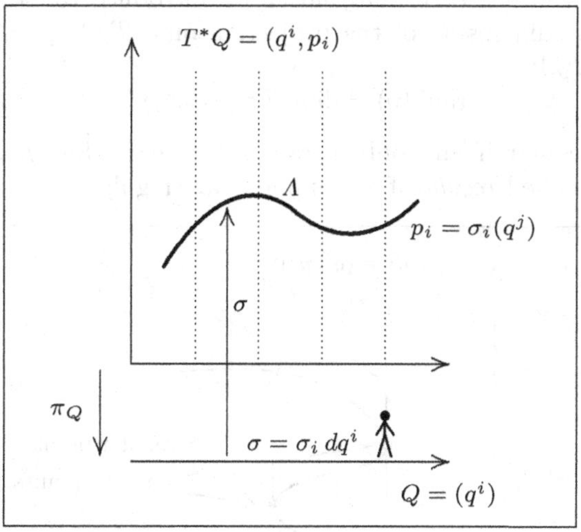

Fig. 4.2 The image of a one-form as a section of a cotangent bundle

4.5 Lagrangian singularities and caustics

Let $\Lambda \subset T^*Q$ be a Lagrangian submanifold of a cotangent bundle. A *regular point* $p \in \Lambda$ is a point where Λ is *transversal to the fibers* , that is, where the tangent space $T_p\Lambda$ is complementary to the space V_p of the vertical vectors at p:

$$T_p\Lambda \cap V_p = 0, \;\; T_p\Lambda + V_p = T_p(T^*Q).$$

Observe that these two conditions, which express the complementarity of the two subspaces $T_p\Lambda$ and V_p, are in fact equivalent, because the subspaces $T_p\Lambda$ and V_p are both Lagrangian subspaces of $T_p(T^*Q)$,

$$\begin{aligned} T_p\Lambda \cap V_p = 0 \iff & (T_p\Lambda \cap V_p)^{\S} = T_p(T^*Q) \\ \iff & (T_p\Lambda)^{\S} + (V_p)^{\S} = T_p(T^*Q) \\ \iff & T_p\Lambda + V_p = T_p(T^*Q). \end{aligned}$$

A nonregular point is called *singular* or *critical.* A singular point of a Lagrangian submanifold is also called *Lagrangian singularity* or *catastrophe.* The set $\Gamma(\Lambda) \subset Q$ of the points of Q on which are based all the singular points is called the *caustic* of Λ.

A point p is regular if and only if the tangent space $T_p\Lambda$ does not contain vertical vectors except the zero-vector. A point is regular if and only if the restriction $\pi\colon \Lambda \to Q$ of the cotangent fibration $\pi_Q\colon T^*Q \to Q$ to Λ is a submersion at p.

To measure the degree of singularity we introduce the *rank* of a point $p \in \Lambda$: it is the dimension of the projection onto T_qQ, $q = \pi_Q(p)$, of the tangent space $T_p\Lambda$:

$$\operatorname{rank}(p) = \dim\left(T\pi_Q(T_p\Lambda)\right).$$

A point p is regular if and only if $\operatorname{rank}(p) = n = \dim Q$. A Lagrangian submanifold is called *regular* if all its points are regular.

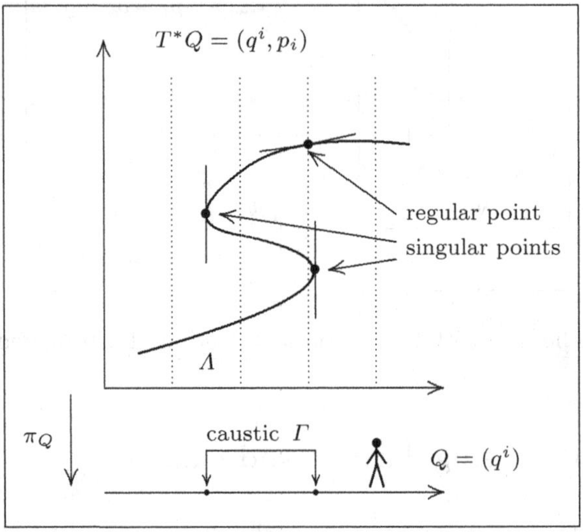

Fig. 4.3 Singular points and caustic

As for any submanifold, a Lagrangian submanifold can be represented, at least locally, by *parametric equations* or by *implicit equations.* For a Lagrangian submanifold there is, however, a third local representation, by means of *generating families.* This is examined in the next section. In this section we consider the particular case of a representation by means of *generating functions.*

4.5.1 Parametric equations

A system of $2n$ parametric equations in n parameters (u^k),

$$\begin{cases} q^i = q^i(u^k), \\ p_i = p_i(u^k), \end{cases} \tag{4.14}$$

represents a *local immersion* $\iota\colon \Lambda \to T^*Q$ of an n-dimensional manifold Λ, with coordinates (u^k), if and only if

$$\operatorname{rank}\left[\frac{\partial q^i}{\partial u^k}\,\middle|\,\frac{\partial p_i}{\partial u^k}\right] = n \qquad (\max). \tag{4.15}$$

Indeed, the tangent map $T\iota$ is represented by equations

$$\begin{cases} \dot{q}^i = \dfrac{\partial q^i}{\partial u^k}\,\dot{u}^k, \\ \dot{p}_i = \dfrac{\partial p_i}{\partial u^k}\,\dot{u}^k, \end{cases}$$

and this is a linear injective map at each point if and only if the rank of the matrix (4.15) is maximal ($\dot{q}^i = 0$ and $\dot{p}_i = 0$ must imply $\dot{u}^k = 0$). The submanifold Λ is Lagrangian if and only if

$$\frac{\partial q^i}{\partial u^k}\frac{\partial p_i}{\partial u^j} - \frac{\partial q^i}{\partial u^j}\frac{\partial p_i}{\partial u^k} = 0. \tag{4.16}$$

Indeed, these equations are equivalent to the isotropy condition $\iota^* d\theta_Q = 0$. The left-hand sides of (4.16) are called *Lagrangian brackets*. In this representation, the rank of a point is given by

$$\operatorname{rank}(p) = \operatorname{rank}\left[\frac{\partial q^i}{\partial u^k}\right]_p,$$

where the evaluation at the point p of the matrix means its evaluation at those values of the parameters u^k corresponding to the point p. Indeed, the tangent map $T\pi\colon T\Lambda \to TQ$ is described by equations $\dot{q}^i = \partial q^i/\partial u^k\,\dot{u}^k$. Hence,

$$p \text{ regular} \iff \det\left[\frac{\partial q^i}{\partial u^k}\right]_p \neq 0.$$

This proves the following.

Theorem 4.2. *If a Lagrangian submanifold is represented by parametric equations* (4.14) *then the set of its singular points is determined by equation*

$$\boxed{\det\left[\frac{\partial q^i}{\partial u^k}\right]_\Lambda = 0} \tag{4.17}$$

4.5.2 Implicit equations

A submanifold $\Lambda \subset T^*Q$ of codimension n (thus, of dimension n) can be represented (at least locally) by n independent equations

$$\Lambda^i(q, p) = 0. \tag{4.18}$$

This means that

$$\operatorname{rank} \left[\frac{\partial \Lambda^i}{\partial q^k} \;\middle|\; \frac{\partial \Lambda^i}{\partial p_k} \right] = n \qquad \text{(max)}$$

at each point of Λ that is, for each set of values of the coordinates satisfying Eqs. (4.18). The submanifold Λ is Lagrangian if and only if it is coisotropic: that is, if and only if $\{\Lambda^i, \Lambda^j\}|\Lambda = 0$. In this representation,

$$\operatorname{rank}(p) = \operatorname{rank} \left[\frac{\partial \Lambda^i}{\partial p_k} \right]_p . \tag{4.19}$$

Indeed, the tangent subbundle $T\Lambda \subset TT^*Q$ is described by equations

$$\frac{\partial \Lambda^i}{\partial q^k} \dot{q}^k + \frac{\partial \Lambda^i}{\partial p_k} \dot{p}_k = 0,$$

so that at any point p the dimension of the space of the vertical vectors, for which $\dot{q}^i = 0$, is given by the corank of the matrix (4.19) at that point; this is the codimension of the space $T_p\pi(T_p\Lambda)$. Hence,

$$p \text{ regular} \iff \det \left[\frac{\partial \Lambda^i}{\partial p_k} \right]_p \neq 0. \tag{4.20}$$

This proves the following.

Theorem 4.3. *If a Lagrangian submanifold is represented by implicit equations* 4.18) *then the set of its singular points is determined by equation*

$$\boxed{\det \left[\frac{\partial \Lambda^i}{\partial p_k} \right]_\Lambda = 0} \tag{4.21}$$

4.5.3 Generating functions

In a neighborhood of a regular point p a Lagrangian submanifold can be described by equations of the kind

$$p_i = \frac{\partial G}{\partial q^i}, \tag{4.22}$$

where $G(q)$ is a function in a neighborhood of the point $q = \pi_Q(p)$. Indeed, due to condition (4.20), Eqs. (4.18) are locally solvable with respect to p_i, $p_i = S_i(q^j)$. Being Λ Lagrangian, the form $S = S_i\, dq^i$ is closed, thus locally exact.

However, there are cases in which a representation of the kind (4.22) holds also in a neighborhood of a singular point. A simple example is the following.

Example 4.1. Let us consider $Q = \mathbb{R} = (q)$, $T^*Q = \mathbb{R}^2 = (q, p)$, and Λ the curve $q = p^3$. All points of Λ are regular except $(q, p) = (0, 0)$. Since $p = q^{1/3}$, this Lagrangian submanifold is the image of the one-form $S = q^{1/3}\, dq$; thus, it is generated by the function $G(q) = \frac{3}{4}\, q^{4/3}$. Note that this function is only C^1 (it does not admits the second derivative at the point $q = 0$). However, as we show in Example 4.2, this Lagrangian submanifold admits a C^∞ "generating family". ♢

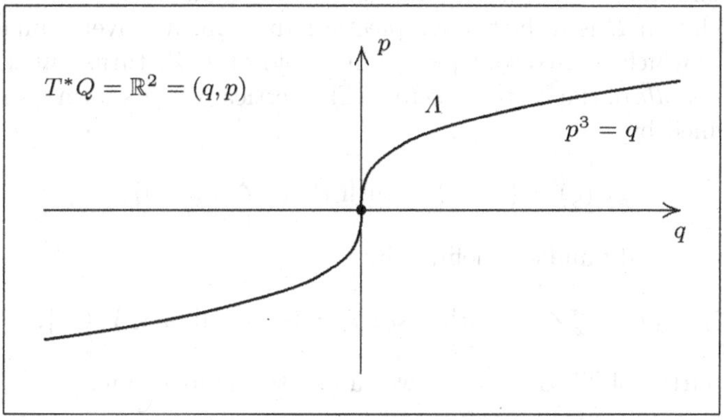

Fig. 4.4 Example 4.1

4.6 Generating families and Morse families

Let there be given a fibration $\zeta\colon Z \to Q$ over a manifold Q and a smooth function $G\colon Z \to \mathbb{R}$. With the fibration ζ we associate a relation $\widehat{R}\colon T^*Q \leftarrow T^*Z$ defined by

$$(p, f) \in \widehat{R} \iff \begin{cases} (p, f) \in T^*_q Q \times T^*_z Z, \ \ (q, z) \in R \ \text{ and} \\ \langle v, p \rangle = \langle w, f \rangle, \ \ \text{for all } v \in T_q Q \text{ and } w \in T_z Z \\ \text{such that } T\zeta(w) = v. \end{cases} \tag{4.23}$$

With the function G we generate a Lagrangian submanifold $\Lambda = dG(Z)$ of T^*Z.

Definition 4.3. The image of Λ by the relation $\widehat{R}$,

$$\Lambda_\circ = \widehat{R} \circ \Lambda \subset T^*Q,$$

is the *Lagrangian set* in T^*Q generated by the *generating family* (ζ, G). ♡

Due to (4.23) the explicit definition of $\Lambda_\circ$ is

$$\boxed{p \in \Lambda \iff \begin{cases} p \in T_q^*Q, \quad \text{and} \\ \langle v, p\rangle = \langle w, dG\rangle, \quad \text{for all } v \in T_qQ \text{ and } w \in T_zZ \\ \text{such that } q = \zeta(z) \text{ and } T\zeta(w) = v. \end{cases}} \tag{4.24}$$

This definition needs an explanation. As we show in Sect. 5.1 and Sect. 5.2, the relation $\widehat{R}$ is in fact a symplectic reduction. Its inverse image, $C = \widehat{R}^\top \circ T^*Q$, which is a coisotropic submanifold of T^*Z, turns out to be the *conormal bundle* of the *vertical vectors.* The vertical vectors form a subbundle of TZ defined by

$$V(\zeta) = \{w \in TZ \ \text{ such that } T\zeta(w) = 0\},$$

and the conormal bundle is defined by

$$V^\circ(\zeta) = \{f \in T_z^*Z \ \text{ such that } \langle w, f\rangle = 0, \ \text{ for all } w \in V_z(\zeta)\}. \tag{4.25}$$

Then, by virtue of Theorem 3.18, we can state the following.

Theorem 4.4. *If Λ and $C = V^\circ(\zeta)$ have a transverse intersection, then $\Lambda_\circ$ is a (may be immersed) Lagrangian submanifold of T^*Q.*

Although it is a special case, the transverse intersection plays an important role in this context and deserves to be highlighted in a definition.

Definition 4.4. If Λ and $C = V^\circ(\zeta)$ have a transverse intersection then the generating family (ζ, G) is called a *Morse family.*[1] ♡

---

[1] For a general survey on the notion of "generating family", or "generating function", and its applications to analytical mechanics see (Tulczyjew 1974), (Weinstein 1977), (Tulczyjew 1977b), (Arnold 1986) and (Libermann and Marle 1987). The definition given here differs from that given for instance in (Arnold 1986) and in (Chaperon 1995), where the term "generating family" is introduced for what here, or in (Weinstein 1977), is called a "Morse family". In (Chaperon 1995) Sect. 1.3, the term "phase function" is used for a function on $Z = Q \times E$, where E is a finite-dimensional vector space. As remarked in (Arnold 1986) the notion of generating family already appears in the works of Jacobi and Lie and in (Whittaker 1927). The expression *generating family* or *Morse family* is suggested by the fact that G is regarded as a family of functions parametrized by the points $q \in Q$.

A fundamental fact is that *any Lagrangian submanifold of a cotangent bundle is locally generated by Morse families*:

Theorem 4.5. *Maslov–Hörmander theorem. If $\Lambda \subset T^*Q$ is a Lagrangian submanifold, then for each $p \in \Lambda$ there exists a Morse family generating an open neighborhood $U \subseteq \Lambda$ containing p.*[2]

Remark 4.2. A hard, very intriguing and alive problem is the study of the existence of *global generating families.* Although this argument goes beyond the scope of this book, notable examples of global generating functions are discussed in Chaps. 7, 8, and 10. ♢

Remark 4.3. In almost all examples of generating families discussed in this book the fibration $\zeta \colon Z \to Q$ is trivial: $Z = Q \times U$ and $\zeta \colon Q \times U \to Q$ is the canonical projection. Then the manifold U is called the *supplementary manifold* and Q the *configuration manifold.* For a generating family

$$G \colon Q \times U \to \mathbb{R}$$

we sometimes use the shorthand notation

$$G(Q; U)$$

in which a semicolon separates the configuration manifold Q from the supplementary manifold U. Similarly, any local coordinate representation of a generating family $G(Q; U)$ is denoted by

$$G(q^i; u^\alpha),$$

where (q^i) are coordinates on Q and (u^α) are coordinates on U. The (u^α) are called *extra* or *supplementary variables.* ♢

Example 4.2. The Lagrangian submanifold of Example 4.1 is generated by the global C^∞ Morse family $G(q; u) = uq - \frac{1}{4}\, u^4$. ♢

Remark 4.4. In considering generating families that are Morse families, as is commonly done in the literature, we lose the possibility of dealing with many physically interesting applications. For instance, there are *systems of rays* or sets of *equilibrium states* of static systems that are Lagrangian sets, and not Lagrangian submanifolds (see Chaps. 5 and 6); there are global Hamilton principal functions that are not Morse families (Chap. 8),[3] ♢

Before going on, we must introduce an additional object related to a generating family.

---

[2] Go to Sect. 9.5.

[3] See also (Cardin 2002), where such an example arises in the construction of global solutions of the Cauchy problem for a t-dependent Hamilton–Jacobi equation).

Definition 4.5. The *critical set* of a generating family (ζ, G) is the subset $\Xi \subset Z$ of the stationary points of G along the fibers of ζ (i.e., of those points where the differential of the function G restricted to the fibers vanishes). ♡

Remark 4.5. Let $\pi_Z \colon T^*Z \to Z$ be the cotangent fibration over Z. Then

$$\Xi = \pi_Z(\Lambda \cap C). \tag{4.26}$$

The Lagrangian submanifold Λ is the image of a section of $\pi_Z \colon T^*Z \to Z$. Then $\Lambda \cap C$ is a submanifold if and only if the critical set Ξ is a submanifold of Z. ◇

4.7 Coordinate representation of generating families

Let us consider a local coordinate representation $G(q^i; u^\alpha)$ of a generating family $G(Q; U)$ and denote by $(p_i; \pi_\alpha)$ the canonical momenta associated with $(q^i; u^\alpha)$.

Theorem 4.6. *The Lagrangian set $\Lambda_\circ$ is described by equations*

$$\Lambda_\circ \colon \begin{cases} p_i = \dfrac{\partial G}{\partial q^i}, \\ 0 \;= \dfrac{\partial G}{\partial u^\alpha}, \end{cases} \tag{4.27}$$

in the sense that a point $p \in T^*Q$ belongs to $\Lambda_\circ$ if and only if its coordinates (q^i, p_i) satisfy equations (4.29) for some value of the (u^α).

Proof. The Lagrangian submanifold Λ is described by equations

$$\Lambda \colon \begin{cases} p_i \;= \dfrac{\partial G}{\partial q^i}, \\ \pi_\alpha = \dfrac{\partial G}{\partial u^\alpha}. \end{cases} \tag{4.28}$$

In (4.24) equation $\langle v, p\rangle = \langle w, dG\rangle$ and condition $T\zeta(W) = v$ become $p_i v^i = \pi_i w^i + \pi_\alpha w^\alpha$ and $w^i = v^i$, respectively. It follows that

$$p_i = \pi_i, \quad \pi_\alpha = 0,$$

and (4.27) is proved. □

Remark 4.6. The critical set Ξ is described by equations

$$\frac{\partial G}{\partial u^\alpha} = 0. \quad \diamondsuit \tag{4.29}$$

Remark 4.7. Equations (4.29) are equivalent to the "differential equation"

$$p_i\, dq^i = dG, \tag{4.30}$$

or to the "variational equation"

$$p_i\, \delta q^i = \delta G. \qquad \diamondsuit \tag{4.31}$$

Theorem 4.7. *A generating family is a Morse family, that is, the submanifolds Λ and C have a transverse intersection, if and only if the $r \times (n+r)$ matrix*

$$\big[\boldsymbol{G}_{uq} \,\big|\, \boldsymbol{G}_{uu}\big] = \left[\frac{\partial^2 G}{\partial u^\alpha \partial q^i} \,\middle|\, \frac{\partial^2 G}{\partial u^\alpha \partial u^\beta}\right] \tag{4.32}$$

has maximal rank ($= n$) at each point of the critical set Ξ.

Proof. Due to Eqs. (4.28), the spaces $T_x\Lambda$ are defined by equations

$$\begin{cases} \dot{p}_i - \dfrac{\partial^2 G}{\partial q^i \partial q^j}\, \dot{q}^j - \dfrac{\partial^2 G}{\partial q^i \partial u^\alpha}\, \dot{u}^\alpha & = 0, \\ \dot{\pi}_\alpha - \dfrac{\partial^2 G}{\partial u^\alpha \partial q^j}\, \dot{q}^j - \dfrac{\partial^2 G}{\partial u^\alpha \partial u^\beta}\, \dot{u}^\beta & = 0. \end{cases}$$

On the other hand, the spaces $(T_xC)^\S$, whose vectors are tangent to the fibers of $\widehat{R}$, are characterized by the conditions $\dot{q}^i = \dot{p}_i = \dot{\pi}_\alpha = 0$. Then the equations of $T_x\Lambda \cap (T_xC)^\S$ are

$$\begin{cases} \dfrac{\partial^2 G}{\partial q^i \partial u^\alpha}\, \dot{u}^\alpha & = 0, \\ \dfrac{\partial^2 G}{\partial u^\alpha \partial u^\beta}\, \dot{u}^\beta & = 0, \end{cases} \tag{4.33}$$

The transversality condition is equivalent to $T_x\Lambda \cap (T_xC)^\S = 0$ (Remark 3.6). Then Eqs. (4.33) must have $\dot{u}^\alpha = 0$ as a unique solution. This is equivalent to saying that the matrix (4.32) has maximal rank. □

4.8 Equivalence and reduction of generating families

We can say that two generating families (or Morse families) are *equivalent* if they generate the same Lagrangian set (or the same Lagrangian submanifold).[4] Two generating families differing by an additive constant are obviously equivalent.

[4] For the equivalence theory of Morse families, and related references, see (Arnold 1986), (Libermann and Marle 1987), (Viterbo 1992), and (Théret 1999).

In some cases it is possible to reduce the dimension of the supplementary manifold (i.e., the number of the supplementary variables) of a given generating family G and get an equivalent *reduced generating family.* This depends on the critical set of G. For instance, if the critical set is the image of a section $\xi\colon Q \to \mathcal{Z}$ of ζ, then Λ is generated (in the ordinary sense) by the function $G \circ \xi\colon Q \to \mathbb{R}$. This is, of course, an extreme case. In general, we can remove some of the extra variables (u^α) when, by means of Eqs. (4.29) of the critical set, they can be expressed as functions of the coordinates and of the remaining extra variables. Note that the extra variables can be totally removed, so that one gets an ordinary generating function, if and only if

$$\det \left[\frac{\partial^2 G}{\partial u^\alpha \partial u^\beta} \right]_\Xi \neq 0.$$

This is the case of a regular Lagrangian submanifold (no caustic), in accordance with the next Theorem 4.8.

4.9 The caustic of a Lagrangian submanifold generated by a Morse family

Theorem 4.8. *If a Lagrangian submanifold $\Lambda_\circ \subset T^*Q$ is generated by a Morse family $G\colon Q \times U \to \mathbb{R}$, then the caustic $\Gamma \subseteq Q$ of $\Lambda_\circ$ is the set described by equations*

$$\boxed{\det \left[\frac{\partial^2 G}{\partial u^\alpha \partial u^\beta} \right] = 0, \quad \frac{\partial G}{\partial u^\alpha} = 0} \tag{4.34}$$

A point $q \in Q$ belongs to the caustic Γ if and only if its coordinates (q^i) satisfy equations (4.34) *for some values of the extra variables (u^α).*

Equations (4.34) can be summarized by the single equation

$$\boxed{\det \left[\frac{\partial^2 G}{\partial u^\alpha \partial u^\beta} \right]_\Xi = 0} \tag{4.35}$$

where Ξ is the critical set.

Proof. A point $p \in \Lambda_\circ$ is regular when the tangent space $T_p\Lambda_\circ$ has no vertical vectors except the zero-vector. This means that at the point p the condition $\dot{q}^i = 0$ must imply $\dot{p}_i = 0$. From Eqs. (4.27) of $\Lambda_\circ$ we derive the equations of $T\Lambda_\circ$:

$$* \begin{cases} p_i = \dfrac{\partial G}{\partial q^i}, \\ 0 = \dfrac{\partial G}{\partial u^\alpha}, \end{cases} \qquad ** \begin{cases} \dot{p}_i - \dfrac{\partial^2 G}{\partial q^i \partial q^j}\, \dot{q}^j - \dfrac{\partial^2 G}{\partial q^i \partial u^\alpha}\, \dot{u}^\alpha = 0, \\ \dfrac{\partial^2 G}{\partial u^\alpha \partial q^j}\, \dot{q}^j + \dfrac{\partial^2 G}{\partial u^\alpha \partial u^\beta}\, \dot{u}^\beta \qquad = 0. \end{cases} \tag{4.36}$$

If we put $\dot{q}^i = 0$ in the second set $**$ of these equations then we get, in matrix notation,

$$\begin{cases} \dot{\boldsymbol{p}} - \boldsymbol{G}_{qu}\, \dot{\boldsymbol{u}} = 0, \\ \boldsymbol{G}_{uu}\, \dot{\boldsymbol{u}} = 0. \end{cases}$$

(i) If $\det \boldsymbol{G}_{uu} \neq 0$ then $\dot{\boldsymbol{u}} = 0$ and consequently $\dot{\boldsymbol{p}} = 0$: regular point. (ii) If $\det \boldsymbol{G}_{uu} = 0$ then there exists a vector $\dot{\boldsymbol{u}} \neq 0$ satisfying the second equation. This means $\dot{\boldsymbol{u}} \in \ker \boldsymbol{G}_{uu}$. Suppose $\boldsymbol{G}_{qu}(\dot{\boldsymbol{u}}) = 0$. This means that $\dot{\boldsymbol{u}}$ is also in the kernel of $\boldsymbol{G}_{qu}$. As a consequence, $\dot{\boldsymbol{u}}$ is in the kernel of the whole matrix $[\boldsymbol{G}_{qu}, \boldsymbol{G}_{uu}]$. But this matrix has maximal rank by assumption. Then $\dot{\boldsymbol{u}} = 0$: absurd. Hence, $\dot{\boldsymbol{p}} = \boldsymbol{G}_{qu}(\dot{\boldsymbol{u}}) \neq 0$: singular point. The second equation in $*$ tell us that $\det \boldsymbol{G}_{uu}$ must be computed on the critical set Ξ. The first equation $*$ simply gives the values of the coordinates (p_i) of the singular points. □

4.10 Generating families of symplectic relations

The image of the differential dG of a smooth function $G\colon Q_2 \times Q_1 \to \mathbb{R}$, interpreted as a section of the cotangent fibration $T^*(Q_2 \times Q_1) \to Q_2 \times Q_1$, is a Lagrangian submanifold R' of $T^*(Q_2 \times Q_1)$ with respect to the canonical symplectic form $\theta_{Q_2 \times Q_1}$. But, in accordance with what we have established in Sect. 3.5, this is not a symplectic relation because it is not Lagrangian with respect to the symplectic form $\omega_{Q_2} \ominus \omega_{Q_1}$. In order to get a symplectic relation we must apply the natural isomorphism $T^*(Q_2 \times Q_1) \to T^*Q_2 \times T^*Q_1$, so that $\theta_{Q_2 \times Q_1}$ can be identified with $\omega_{Q_2} \oplus \omega_{Q_1}$, and then apply the symplectic transformation

$$\iota\colon T^*Q_2 \times T^*Q_1 \to T^*Q_2 \times T^*Q_1\colon (p_2, p_1) \mapsto (p_2, -p_1) \tag{4.37}$$

so that

$$\iota^*(\omega_{Q_2} \ominus \omega_{Q_1}) = \omega_{Q_2} \oplus \omega_{Q_1}.$$

In this way we get a Lagrangian submanifold $R = \iota(R')$ with respect to the symplectic form $\omega_{Q_2} \ominus \omega_{Q_1}$, that is, a symplectic relation R from T^*Q_1 to T^*Q_2.

Let us denote by q_1 and q_2 two generic coordinate systems on Q_1 and Q_2, and by p_1 and p_2 the corresponding momenta on T^*Q_1 and T^*Q_2. Here, we can avoid the use of indices for labeling the coordinates so that we can write simpler formulae. For instance,

$$\boxed{\omega_{Q_2} \ominus \omega_{Q_1} = dp_2 \wedge dq_2 - dp_1 \wedge dq_1} \tag{4.38}$$

instead of $\omega_{Q_2} \ominus \omega_{Q_1} = d(p_2)_i \wedge dq_2^i - d(p_1)_\iota \wedge dq_1^\iota$. Conclusion:

Theorem 4.9. *Any smooth real function $G(q_2, q_1)$ on $Q_2 \times Q_1$ generates a symplectic relation $R\colon T^*Q_2 \leftarrow T^*Q_1$ by means of equations*

$$\boxed{p_1 = -\frac{\partial G}{\partial q_1}, \quad p_2 = \frac{\partial G}{\partial q_2}} \tag{4.39}$$

But we know that not all Lagrangian submanifolds are generated in this way. Hence, we are necessarily led to introduce a more general notion.

Definition 4.6. Let $G\colon Q_1 \times Q_2 \times U \to \mathbb{R}$ be a smooth function. The subset $R \subset T^*Q_2 \times T^*Q_1$ defined by equations

$$\boxed{p_1 = -\frac{\partial G}{\partial q_1}, \quad p_2 = \frac{\partial G}{\partial q_2}, \quad 0 = \frac{\partial G}{\partial u}} \tag{4.40}$$

is the *symplectic relation $R\colon T^*Q_2 \leftarrow T^*Q_1$ generated by the family G.* ♡

Note that this extension of the notion of symplectic relation is possible because we are in the category of the cotangent bundles.

Remark 4.8. Equations (4.40) should read as follows: *two points belong to the relation if their canonical coordinates satisfy these equations for some values of the coordinates u.* Equations (4.40) are equivalent to the differential equation

$$\boxed{p_2\, dq_2 - p_1\, dq_1 = dG} \tag{4.41}$$

◇

Remark 4.9. In this context we may recall what was said in Remark 4.3: for a generating family $G\colon Q_1 \times Q_2 \times U \to \mathbb{R}$ of a symplectic relation we use the shorthand notation

$$G(Q_2 \times Q_1; U)$$

or, for any coordinate representation,

$$G(q_2, q_1; u). \qquad \diamondsuit$$

4.11 Generating families of symmetric relations

Theorem 4.10. *If a symplectic relation $R\colon T^*Q_2 \leftarrow T^*Q_1$ is generated by $G(Q_2 \times Q_1; U)$, then the transpose relation $R^\top : T^*Q_1 \leftarrow T^*Q_2$ is generated by the family*

$$G^{\top}(Q_1 \times Q_2; U)$$

defined by

$$G^{\top}(q_1, q_2; u) = -G(q_2, q_1; u). \tag{4.42}$$

Proof. If R is described by Eqs. (4.40) then $R^{\top}$ is described by similar equations

$$p_2 = -\frac{\partial G^{\top}}{\partial q_2}, \quad p_1 = \frac{\partial G^{\top}}{\partial q_1}, \quad 0 = \frac{\partial G^{\top}}{\partial u}. \tag{4.43}$$

The relations R and $R^{\top}$ must be described by equivalent equations, because these two relations differ only by the order in the pairs. Equations (4.40) and (4.43) coincide if $S^{\top}$ is defined as in (4.42). □

It follows that

Theorem 4.11. *A symplectic relation $D\colon T^*Q \leftarrow T^*Q$ generated by $G(Q \times Q; U)$ is symmetric,*

$$D^{\top} = D,$$

if and only if G is skew-symmetric, up to an additive constant, on the pairs (q, q') belonging to the critical set,

$$\boxed{G(q, q'; u) = -G(q', q; u) + c, \quad (q, q'; u) \in \Xi, \quad c \in \mathbb{R}} \tag{4.44}$$

Proof. If D is symmetric then $G(q, q', u) = -G(q', q, u) + c$, where c is a constant. If we replace G by $\bar{G} = G - c/2$, then $\bar{G}(q, q'; u) + \bar{G}(q', q; u) = G(q, q'; u) + G(q', q; u) - c = 0$. □

Remark 4.10. In all the following applications the generating families of symmetric relations are skew-symmetric *in the proper sense*, that is, with $c = 0$:

$$G(q, q'; u) + G(q', q; u) = 0. \qquad \diamond$$

4.12 The composition of generating families

Theorem 4.12. *If two symplectic relations $R_1\colon T^*Q_1 \leftarrow T^*Q_0$ and $R_2\colon T^*Q_2 \leftarrow T^*Q_1$ are generated by the families $G_1(Q_1 \times Q_0; U_1)$ and $G_2(Q_2 \times Q_1; U_2)$, then the relation $R_2 \circ R_1 \leftarrow T^*Q_2 \times T^*Q_0$ is generated by the family*

$$G_{21}(Q_2 \times Q_0; Q_1 \times U_2 \times U_1)$$

defined by

$$\boxed{G_{21}(q_2, q_0; q_1, u_2, u_1) = G_2(q_2, q_1; u_2) + G_1(q_1, q_0; u_1)} \tag{4.45}$$

We denote this composition rule by[5]

$$\boxed{G_{21} = G_2 \oplus G_1} \tag{4.46}$$

Remark 4.11. In the generating family G_{21} the manifold Q_1 plays the role of supplementary manifold, together with U_1 and U_2. ♢

Proof. The two relations are respectively described by equations

$$R_1 : \begin{cases} p_0 = -\dfrac{\partial G_1}{\partial q_0}, \\ p_1 = \dfrac{\partial G_1}{\partial q_1}, \\ 0 = \dfrac{\partial G_1}{\partial u_1}. \end{cases} \qquad R_2 : \begin{cases} p_1 = -\dfrac{\partial G_2}{\partial q_1}, \\ p_2 = \dfrac{\partial G_2}{\partial q_2}, \\ 0 = \dfrac{\partial G_2}{\partial u_2}. \end{cases}$$

In composing the two relations the two sets of coordinates p_1 must coincide. Thus, the relation $R_2 \circ R_1$ is described by equations

$$\begin{cases} p_0 = -\dfrac{\partial G_1}{\partial q_0}, \\ p_2 = \dfrac{\partial G_2}{\partial q_2}, \end{cases} \qquad \begin{cases} 0 = \dfrac{\partial G_2}{\partial q_1} + \dfrac{\partial G_1}{\partial q_1}, \\ 0 = \dfrac{\partial G_2}{\partial u_2}, \\ 0 = \dfrac{\partial G_1}{\partial u_1}. \end{cases}$$

These equations are equivalent to the single equation

$$\boxed{p_2\, dq_2 - p_0\, dq_0 = d(G_2 + G_1)} \tag{4.47}$$

This proves the composition rule (4.45). □

Remark 4.12. Equation (4.47), as an equivalent form of the composition rule (4.45), is very useful in practical applications. ♢

[5] The composition rule of "generating forms" of linear symplectic relations has been introduced in (Ławruk et al. 1975) and (Benenti and Tulczyjew 1981), and extended to generating families of symplectic relations in (Benenti 1988).

Chapter 5
Canonical Lift on Cotangent Bundles

Abstract There exists an operation, that we call *canonical lift* and denote by a "hat" $\widehat{\ }$, which creates "symplectic objects" on a cotangent bundle T^*Q starting from "objects" on the manifold Q (vector fields, maps, submanifolds, etc.). It plays an important role in the theory of symplectic relations and in its applications. The basic lift, from which all other canonical lifts can be derived, is that of a submanifold.

5.1 Canonical lift of submanifolds

Definition 5.1. The *canonical lift of a submanifold* $\Sigma \subseteq Q$ is the set

$$\widehat{\Sigma} = T^\circ \Sigma \subset T^*Q$$

of the covectors annihilating the vectors tangent to Σ,

$$p \in \widehat{\Sigma} \iff \begin{cases} p \in T_q^*Q, & q \in \Sigma, \\ \langle v, p \rangle = 0, & \text{for all } v \in T_q\Sigma. \end{cases} \qquad \heartsuit \qquad (5.1)$$

Remark 5.1. The set $\widehat{\Sigma}$ has a mechanical interpretation: if Σ is a *smooth constraint* imposed on the configuration manifold Q of a holonomic system, then $\widehat{\Sigma}$ is the set of the *reactive forces*, whose virtual work is zero. If we interpret these forces as vectors (by means of a metric tensor) then $\widehat{\Sigma}$ is the set of all vectors orthogonal to Σ. Special remarkable cases are

$$\begin{cases} \Sigma = q \in Q \quad \text{(a point of } Q) \quad \longmapsto \quad \widehat{q} = T_q^*Q \quad \text{(the fiber over } q) \\ \Sigma = Q \quad \longmapsto \quad \widehat{Q} = Q \quad \text{(interpreted as the zero-section of } T^*Q). \end{cases}$$

The *zero-section* of T^*Q is the set of all zero-covectors; thus it is identified with Q itself. $\diamondsuit$

Definition 5.2. Let $\Sigma \subseteq Q$ be a submanifold and $F\colon \Sigma \to \mathbb{R}$ a smooth function. The *canonical lift of a submanifold with function* is the set $\widehat{(S,F)} \subset T^*Q$ defined by

$$p \in \widehat{(\Sigma, F)} \iff \begin{cases} p \in T_q^*Q, \quad q \in \Sigma, \\ \langle v, p\rangle = \langle v, dF\rangle \quad \text{for all } v \in T_q\Sigma. \end{cases} \qquad \heartsuit \qquad (5.2)$$

This is the set of the covectors p whose pairing $\langle v, p\rangle$ with any vector v tangent to Σ is equal to the derivative of the function F with respect to v. In this definition, F can be a function on the whole Q or on an open neighborhood of Σ. Indeed, only the restriction of F to Σ is involved.

Remark 5.2. The second line of (5.2) is equivalent to $p - d_qF \in \widehat{\Sigma}$. Note that $\widehat{(\Sigma, c)} = \widehat{\Sigma}$, $c =$ constant. If F represents a potential energy, then $\widehat{(S,F)}$ is the set of *equilibrium states.* $\diamondsuit$

Theorem 5.1. *The canonical lifts $\widehat{\Sigma}$ and $\widehat{(\Sigma, F)}$ are Lagrangian submanifolds.*[1]

This can be proved in a direct way[2] or by using Morse families, as shown by the following

Theorem 5.2. *If Σ is defined by equations*

$$\Sigma_\alpha(q) = 0, \quad \alpha = 1, \dots, r, \qquad (5.3)$$

then $\widehat{(\Sigma, F)}$ is generated by the Morse family $G\colon Q \times \mathbb{R}^r \to \mathbb{R}$ defined by

$$\boxed{G(q; \lambda^\alpha) = \lambda^\alpha\, \Sigma_\alpha(q) + F(q)} \qquad (5.4)$$

Proof. The critical set Ξ is determined by equations $\partial G/\partial \lambda^\alpha = \Sigma_\alpha = 0$, thus it coincides with $\Sigma \times \mathbb{R}^r$. The maximal rank condition is fulfilled:

---

[1] Lagrangian submanifolds of this kind have been introduced in (Tulczyjew 1977b). The definition of the canonical lift $\widehat{\Sigma}$ can be extended to any subset $\Sigma \subset Q$, by a suitable definition of the tangent $T\Sigma$ of a subset given in (Tulczyjew 1989).

[2] Let (q^i) be local coordinates on Q adapted to Σ. This means that Σ is locally described by equations $q^\alpha = 0$ ($\alpha = 1, \dots, r$), $r = \operatorname{codim}(\Sigma)$. Then, $T\Sigma$ is described by equations $q^\alpha = 0$, $\dot{q}^\alpha = 0$, and the condition $\langle v, p\rangle = \langle v, dF\rangle$, for all $v \in T_q\Sigma$ becomes $\dot{q}^a\,(p_a - \partial_a F) = 0$, for all $(\dot{q}^a) \in \mathbb{R}^{n-r}$ (where $a = r+1, \dots, n$). Thus, $\Lambda = \widehat{(\Sigma, F)}$ is a submanifold of dimension n described by the n equations $q^\alpha = 0$, $p_a = \partial_a F$. Because of the dimension, the isotropy follows from coisotropic condition $\{q^\alpha, p_a - \partial_a F\} = 0$.

$$\operatorname{rank}\left[\frac{\partial^2 G}{\partial\lambda^\alpha \partial q^i}\,\middle|\,\frac{\partial^2 G}{\partial\lambda^\alpha\partial\lambda^\beta}\right]_\Xi = \operatorname{rank}\left[\frac{\partial\Sigma_\alpha}{\partial q^i}\,\middle|\,0\right]_\Xi = \operatorname{rank}\left[\frac{\partial\Sigma_\alpha}{\partial q^i}\right]_\Xi = r.$$

Thus, G is a Morse family and generates a Lagrangian submanifold Λ by equations

$$\begin{cases} p_i = \dfrac{\partial G}{\partial q^i} = \lambda^\alpha\, \partial_i \Sigma_\alpha + \partial_i F, \\ 0 = \dfrac{\partial G}{\partial \lambda^\alpha} = \Sigma_\alpha. \end{cases}$$

The vectors v tangent to Σ are characterized by equations

$$\frac{\partial \Sigma_\alpha}{\partial q^i}\,\dot{q}^i = 0,$$

thus,

$$\langle v, p\rangle = \dot{q}^i\, p_i = \dot{q}^i\,(\lambda^\alpha\,\partial_i\Sigma_\alpha + \partial_i F) = \dot{q}^i\,\partial_i F = \langle v, dF\rangle$$

for all $p \in \Lambda$. This shows that $\Lambda = \widehat{\Sigma}$. □

Remark 5.3. The supplementary variables (λ^α) in (5.3) play the role of *Lagrangian multipliers.* We say that $\widehat{(\Sigma, F)}$ is the *Lagrangian submanifold generated by the function F on the constraint Σ.* The Lagrangian submanifold $\Lambda = \widehat{(\Sigma, F)}$ projects onto Σ,

$$\pi_Q\widehat{(\Sigma, F)} = \Sigma,$$

and the restriction $\pi\colon \Lambda \to \Sigma$ of π_Q to $\Lambda = \widehat{(\Sigma, F)}$ is a surjective submersion. Hence, all points have constant rank equal to $\dim S$ and the caustic is S, unless Σ is an open subset of Q; in this case, Λ is regular. The canonical lift $\Lambda = \widehat{(\Sigma, F)}$ is a Lagrangian submanifold of a special kind, which we call *exact*; see Sect. 9.2. Indeed, from (5.2) it follows that for all $w \in T_p\Lambda$, $\langle w, \theta_Q\rangle = \langle v, p\rangle = \langle v, dF\rangle = \langle w, \pi^* dF\rangle$, where $v = T\pi(w)$. This means that

$$\theta_Q|\widehat{(\Sigma, F)} = d\pi^* F. \tag{5.5}$$

This shows that the pullback of the Liouville one-form to Λ is exact. For $F = 0$ (or constant) we have in particular $\theta_Q|\widehat{\Sigma} = 0$. ◇

5.2 Canonical lift of relations

Definition 5.3. The *canonical lift of a smooth relation $R \leftarrow Q_2 \times Q_1$* is the symplectic relation $\widehat{R} \leftarrow T^*Q_2 \times T^*Q_1$ defined by

$$(p_2,p_1)\in\widehat{R} \iff \begin{cases}(p_2,p_1)\in T^*_{q_2}Q_2\times T^*_{q_1}Q_1,\quad (q_2,q_1)\in R,\\ \text{and}\\ \langle v_2,p_2\rangle=\langle v_1,p_1\rangle \text{ for all } (v_2,v_1)\in T_{(q_2,q_1)}R.\end{cases} \tag{5.6}$$

♡

Remark 5.4. It must be emphasized that this is *not* the "true" canonical lift $\widehat{R}$ of the submanifold R which, according to (5.1), should be the set

$$\widehat{R}=\Big\{\bar{p}\in T^*_{(q_2,q_1)}(Q_2\times Q_1) \text{ such that } (q_2,q_1)\in R,$$
$$\text{and } \langle\bar{v},\bar{p}\rangle=0 \text{ for all } \bar{v}\in T_{(q_2,q_1)}R\Big\}.$$

This is a Lagrangian submanifold of $\widehat{R}\subset T^*(Q_2\times Q_1)$ with respect to the canonical symplectic form $d\theta_{Q_2\times Q_1}$. Hence, it is *not* a symplectic relation from T^*Q_1 to T^*Q_2. In order to get a symplectic relation from T^*Q_1 to T^*Q_2 we use the natural identification $T^*(Q_2\times Q_1)\simeq T^*Q_2\times T^*Q_1$ and the symplectomorphism (4.37),

$$\iota\colon T^*Q_2\times T^*Q_1\to T^*Q_2\times T^*Q_1\colon (p_2,p_1)\mapsto(p_2,-p_1).$$

Then we find (5.6). We use the same symbol $\widehat{R}$ for simplicity, inasmuch as there is no danger of confusion. Indeed, if we consider a submanifold $S\subseteq Q$ as a *zero-relation* $\Sigma\subseteq Q\times 0$, then the canonical lift of Σ as a relation, defined in this section, is just the symplectic zero-relation in $T^*Q\times 0$ associated with the Lagrangian submanifold $\widehat{R}\subset T^*Q$ defined in the preceding section. ◇

In a similar way we can introduce the canonical lift of a relation $R\subseteq Q_2\times Q_1$ endowed with a function $F\colon R\to\mathbb{R}$ or $F\colon Q_2\times Q_1\to\mathbb{R}$. It is the following smooth symplectic relation.

Definition 5.4. The *canonical lift of a smooth relation* $R\colon Q_2\leftarrow Q_1$ *endowed with a function* F is the symplectic relation

$$\widehat{(R,F)}\colon T^*Q_2\leftarrow T^*Q_1$$

defined by

$$(p_2,p_1)\in\widehat{(R,F)} \iff \begin{cases}(p_2,p_1)\in T_{(q_2,q_1)}(Q_2\times Q_1),\quad (q_2,q_1)\in R\\ \text{and}\\ \langle v_2,p_2\rangle-\langle v_1,p_1\rangle=\langle (v_2,v_1),dF\rangle \text{ for all } (v_2,v_1)\in T_{(q_2,q_1)}R.\end{cases} \quad ♡ \tag{5.7}$$

Remark 5.5. From this definition it follows that (see Eq. (5.5))

$$\theta_{Q_2} \ominus \theta_{Q_1} | \widehat{(R,F)} = \pi^* dF,$$

where $\pi \colon \widehat{(R,F)} \to R$ is the surjective submersion associated with $\pi_{Q_2} \times \pi_{Q_1}$. Hence, in particular,

$$\theta_{Q_2} \ominus \theta_{Q_1} | \widehat{R} = 0. \qquad \diamondsuit$$

We can apply Theorem 5.2 to build up a Morse family for $\widehat{R}$.

Theorem 5.3. *If the relation $R \colon Q_2 \leftarrow Q_1$ is represented by equations*

$$R_\alpha(q_2, q_1) = 0, \tag{5.8}$$

where q_1 and q_2 are local coordinates on Q_2 and Q_1, then $\widehat{R}$ is generated by the Morse family

$$G(q_2, q_1; \lambda^\alpha) = \lambda^\alpha R_\alpha(q_2, q_1) \tag{5.9}$$

through the equations

$$\begin{cases} p_1 = -\dfrac{\partial G}{\partial q_1} = -\lambda^\alpha \dfrac{\partial R_\alpha}{\partial q_1}, \\ p_2 = \dfrac{\partial G}{\partial q_2} = \lambda^\alpha \dfrac{\partial R_\alpha}{\partial q_2}, \\ 0 = \dfrac{\partial G}{\partial \lambda^\alpha} = R_\alpha(q_2, q_1). \end{cases} \tag{5.10}$$

Example 5.1. Take $Q_1 = \mathbb{R}$, $(q_1 = x)$, $Q_2 = \mathbb{R}$, $(q_2 = y)$ and the relation R defined by the implicit equation $x^2 + y^2 - 1 = 0$. Apply (5.9) and (5.10) to

$$G(y, x; \lambda) = \lambda\, (x^2 + y^2 - 1).$$

Then,

$$\begin{cases} p_x = -\dfrac{\partial G}{\partial x} = -2\lambda x, \\ p_y = \dfrac{\partial G}{\partial y} = 2\lambda y, \\ 0 = \dfrac{\partial G}{\partial \lambda} = x^2 + y^2 - 1. \end{cases} \tag{5.11}$$

The components of the vectors tangent to the unit circle R at a point (x, y) are of the kind

$$v_x = \alpha\, y, \quad v_y = -\alpha\, x, \quad \alpha \in \mathbb{R},$$

and the first two Eqs. (5.11) tell us that $\langle v_x, p_x \rangle = \langle v_y, p_y \rangle$, in accordance with (5.6), the definiton of $\widehat{R}$. $\diamondsuit$

5.3 Canonical lift of diagonal relations

Definition 5.5. The *diagonal relation of a submanifold* $\Sigma \subseteq Q$ is the relation $\Delta_\Sigma \subset Q \times Q$ defined by

$$(q, q') \in \Delta_\Sigma \iff q = q' \in \Sigma. \qquad \heartsuit$$

By applying (5.6) we find that the canonical lift $\widehat{\Delta}_\Sigma$ of a diagonal relation is defined by

$$(p_2, p_1) \in \widehat{\Delta}_\Sigma \iff$$

$$\begin{cases} (p_2, p_1) \in T^*_{q_2}Q \times T^*_{q_1}Q, \ \ (q_2, q_1) \in \Delta_\Sigma, \\ \text{and} \\ \langle v_2, p_2 \rangle = \langle v_1, p_1 \rangle \ \text{ for all } (v_2, v_1) \in T_{(q_2,q_1)}\Delta_\Sigma. \end{cases} \tag{5.12}$$

However, we observe that a tangent vector $(v_2, v_1) \in T_{(q_2,q_1)}\Delta_\Sigma$ is an equivalence class of a curve γ on Δ_Σ and such a curve is necessarily of the form $t \mapsto \gamma(t) = (q(t), q(t))$. Thus, the pair $(v_2, v_1) \in T_{(q_2,q_1)}\Delta_\Sigma$ is represented by a unique vector $v \in T_q\Sigma$. Then,

$$\widehat{\Delta}_\Sigma = \{(p_2, p_1) \in T_{(q_2,q_1)}(Q \times Q),\, q_2 = q_1 \in \Sigma,\, \langle v, p_2 - p_1 \rangle = 0, \ \text{for all } v \in T_{q_1}\Sigma\}$$

and, since the second condition in (5.12) means that $\langle v, p_2 - p_1 \rangle = 0$, for all $v \in T_q\Sigma$, we can state for $\widehat{\Delta}_\Sigma$ the following.

Definition 5.6. The *canonical lift of the diagonal relation of a submanifold* $\Sigma \subseteq Q$ is the relation $\widehat{\Delta}_\Sigma \subset T^*Q \times T^*Q$ defined by

$$\boxed{(p_2, p_1) \in \widehat{\Delta}_\Sigma \iff \begin{cases} \pi_Q(p_1) = \pi_Q(p_2) = q \in \Sigma, \\ \text{and} \\ \langle v, p_2 - p_1 \rangle = 0 \quad \text{for all } v \in T_q\Sigma. \end{cases}} \qquad \heartsuit \tag{5.13}$$

An extension of Definition 5.6, justified by Eq. (5.7), is given by

Definition 5.7. The *canonical lift of the diagonal relation of a submanifold* $\Sigma \subseteq Q$ *endowed with a function* $F \colon \Sigma \to \mathbb{R}$ is the relation $\widehat{\Delta}_{\Sigma,F} \subseteq T^*Q \times T^*Q$ defined by

$$\boxed{\begin{aligned}&(p_2, p_1) \in \widehat{\Delta}_{\Sigma,F} \iff \\ &\begin{cases} \pi_Q(p_1) = \pi_Q(p_2) = q \in \Sigma \\ \text{and} \\ \langle v, p_2 - p_1 - d_q F \rangle = 0 \quad \text{for all } v \in T_q\Sigma \end{cases}\end{aligned}} \quad \heartsuit \tag{5.14}$$

The generating families of these two canonical lifts are given by the following theorem.

Theorem 5.4. *If Σ is given by s equations $\Sigma_\alpha(q) = 0$, then $\widehat{\Delta}_\Sigma$ is generated by the Morse family*[3]

$$G_\Sigma(q_2, q_1; \lambda^\alpha, \lambda_i) = \lambda^\alpha \, \Sigma_\alpha(q_1) + \lambda_i \, (q_2^i - q_1^i), \tag{5.15}$$

and $\widehat{\Delta}_{\Sigma,F}$ by

$$G_{\Sigma,F}(q_2, q_1; \lambda^\alpha, \lambda_i) = F(q_1) + \lambda^\alpha \, \Sigma_\alpha(q_1) + \lambda_i \, (q_2^i - q_1^i), \tag{5.16}$$

where $F(q^1)$ is any extension of F from Σ to Q.

Proof. Δ_Σ is described by the following $s+1$ equations

$$\Sigma_\alpha(q_1) = 0, \quad q_2 - q_1 = 0.$$

Then (5.15) is a special case of (5.9) in Theorem 5.3. Regarding (5.16), the associated differential equations are (we set $G = G_{\Sigma,F}$ for simplicity)

$$\begin{aligned} p_{1i} &= -\frac{\partial G}{\partial q_1^i} = -\frac{\partial F}{\partial q_1^i} - \lambda^\alpha \frac{\partial \Sigma_\alpha}{\partial q_1^i} + \lambda_i, \quad & 0 &= \frac{\partial G}{\partial \lambda^\alpha} = \Sigma_\alpha, \\ p_{2i} &= \frac{\partial G}{\partial q_2^i} = \lambda_i, & 0 &= \frac{\partial G}{\partial \lambda_i} = q_2^i - q_1^i. \end{aligned}$$

Then, for any vector v^i we have

$$v^i (p_{2i} - p_{1i}) = v^i \frac{\partial F}{\partial q_1^i} + \lambda^\alpha \, v^i \frac{\partial \Sigma_\alpha}{\partial q_1^i}.$$

But the vectors tangent to Σ are characterized by equations $v^i \, \partial \Sigma_\alpha / \partial q^i = 0$. So, for them we have $v^i (p_{2i} - p_{1i}) = v^i \, \partial F / \partial q_1^i$, and we get Eq. (5.14). □

[3] Here, q, q_1, q_2 denote a generic coordinate system on $Q = Q_1 = Q_2$.

5.4 Canonical lift of reductions and diffeomorphisms

The definition of canonical lift of a relation, Sect. 5.2, can be applied to (the graph of) a map, in particular to a diffeomorphism or to a surjective submersion. These two last extensions are special cases of the canonical lift of a reduction. Let $R \subset Q_2 \times Q_1$ be a smooth reduction, that is, the graph of a surjective submersion $\rho\colon A \to Q_2$ from a submanifold $A \subseteq Q_1$ onto Q_2. Let $V(\rho) \subset TQ_1$ be the subbundle of the vertical vectors, that is, of the vectors tangent to the fibers of ρ,

$$V(\rho) = \{v \in TQ_1 \text{ such that } T\rho(v)\} = 0$$

and $V^\circ(\rho) \subset T^*Q_1$ the subbundle of the covectors annihilating the vertical vectors,

$$V^\circ(\rho) = \{p \in T^*Q_1 \text{ such that } \langle v, p\rangle = 0, \text{ for all } v \in V(\rho) \cap T_qQ_1,\ q = \pi_{Q_1}(p)\}.$$

It can be proved that[4]

Theorem 5.5. *The canonical lift of a reduction $R\colon Q_2 \leftarrow Q_1$, with underlying submersion $\rho\colon A \to Q_2$, is a symplectic reduction $\widehat{R}\colon T^*Q_2 \leftarrow T^*Q_1$, whose inverse image $C = \widehat{R}^\top \circ T^*_AQ_2 \subset T^*Q_1$ is the coisotropic submanifold $C = V^\circ(\rho)$ made of the covectors annihilating the vectors tangent to the fibers of ρ. The underlying surjective submersion $\widehat{\rho}\colon C \to T^*Q_2$ is defined by equation*

$$\langle T\rho(v), \widehat{\rho}(p)\rangle = \langle v, p\rangle, \tag{5.17}$$

where $v \in T_qQ_1$ and $q = \pi_{Q_1}(p)$. As a consequence,

$$\widehat{\rho}^*\theta_{Q_2} = \theta_{Q_1}|T^*_AQ_1.$$

Theorem 5.6. *The composition of two reductions $S \circ R$ is a reduction and*

$$\boxed{\widehat{S \circ R} = \widehat{\Sigma} \circ \widehat{R}} \tag{5.18}$$

We also observe that $\widehat{\mathrm{id}}_Q = \mathrm{id}_{T^*Q}$. Then the last theorem shows that *the canonical lift is a covariant functor from the category of smooth reductions to the category of symplectic reductions.*

Remark 5.6. Canonical lift of a diffeomorphism. A diffeomorphism $\rho\colon Q_1 \to Q_2$ is a special case of reduction. By (5.17) we can see that its canonical lift $\widehat{\rho}\colon T^*Q_1 \to T^*Q_2$ is the symplectomorphism defined by

$$\langle v, \widehat{\rho}(p)\rangle = \langle T\rho^{-1}(v), p\rangle. \tag{5.19}$$

[4] (Benenti 1983b).

It preserves the Liouville forms: $\widehat{\rho}^* \theta_{Q_2} = \theta_{Q_1}$. Note that the pair $(\rho, \widehat{\rho})$ is a fiber-bundle isomorphism $\pi_{Q_2} \circ \widehat{\rho} = \rho \circ \pi_{Q_1}$. ♢

5.5 Canonical lift of vector fields

Definition 5.8. The *canonical lift of a vector field* X on a manifold Q is the Hamiltonian vector field on T^*Q generated by the function P_X,

$$i_{\widehat{X}} d\theta_Q = -dP_X. \qquad \heartsuit$$

If X^i are the components of X in a coordinate system (q^i) then the components $\widehat{X}^i = \langle \widehat{X}, dq^i \rangle$ and $\widehat{X}_i = \langle \widehat{X}, dp_i \rangle$ in the canonical coordinates (q^i, p_i) are

$$\widehat{X}^i = X^i, \quad \widehat{X}_i = -\frac{\partial X^j}{\partial q^i} p_j. \tag{5.20}$$

The canonical lift of vector fields has the following properties.

- The vector field $\widehat{X}$ is projectable onto X; that is,

$$T\pi_Q \circ \widehat{X} = X \circ \pi_Q.$$

- The restriction of $\widehat{X}$ to the zero-section of T^*Q coincides with X.
- The following equations hold.

$$i_{\widehat{X}} \theta_Q = P_X, \quad d_{\widehat{X}} \theta_Q = 0.$$

- The map $X \mapsto \widehat{X}$ is a Lie-algebra homomorphism,

$$(aX + bY)\widehat{\ } = a\widehat{X} + b\widehat{Y} \quad (a, b \in \mathbb{R}), \quad [\widehat{X}, \widehat{Y}] = [X, Y]\widehat{\ }.$$

- For each smooth function F on T^*Q,

$$d_{\widehat{X}} F = \{P_X, F\}.$$

The following theorem shows that the above definition of canonical lift of vector fields is strictly related to the basic definition of the canonical lift of relations.

Theorem 5.7. *If X is a complete vector field with one-parameter group $\varphi_t^X : Q \to Q$, $t \in \mathbb{R}$, then its canonical lift $\widehat{X}$ is complete and its one-parameter group*

$$\varphi_t^{\widehat{X}} : T^*Q \to T^*Q$$

is the canonical lift of φ_t^X,

$$\varphi_t^{\widehat{X}} = \widehat{(\varphi_t^X)}.$$

Proof. Let us put $\varphi_t^X = \varphi_t$ for simplicity. Due to the definition of canonical lift of a diffeomorphism, formula (5.19), we can write

$$\langle v, \widehat{\varphi}_t(p_0)\rangle = \langle T\varphi_t^{-1}(v), p_0\rangle \tag{5.21}$$

for all $p_0 \in T^*_{q_0}Q$ and $v \in T_qQ$, with $q = \varphi_t(q_0)$. Due to the functorial properties of T, $\widehat{\varphi}_t$ is a one-parameter group of transformations on T^*Q. Let $\widehat{X}$ be the corresponding (complete) vector field, and let (X^i, X_i) be its components in standard canonical coordinates $(q,p) = (q^i, p_i)$. Let us consider a local coordinate representation of $\widehat{\varphi}_t$,

$$q^i = \varphi^i(t, q_0), \quad p_i = \varphi_i(t, q_0, p_0).$$

Then (see Sect. 1.7),

$$X^i(q_0, p_0) = \dot{\varphi}^i(0, q_0), \quad X_i(q_0, p_0) = \dot{\varphi}_i(0, q_0, p_0). \tag{5.22}$$

On the other hand, Eq. (5.21) is equivalent to equation

$$v^i \varphi_i(t, q_0, p_0) = p_{0i}\, \frac{\partial \varphi^i(-t, q_0)}{\partial q_0^j}\, v^j,$$

for all (v^i), thus to equation

$$\varphi_i(t, q_0, p_0) = p_{0j}\, \frac{\partial \varphi^j(-t, q_0)}{\partial q_0^i}. \tag{5.23}$$

From (5.22) and (5.23) it follows that

$$X_i(q_0, p_0) = p_{0j}\, \left.\frac{\partial \dot{\varphi}^j(-t, q_0)}{\partial q_0^i}\right|_{t=0} = -\, p_{0j}\, \frac{\partial X^j}{\partial q_0^i} = -\, \frac{\partial P_X}{\partial q_0^i}.$$

Due to (5.20), this is sufficient to prove that the vector field (X^i, X_i) is the canonical lift of $X = (X^i)$. □

5.6 Symplectic relations generated by a submanifold

It is useful to take a look of the symplectic relations that can be generated by a submanifold Σ of a manifold Q.

At an early stage, a submanifold $\Sigma \subseteq Q$ generates three smooth relations.

- The *zero-relation*

$$\Sigma \times \{0\} \subset Q \times \{0\}, \quad (q, 0) \in \Sigma \times \{0\} \iff q \in \Sigma.$$

- The *injection-relation*

$$R_\Sigma \subset \Sigma \times Q, \quad (q, q') \in R_\Sigma \iff q = q' \in \Sigma.$$

- The *diagonal relation*

$$\Delta_\Sigma \subset Q \times Q, \quad (q, q') \in \Delta_\Sigma \iff q = q' \in \Sigma.$$

By their canonical lifts we get three symplectic relations between cotangent bundles,

$$\begin{cases} \widehat{\Sigma} \subset T^*Q \times \{0\}, \\ \widehat{R}_\Sigma \subset T^*\Sigma \times T^*Q, \\ \widehat{\Delta}_\Sigma \subset T^*Q \times T^*Q. \end{cases}$$

The first relation is the canonical lift of Σ interpreted as a zero-relation. The third one has been examined in Sect. 5.3. It is an interesting fact that[5]

Theorem 5.8. *The symplectic relations $\widehat{R}_\Sigma$ and $\widehat{\Delta}_\Sigma$ are, respectively, the reduction relation and the characteristic relation of the coisotropic submanifold*

$$C_\Sigma = T^*_\Sigma Q = \{p \in T^*Q \text{ such that } \pi_Q(p) \in \Sigma\}$$

made of the covectors based on points of Σ,

$$R_{C_\Sigma} = \widehat{R}_\Sigma, \quad D_{C_\Sigma} = \widehat{\Delta}_\Sigma.$$

Indeed, the characteristics of C_Σ are the equivalence classes of the equivalence relation defined by

$$p \sim p' \iff \begin{cases} \pi_Q(p) = \pi_Q(p') = q \in \Sigma, \\ \text{and} \\ \langle v, p - p' \rangle = 0 \text{ for all } v \in T_q\Sigma. \end{cases}$$

Remark 5.7. The canonical lift $\widehat{\Sigma}$ is invariant under the characteristic relation D_{C_Σ},

$$D_{C_\Sigma} \circ \widehat{\Sigma} = \widehat{\Sigma}.$$

Indeed,

$$\begin{aligned} D_{C_\Sigma} \circ \widehat{\Sigma} &= \{p \in T^*Q \text{ such that there exists a } p' \in T^\circ\Sigma \\ &\qquad \text{with } (p, p') \in D_{C_\Sigma}\} \\ &= \{p \in T^*Q \text{ such that there exists a } p' \in T^\circ\Sigma \\ &\qquad \text{with } p - p' \in T^\circ\Sigma\} = T^\circ\Sigma = \widehat{\Sigma}. \end{aligned}$$

---

[5] (Benenti 1988).

The same holds for the canonical lift with a function:

$$D_{C_\Sigma} \circ \widehat{(\Sigma, F)} = \widehat{(\Sigma, F)}. \tag{5.24}$$

This fact can be interpreted as follows. The Lagrangian submanifold $\widehat{(\Sigma, F)}$ is the geometrical solution of the Hamilton–Jacobi $C = T^*_\Sigma Q$ determined by the initial data (Σ, F). ♢

Remark 5.8. The characteristics of $C_\Sigma = T^*_\Sigma Q$ are *vertical submanifolds* (i.e., their tangent vectors are vertical) and the rays are the points of Σ (the "rays" are the projections on Q of the characteristics; see Sect. 6.3). ♢

Remark 5.9. We can consider Σ as the zero-section of $T^*\Sigma$. Then, Σ is a Lagrangian submanifold of $T^*\Sigma$. Hence, its inverse image $R^\top_{C_\Sigma} \circ \Sigma$ is a Lagrangian submanifold of T^*Q. This Lagrangian submanifold coincides with the canonical lift of Σ,

$$R^\top_{C_\Sigma} \circ \Sigma = \widehat{\Sigma}.$$

Indeed, because $R_{C_\Sigma} = \widehat{R}_\Sigma$ and a covector $p' \in \Sigma \subset T^*\Sigma$ is a zero-covector, we have

$$\begin{aligned} R^\top_{C_\Sigma} \circ \Sigma &= \{p \in T^*Q \text{ such that there exists a } p' \in \Sigma \\ &\quad \text{with } (p', p) \in \widehat{R}_\Sigma\} \\ &= \{p \in T^*Q \text{ such that there exists a } p' \in \Sigma \\ &\quad \text{with } \pi_Q(p') = \pi_Q(p) = q, \\ &\quad \text{and } \langle v, p - p' \rangle = 0 \text{ for all } v \in T_q\Sigma\} \\ &= \{p \in T^*Q \text{ such that } \langle v, p \rangle = 0 \text{ for all } v \in T_q\Sigma\} = \widehat{\Sigma}. \end{aligned}$$

In a similar way it can be proved that

$$R^\top_{C_\Sigma} \circ dF(\Sigma) = \widehat{(\Sigma, F)}, \tag{5.25}$$

$dF(\Sigma) \subset T^*\Sigma$ is the Lagrangian submanifold generated by $F\colon \Sigma \to \mathbb{R}$. ♢

Chapter 6
The Geometry of the Hamilton–Jacobi Equation

Abstract A coisotropic submanifold of a cotangent bundle gives rise to several geometric objects that allow an appropriate and quite general discussion of the Hamilton–Jacobi equations. For example, the concept of "solution" appears to have two meanings: from a geometrical viewpoint, it is a Lagrangian submanifold of C (or, possibly, a Lagrangian set contained in C), and, from an analytical viewpoint, it is a generating family satisfying a certain system of first-order PDE. One of the main problems related to a Hamilton–Jacobi equation is how to generate a (possibly unique) maximal solution from suitable *initial conditions* (*Cauchy problem*). We illustrate a geometrical construction of such a solution, by using the composition rule of symplectic relations, then we can transform this geometrical construction into an analytical method. Furthermore, other classical notions of geometrical optics, such as the system of rays and caustic of a system of rays, are more easily intelligible and manageable in a geometrical context.

6.1 The Hamilton–Jacobi equation

Definition 6.1. A *Hamilton–Jacobi equation* is a coisotropic submanifold C of a cotangent bundle T^*Q. A *geometrical solution* is a Lagrangian submanifold or a Lagrangian set Λ contained in C. ♡

In particular we say that Λ is

- a *smooth solution* when it is a Lagrangian submanifold;
- a *regular solution* when it is a Lagrangian submanifold without singular points;
- a *nonsmooth solution* when it is a Lagrangian set.

From an analytical viewpoint all these cases are included in a unique definition.

Definition 6.2. A *solution* of a Hamilton–Jacobi equation $C \subset T^*Q$ is a family $G(Q;U)$ generating a Lagrangian submanifold, or a Lagrangian set, contained in C. ♡

Remark 6.1. Let us see why in the above definitions we have to consider a coisotropic submanifold. Let us call *integrable* a submanifold $C \subset T^*Q$ such that at each point $p_\circ \in C$ there exists a Lagrangian submanifold Λ contained in C and containing $p_\circ$. Then we can state the following.

Theorem 6.1. *A Hamilton–Jacobi equation $C \subset T^*Q$ is integrable if and only if it is a coisotropic submanifold.*

Proof. (i) Assume that C is integrable. From $T_p\Lambda \subset T_pC$ it follows that $T_p^\S C \subset T_p^\S \Lambda = T_p\Lambda$, since Λ is Lagrangian. Thus, $T_p^\S C \subset T_pC$, and C is coisotropic. (ii) Assume that C is coisotropic. Take a point $p \in C$. There always exists a neighborhood $C_p \subset C$ of p, such that the reduced set M/C_p (here, $M = T^*Q$) is a manifold, thus a symplectic manifold, and a symplectic reduction $R = R_{C_p}$ is defined from M to this manifold. Consider the reduced point $\gamma = R \circ \{p\}$. We observe that for each point of a symplectic manifold there always exists a Lagrangian submanifold containing that point.[1] Consider a Lagrangian submanifold Λ_γ containing γ. The inverse image $\Lambda = R^\top \circ \Lambda_\gamma$ is a Lagrangian submanifold containing p. □

Note that this theorem is at the level of the category of symplectic manifolds: the cotangent bundle structure is not involved.[2] ♢

In a canonical coordinate system $(q^i, p_i) = (q, p)$ a Hamilton–Jacobi equation $C \subset T^*Q$ can be represented by a system of independent equations

$$C_a(q,p) = 0, \quad a = 1, \ldots, k. \tag{6.1}$$

The coisotropy of C is characterized by equations

$$\{C_a, C_b|C = 0.$$

Then, according to Definition 6.2, a "solution" is a generating family $G(q^i; u^\alpha)$ satisfying the following equations:

$$\boxed{\begin{aligned} &C_a\left(q, \frac{\partial G}{\partial q}\right) = 0, \quad a = 1, \ldots, k, \\ &\frac{\partial G}{\partial u^\alpha} = 0, \qquad\qquad \alpha = 1, \ldots, r, \end{aligned}} \tag{6.2}$$

[1] This follows from the existence of local canonical coordinates (Darboux theorem). Indeed, if $\omega = dy_i \wedge dx^i$, where (x^i, y_i) are canonical coordinates such that $x^i(p) = 0$, then equations $x^i = 0$ define a Lagrangian submanifold containing p.

[2] In (Abraham and Marsden 1978) a coisotropic submanifold is also called an *integrable submanifold.*

for some values of the parameters $u = (u^\alpha)$.

Remark 6.2. When C has codimension 1, we find the ordinary Hamilton–Jacobi equation, that is, a single first-order PDE,

$$C\left(q^i, \frac{\partial G}{\partial q^i}\right) = 0, \tag{6.3}$$

where $C(q, p)$ is a smooth function in $2n$ variables $(q, p) = (q^i, p_i)$, and equations

$$p_i = \frac{\partial G}{\partial q^i}.$$

describe a regular Lagrangian submanifold contained in C.[3] ◇

6.2 Examples of Hamilton–Jacobi equations

Example 6.1. The terminology we use is taken from geometrical optics, inasmuch as one of the most important examples of Hamilton–Jacobi equations is the *eikonal equation*,

$$\boxed{g^{ij}(q)\, p_i\, p_j - n^2(q) = 0} \tag{6.4}$$

determined by a positive-definite contravariant metric tensor g^{ij} on a manifold Q and by a function $n\colon Q \to \mathbb{R}$. A special but fundamental case is that of an *isotropic medium*, where $Q = \mathbb{R}^3$ is the Euclidean three-space and n is the *refraction index*,

$$n = \frac{c}{v},$$

where v is the velocity of the light in the medium. A homogeneous medium is characterized by $n =$ constant, the vacuum by $n = 1$.[4] ◇

[3] The common geometrical interpretation of the "Hamilton–Jacobi equation" is a hypersurface (i.e., a submanifold of codimension 1) of a cotangent bundle or of a contact manifold, see (Vinogradov and Kupershmidt 1977) and (Arnold 1980). The fundamental elements of the geometrical theory of the Hamilton–Jacobi equation, interpreted as a submanifold $C \subset T^*Q_n$ of any dimension $k < n$, are given in the short note (Tulczyjew 1975). Some of these elements have been developed in (Benenti and Tulczyjew 1980) and (Benenti 1983a,b).

[4] In the gravitational lensing theory the effective refraction index is $n(\boldsymbol{x}) = 1 - 2\,U(\boldsymbol{x})/c^2$, where $U(\boldsymbol{x})$ is the Newtonian potential of the mass distribution $\rho(\boldsymbol{x})$,

$$U(\boldsymbol{x}) = -\,G \int \frac{\rho(\boldsymbol{x}')}{|\boldsymbol{x} - \boldsymbol{x}'|}\, d^3x'.$$

See, for example, the article of N. Straumann in (Straumann et al. 1998).

Example 6.2. The Hamilton–Jacobi equation of a holonomic time-independent and conservative dynamical system, for a fixed value of the total energy $E \in \mathbb{R}$,

$$\tfrac{1}{2}\, g^{ij}(q)\, p_i\, p_j + V(q) - E = 0. \tag{6.5}$$

The reduced symplectic manifold R_C is the *manifold of the orbits* of total energy E (Souriau 1970). ♢

Example 6.3. The Hamilton–Jacobi equation of the Kepler motions in the Euclidean space $\mathbb{R}^n$ with a fixed value of the energy E, see Eq. (6.5), Sect. 3.10. ♢

Example 6.4. The Hamilton–Jacobi equation of a holonomic time-independent conservative system

$$\tfrac{1}{2}\, g^{ij}(q)\, p_i\, p_j + V(q) + p_0 = 0, \tag{6.6}$$

in the cotangent bundle of the extended configuration manifold $\mathbb{R} \times Q$, where $\mathbb{R} = (t) = (q^0)$ is the time-axis. This way of considering classical dynamics is called *homogeneous formalism*: time is considered as a Lagrangian coordinate. It can be extended to time-dependent holonomic systems,

$$\tfrac{1}{2}\, g^{ij}(t, q)\, p_i\, p_j + V(t, q) + p_0 = 0. \tag{6.7}$$

♢

Example 6.5. The Hamilton–Jacobi equation associated with a vector field $X = (X^i)$ on a manifold Q,

$$X^i p_i = 0, \tag{6.8}$$

whose solutions are the first integrals of X. In order to avoid singularities, it is convenient to consider its extension to the cotangent bundle of $\mathbb{R} \times Q$,

$$X^i p_i + p_0 = 0. \qquad ♢ \tag{6.9}$$

Example 6.6. The Hamilton–Jacobi equation associated with a completely integrable distribution,

$$X^i_\alpha p_i = 0, \tag{6.10}$$

where $X_\alpha = (X^i_\alpha)$ are $r \leq n$ independent vector fields spanning the distribution. In this case the coisotropic submanifold C has codimension $r \geq 1$ (see Sect. 9.1). ♢

Example 6.7. If $\Sigma \subset Q$ is a submanifold then $C = T^*_\Sigma$ in a Hamilton–Jacobi equation; see Remarks 5.7 and 5.8. ♢

6.3 Characteristics and rays

Due to Theorem 3.11, if a coisotropic submanifold C is represented by $k \leq n$ independent equations $C_a = 0$, then the functions C_a generate characteristic vector fields $\boldsymbol{X}_a$. The corresponding Hamilton equations are[5]

$$\begin{cases} \dot{q}^i = \partial^i C_a, \\ \dot{p}_i = -\,\partial_i C_a. \end{cases}$$

These vectors are pointwise independent because the rank of the $2n \times k$ matrix

$$\left[\partial^i C_a \mid \partial_i C_a\right]$$

is maximal. By linear combinations $\boldsymbol{X} = \lambda^a\, \boldsymbol{X}_a$ these fields span the characteristic distribution. Thus, the differential system associated with a characteristic vector field is of the kind

$$\begin{cases} \dot{q}^i = \lambda^a\, \partial^i C_a, \\ \dot{p}_i = -\,\lambda^a\, \partial_i C_a, \end{cases}$$

where λ^a are k arbitrary functions.

Definition 6.3. The *rays* of a Hamilton–Jacobi equation $C \subset T^*Q$ are the projections onto the configuration manifold Q of the characteristics of C. ♡

Remark 6.3. The characteristics of C project onto (immersed) submanifolds of Q of dimension equal to the codimension of C if the characteristics are transversal to the fibers,[6]

$$T^{\S}C \cap V(T^*Q) = 0.$$

In this case we say that the Hamilton–Jacobi equation C is *regular.* This condition is fulfilled if the rank of the $n \times k$ matrix $[\partial^i C_a]$ is maximal,

$$\operatorname{rank}\left[\partial^i C_a\right] = k. \tag{6.11}$$

Indeed, the vertical vectors are characterized by equations $\dot{q}^i = 0$. A vertical vector in $T^{\S}C$ is the zero-vector if equations $\lambda^a\, \partial^i C_a = 0$ imply $\lambda^a = 0$ thus, $\dot{p}_i = 0$. This happens if the matrix (6.11) has maximal rank. ◇

---

[5] Notation:

$$\partial^i = \frac{\partial}{\partial p_i}, \quad \partial_i = \frac{\partial}{\partial q^i}.$$

[6] The converse is not true in general. Let us consider, for instance, the Lagrangian submanifold of Example 4.1. It is a coisotropic submanifold with only one characteristic, the submanifold itself, which is not transversal to the fiber at the origin, and it projects onto a submanifold, the q-axis.

Remark 6.4. In the case codim $C = 1$, the characteristics are transversal to the fibers if rank$[\partial^i C] = 1$. Then the rays are (one-dimensional) curves. ♢

Remark 6.5. In the case of the eikonal equation, we have $C = |p|^2 - n^2(q)$ so that $\partial^i C = g^{ij} p_j$. It follows that the transversality condition is satisfied for all $(p_i) \neq 0 \in \mathbb{R}^n$, thus, for all $p \in C$, because $n \neq 0$. ♢

Theorem 6.2. *For the vacuum eikonal equation, $g^{ij} p_i p_j = 1$, the rays are oriented geodesics of the Riemannian manifold (Q, g^{ij}). For the eikonal equation $g^{ij} p_i p_j = n^2$, with $n \neq 0$, the rays are the oriented geodesics of the Jacobi metric*

$$\bar{g}^{ij} = \frac{1}{n^2} g^{ij}.$$

Proof. The integral curves on T^*Q of the Hamiltonian dynamical system generated by $H = \frac{1}{2} g^{ij} p_i p_j$ project onto the integral curves (on Q) of the Lagrange equations associated with $L = \frac{1}{2} g_{ij} \dot{q}^i \dot{q}^j$. These integral curves describe motions with constant scalar velocity on geodesic trajectories. □

Example 6.8. The characteristics of the Hamilton–Jacobi equation (6.8), $X^i p_i = 0$, are the unparametrized integral curves of the canonical lift $\widehat{X}$ starting from the points satisfying this equation. Recall that $\widehat{X}$ is the Hamiltonian vector field generated by the Hamiltonian $P_X = X^i p_i$ (see Sect. 5.5). The rays are the unparametrized integral curves of X (i.e., the orbits of X). ♢

Example 6.9. The rays of the Hamilton–Jacobi equation (6.9), $X^i p_i + p_0 = 0$, are parametrized integral curves of X: two integral curves describing the same unparametrized curve differ by the initial point. ♢

Example 6.10. The rays of the Hamilton–Jacobi equation (6.10), $X^i_\alpha p_i = 0$, associated with an integrable distribution, are the integral manifolds of the distribution (see Sect. 9.1 for details). ♢

6.4 Systems of rays and wave fronts

Definition 6.4. A *system of rays* associated with a Hamilton–Jacobi equation $C \subseteq T^*Q$ is the set of the projections on the configuration manifold Q of the characteristics contained in a geometrical solution $\Lambda \subseteq C$. ♡

If C is regular (see Sect. 6.3), then all characteristics project onto submanifolds of dimension equal to the codimension of C; so, if Λ is a smooth solution (i.e., a Lagrangian submanifold) then the corresponding system of rays is made of a set of these submanifolds, with possible points of intersection. In all other cases a system of rays may be a complicated family of subsets of Q.

Let us consider for simplicity the case of a smooth geometrical solution. Assume that Λ is the image of a closed one-form φ on an open domain $U \subset Q$, and that $\varphi \neq 0$ everywhere. Then on the domain U two regular and integrable distributions are defined.

The first distribution $\Delta_W \subset TU$ has the one-form φ as a characteristic form; that is, it is made of the vectors annihilated by φ. The one-form is closed, thus this distribution is completely integrable with integral manifolds of codimension 1. These integral manifolds are called *wave fronts* of the solution Λ. If $\varphi = dG$, then the wave fronts are described by the equations $G =$ constant.

The second distribution $\Delta_R \subset TU$ is the projection onto TU of the characteristic distribution $T^{\S}C$ restricted to Λ. By the *absorption principle* this restriction is well defined: $T^{\S}_{\Lambda}C \subset T\Lambda$. If we assume that on Λ the characteristic distribution is transversal to the fibers (as we have seen, this condition is also satisfied by the eikonal equation), then the distribution R is completely integrable and its integral manifolds form a system of rays (whose dimension coincides with the codimension of the Hamilton–Jacobi equation C). Note that Δ_W and Δ_R have a complementary rank.

The distribution Δ_R is spanned by the projections on Q of the characteristic vector fields $\boldsymbol{X}_a$ restricted to Λ. The dynamical systems corresponding to these projected vector fields are the first set of the Hamilton equations

$$\dot{q}^i = \partial^i C_a(q, p), \tag{6.12}$$

where in the right-hand sides the momenta p are replaced by their expressions in terms of the coordinates q, defined by the components of the one-form $\varphi = \varphi_i \, dq^i$:

$$p_i = \varphi_i(q).$$

Remark 6.6. In the case of the eikonal equation, Eqs. (6.12) become

$$\dot{q}^i = 2 \, g^{ij} \, p_j,$$

and on a regular solution Λ generated by G,

$$\dot{q}^i = 2 \, g^{ij} \, \partial_j G.$$

This shows that the gradient of the generating function G is a vector field spanning the distribution Δ_R. Since G is constant on the wave fronts, we have that *a regular solution of the eikonal equation generates a system of geodesics* (the rays) *orthogonal to a system of hypersurfaces* (the wave fronts).

In fact, this is an equivalence: any orthogonally integrable system of geodesics corresponds to a regular solution of the eikonal equation. This was one of the leading ideas of Hamilton's *Theory of Systems of Rays* (Hamilton 1828). Inasmuch as the wave fronts are orthogonal to a system of geodesics, they are *geodesically parallel*; that is, the ray segments between two given wave fronts have constant length. ♢

Remark 6.7. The wave fronts of the Hamilton–Jacobi equations (6.8) or (6.9), $X^i p_i = 0$, are defined by equations $G =$ constant, where G is any first integral of the vector field X. Since G is constant along the integral curves, in this case any wave front is made of rays. The same property holds for the Hamilton–Jacobi equation (6.10) associated with a completely integrable distribution. ♢

Remark 6.8. The above description of the wave fronts and rays fails in a neighborhood of a singular point: when rays and wave fronts approach the caustic of Λ. An even more complicated situation is that arising from a non-smooth geometrical solution, that is, from a Lagrangian set generated by a solution G (with supplementary variables) that is not a Morse family. In this case wave fronts and caustics are not defined. ♢

Remark 6.9. The vectorial form of the (vacuum) eikonal equation in a Euclidean affine space $Q = \mathbb{R}^n$ is

$$|\boldsymbol{p}|^2 = 1.$$

A system of parallel (oriented) rays is represented by a unit (constant) vector $\boldsymbol{u}$. The generating function of the corresponding Lagrangian submanifold is

$$G(\boldsymbol{x}) = \boldsymbol{x} \cdot \boldsymbol{u}. \tag{6.13}$$

The wave fronts are the $n - 1$-planes orthogonal to $\boldsymbol{u}$. ♢

Remark 6.10. In an Euclidean space, the system of rays originated by a fixed point $\boldsymbol{x}_0$ is generated by the function

$$G_1(\boldsymbol{x}) = |\boldsymbol{x} - \boldsymbol{x}_0|$$

(this *distance function*, as a generating family, is examined in detail in Sect. 6.1) or by the Morse family

$$G_2(\boldsymbol{x}; \boldsymbol{a}) = (\boldsymbol{x} - \boldsymbol{x}_0) \cdot \boldsymbol{a}, \quad \boldsymbol{a} \in \mathbb{S}_{n-1},$$

with supplementary manifold $\mathbb{S}_{n-1}$. The generating function G_1 yields *outgoing rays* only, because

$$\boldsymbol{p} = \frac{\partial G_1}{\partial \boldsymbol{x}} = \frac{\boldsymbol{x} - \boldsymbol{x}_0}{|\boldsymbol{x} - \boldsymbol{x}_0|}. \tag{6.14}$$

Note that it is not differentiable for $\boldsymbol{x} = \boldsymbol{x}_0$, so that the Lagrangian submanifold described by Eq. (6.14) is not defined over the point $\boldsymbol{x}_0$. The generating family G_2 is globally defined and differentiable. The corresponding equations

$$0 = \frac{\partial G_2}{\partial \boldsymbol{a}}, \quad \boldsymbol{p} = \frac{\partial G_2}{\partial \boldsymbol{x}},$$

are equivalent to (see Remark 7.10 below)

$$\begin{cases} \boldsymbol{x} - \boldsymbol{x}_0 \parallel \boldsymbol{a} \quad (\parallel = \text{parallel to}), \\ \boldsymbol{p} = \boldsymbol{a}. \end{cases}$$

Thus, G_2 provides the outgoing as well as the incoming rays. ◇

6.5 The Hamilton principal function

The characteristic relation $D_C \subset T^*Q \times T^*Q$ determined by a coisotropic submanifold $C \subset T^*Q$ is a symplectic relation between cotangent bundles. Thus, at least locally, it is generated by generating families on the product manifold $Q \times Q$. It is a symmetric relation, therefore its generating families are skew-symmetric, in the sense of Theorem 4.11.

Definition 6.5. A *Hamilton principal function* (also called *characteristic function*) of a Hamilton–Jacobi equation $C \subset T^*Q$ is a generating family $S(Q \times Q; A)$ of the characteristic relations $D_C \colon T^*Q \leftarrow T^*Q$. ♡

As we show, a Hamilton principal function can be used for computing (i) all solutions of the Hamilton–Jacobi equation, and (ii) the system of rays associated with any solution.

Remark 6.11. If $S(q^i, q_0^i; a^\alpha)$ is a local representative of a Hamilton principal function, where (q^i) and (q_0^i) are local coordinates on Q and (a^α) local coordinates on A, then D_C is locally described by equations

$$\boxed{p_{0i} = -\frac{\partial S}{\partial q_0^i}, \quad p_i = \frac{\partial S}{\partial q^i}, \quad 0 = \frac{\partial S}{\partial a^\alpha}} \quad \diamond \qquad (6.15)$$

Remark 6.12. If p_0 is a point of C, then the set $D_C \circ \{p_0\}$ is the maximal characteristic containing p_0. If in Eqs. (6.15) the coordinates (q, p) are just the coordinates of p_0, then these equations describe (locally) this characteristic and consequently the corresponding ray. A system of rays corresponding to a solution of the Hamilton–Jacobi equation C can be computed in this way. ◇

Theorem 6.3. *If the coisotropic submanifold C is not a section of the cotangent bundle T^*Q, then the characteristic relation D_C is singular over the diagonal of $Q \times Q$.*

Proof. Assume that D_C is regular at a point $(q, q) \in \Delta_Q$. Hence, it is locally generated by a function $S(q, q_0)$ which is skew-symmetric, Theorem 4.11. In this case Eqs. (6.15) reduce to

$$p_{0i} = -\frac{\partial S}{\partial q_0^i}, \quad p_i = \frac{\partial S}{\partial q^i}.$$

S is skew-symmetric, therefore these equations show that for $q = q_0$ we have $p = p_0$. This means that we have a unique covector $p \in T_q^*Q \cap C$. This is in contradiction to the assumption that C is a section. □

Remark 6.13. If C is a section, then it is a Lagrangian submanifold. If it is connected, then we have only one characteristic (the manifold C itself) and the characteristic relation D_C is defined by

$$(p, p_0) \in D_C \iff p, p_0 \in C.$$

If $G\colon Q \to \mathbb{R}$ is a global generating function of C, then D_C is generated by the global (skew-symmetric) generating function

$$S(q, q_0) = G(q) - G(q_0). \qquad \diamondsuit$$

Remark 6.14. Theorem 6.3 shows that, except for the case considered in Remark 6.13, *a global Hamilton principal function is necessarily a generating family.* In other words, *a global Hamilton principal function cannot be a function on $Q \times Q$ only; that is, a two-point function.* This is a novelty with respect to the classical Hamilton–Jacobi theory, where S is intended to be a function of pairs of points of Q, locally represented by a function $S(q_1, q_0)$ of $2n$ coordinates.[7] Moreover, in the classical theory the Hamilton principal function is defined as an *action integral.* This is due to the following general property. ◇

---

[7] The Hamilton principal function S was introduced by Hamilton as a function depending on $2n+2$ variables (q_1, q_0, t_1, t_0), where q_0 are the initial values of coordinates (at the time t_0) of a holonomic system, and q_1 are their final values (at the time t_1). This function satisfies the Hamilton–Jacobi equations

$$\begin{cases} \dfrac{\partial S}{\partial t_1} + H\left(q_1, \dfrac{\partial S}{\partial q_1}, t_1\right) = 0 \\ \dfrac{\partial S}{\partial t_0} + H\left(q_0, -\dfrac{\partial S}{\partial q_0}, t_0\right) = 0, \end{cases}$$

where $H = H(q, p, t)$ is the (time-dependent) Hamiltonian function of the mechanical system. We have used here the classical notation adopted by Levi-Civita (Levi-Civita and Amaldi 1927) Chap. XI, n. 27. In the homogeneous formalism of Hamiltonian dynamics, time is considered as a coordinate and the n-dimensional configuration manifold Q of the system is replaced by the $n+1$-dimensional *extended configuration manifold* $\mathbb{R} \times Q$. The notion of Hamilton principal function as a generating function of the characteristic relation of a coisotropic submanifold or as a generating function of the symplectic relation $D_t \subset T^*Q \times T^*Q$, between the initial values (at $t_0 = 0$) and the final values (at $t = t_1$) of the coordinates in the motions generated by the Hamiltonian H, has been introduced in (Tulczyjew 1975, 1977b). In Hamiltonian optics other "characteristic functions" are considered (see e.g. (Synge 1962), Chap. II, (Luneburg 1964) p. 100, (Buchdahl 1970) p. 8. What we are considering here is just a generalization of the so-called *point-characteristic function.*

Theorem 6.4. *If we exclude the singular points and assume that the remaining part of D_C is an exact Lagrangian submanifold* (Sect. 9.2), *then a potential function of D_C is the integral*

$$I(p_1, p_0) = \int_{c[p_0, p_1]} \theta_Q, \tag{6.16}$$

where $c[p_0, p_1]$ is any path with extremal points (p_0, p_1) and contained in the characteristic containing these two points.

Proof. From the definition of potential function of a Lagrangian submanifold it follows that a potential function of the Lagrangian submanifold $D_C \subset (T^*Q \times T^*Q, d\theta_Q \ominus d\theta_Q)$ is given by the integral

$$I(p_1, p_0) = \int_c \theta_Q \ominus \theta_Q,$$

taken along any path c over D_C joining a fixed pair $(\bar{p}_1, \bar{p}_0)$ with the moving pair (p_1, p_0). This path can be represented by two curves $c_1(t)$ and $c_0(t)$ on C and defined on the real closed interval $[0, 1]$ such that $c_i(0) = \bar{p}_i$, $c_i(1) = p_i$ $(i = 0, 1)$, 6and $(c_1(t), c_0(t)) \in D_C$. Hence,

$$I(p_1, p_0) = \int_{c_1} \theta_Q - \int_{c_0} \theta_Q = \int_{c_1'} \theta_Q - \int_{c_0'} \theta_Q$$

where $c' = (c_1', c_0')$ is another path having the same property. By choosing $c_0' = c_0$ we see that

$$\int_{c_1} \theta_Q = \int_{c_1'} \theta_Q.$$

This means that θ_Q is exact for the chosen paths from $\bar{p}_1$ to p_1, as well as for those from $\bar{p}_0$ to p_0. Moreover, for each $t \in [0, 1]$ the two points $c_1(t)$ and $c_0(t)$ are the endpoints of a curve $\gamma_t(s)$ defined for $s \in [0, 1]$ with the image on a characteristic. The characteristic are isotropic, therefore θ_Q is exact on all γ_1. It follows that

$$A = \int_{c_1} \theta_Q + \int_{\gamma_1} \theta_Q - \int_{c_0} \theta_Q$$

is a number depending only on the fixed end points $(\bar{p}_1, \bar{p}_0)$. Thus,

$$I(p_1, p_0) = \int_{c_1} \theta_Q - \int_{c_0} \theta_Q = A - \int_{\gamma_1} \theta_Q.$$

A depends only on the fixed points $(\bar{p}_1, \bar{p}_0)$, and the path γ_1 goes from p_1 to p_0, thus the integral (6.16) is a potential function. □

As a consequence of this theorem we have another link with the classical theory.

Theorem 6.5. *For the eikonal equation* $g^{ij}p_ip_j = 1$ *the characteristic relation* D_C *outside the diagonal is locally generated by the distance function*

$$d(q_0, q_1) = \int_{t_0}^{t_1} \sqrt{g_{ij}\dot{q}^i\dot{q}^j}\, dt$$

where the integral is taken along the geodesic $q(t)$ *such that* $q(t_0) = q_0$*, and* $q(t_1) = q_1$.

Proof. The rays are the geodesics, therefore, because of the preceding theorem, a local potential function is given by

$$I(p_1, p_0) = \int_c p_i dq^i = \int_{t_0}^{t_1} p_i\dot{q}^i\, dt = 2\int_{t_0}^{t_1} g^{ij}p_ip_j\, dt,$$

where c is a characteristic from p_0 to p_1 and the integrals are taken along a geodesic $q^i(t)$ such that $\dot{q}^i = 2\,g^{ij}p_j$; see Eq. (6.15. The kinetic energy of the motion $q^i(t)$ is $K = \frac{1}{2}\,v^2 = \frac{1}{2}\,g_{ij}\dot{q}^i\dot{q}^j = 2\,g^{ij}p_ip_j = 2$, so that the scalar constant velocity is $v = ds/dt = 2$. This means that the Euclidean distance is such that $ds = 2\,dt$. Hence the last integral above is just the integral of ds. This shows that the characteristic function projects onto the generating function given by the distance. □

6.6 The Jacobi theorem

A Hamilton principal function can be derived, at least locally, from another *characteristic function* associated with a coisotropic submanifold C.

Definition 6.6. A *complete solution* (or a *complete integral*) of a Hamilton–Jacobi equation $C \subseteq T^*Q$ is a smooth function $W\colon Q \times A \to \mathbb{R}$, where A is a manifold, such that:

- For each $a \in A$, the function

  $$W_a\colon Q \to \mathbb{R}\colon q \mapsto W(q, a)$$

  is a generating function of a Lagrangian submanifold Λ_a contained in C, thus, a regular solution of the Hamilton–Jacobi equation. This means that, if $C_\kappa(q, p) = 0$ are k independent equations of C, then W satisfies the differential equations

  $$\boxed{C_\kappa\left(q, \frac{\partial W}{\partial q}\right) = 0, \quad \kappa = 1, \ldots, k = \operatorname{codim} C} \tag{6.17}$$

for any fixed value of $a \in A$.

- The set $\{\Lambda_a,\, a \in A\}$ forms a Lagrangian foliation of C: for each $p \in C$ there exists a unique $a \in A$ such that $p \in \Lambda_a$. This means that, if (a^α) are local coordinates on A, then the $n \times m$ matrix $\left[\partial^2 W/\partial q \partial a\right]$ has maximal rank,

$$\boxed{\operatorname{rank}\left[\frac{\partial^2 W}{\partial q^i \partial a^\alpha}\right] = m = \dim A} \tag{6.18}$$

This is called the *completeness condition.*

- The canonical projection $\pi\colon C \to A$, which maps a point $p \in C$ to the point $a \in A$ such that $p \in \Lambda_a$, is a smooth map. Because each Λ_a is described by equations

$$\boxed{p_i = \frac{\partial W_a}{\partial q^i}} \tag{6.19}$$

this means that by solving these equations with respect to $a = (a^\alpha)$ the resulting functions $a^\alpha(q,p)$ are differentiable. ♡

Remark 6.15. From this definition it follows that $m = n - k$; that is,

$$\dim A = \dim Q - \operatorname{codim} C.$$

As we show below, the canonical projection π (which is obviously surjective) is a submersion. ◇

Remark 6.16. We can extend this definition by considering generating families $W_a(q; v)$ parametrized by $a \in A$ and defining a Lagrangian foliation of C with singular points. In classical Hamilton–Jacobi theory, W is locally represented by a function of the coordinates q of Q and a set of *constants of integration* (a^α), which represent a point $a \in A$. No extra variables are present, because in the classical theory only ordinary generating functions are considered.[8] ◇

Remark 6.17. When a coisotropic submanifold C of a symplectic manifold (M, ω), in particular, of a cotangent bundle T^*Q, is given, then the main

[8] According to Levi-Civita (Levi-Civita and Amaldi 1927), Ch. X, n. 38, the letter W is used for denoting a complete solution of the time-independent *reduced Hamilton–Jacobi equation* $H - E = 0$, for any fixed value of the energy E. It is a function of the n Lagrangian coordinates q and of n constant parameters $\pi = (\pi^i)$, satisfying the *completeness condition*

$$\det\left[\frac{\partial^2 W}{\partial q \partial \pi}\right] \neq 0$$

Actually, because the energy E becomes a function of these constants, for a fixed value of E, they are not all independent. So, they can be expressed as functions of $n-1$ independent parameters a satisfying the completeness condition (6.18). This is in accordance with Definition 6.6, being in this case $m = n - 1$ because $\operatorname{codim} C = 1$. For a time-independent Hamiltonian H one can think of a complete solution of the Hamilton–Jacobi equation $\partial V/\partial t + H(q, \partial V/\partial q) = 0$ of the form $V = -Et + W$. Then, this equation reduces to $H - E = 0$. In this reduction procedure, Jacobi considered W as depending on E and on further $n-1$ constant parameters a.

problem is to find its characteristics. Indeed, if the characteristics are known, then we can solve the Cauchy problem; that is, we can construct geometrical solutions (Lagrangian submanifolds) $\Lambda \subset C$ starting from initial data (see Sect. 3.9). ♢

For solving this problem the notion of complete integral plays a crucial role, as shown by the following version of what is known as the *theorem of Jacobi*

Theorem 6.6. *Let $W: Q \times A \to \mathbb{R}$ be a complete solution of the Hamilton–Jacobi equation (coisotropic submanifold) $C \subseteq T^*Q$.*[9] *Then,*

- *W generates a symplectic coreduction $R^\top : T^*Q \leftarrow B$ such that $R^\top \circ B = C$, where B is an open submanifold of T^*A.*
- *The characteristics of C are the connected components of the inverse images $R^\top \circ \{b\}$ of the points $b \in B$.*
- *The reduction R is isomorphic to the reduction relation R_C; that is, there exists a symplectomorphism $\varphi: T^*Q/C \to B$ such that $R = \varphi \circ R_C$.*[10]

Proof. (i) Let $(q, a) = (q^i, a^\alpha)$ be local coordinates of $Q \times A$. Let $(q, p; a, b) = (q^i, p_i; a^\alpha, b_\alpha)$ be the corresponding canonical coordinates on $T^*Q \times T^*A$. Let us consider the symplectic relation $R \subset T^*Q \times T^*A$ generated by the function W: it is described by equations

$$\boxed{p_i = \frac{\partial W(q,a)}{\partial q^i}, \quad b_\alpha = -\frac{\partial W(q,a)}{\partial a^\alpha}} \tag{6.20}$$

The Lagrangian submanifolds Λ_a are determined by the first Eqs. (6.20), also see Eq. (6.19). Due to the completeness condition and by the implicit function theorem we get smooth functions

$$a^\alpha = a^\alpha(q,p) \tag{6.21}$$

representing the canonical projection $\pi: C \to A$. By inserting these functions into the second Eqs. (6.20) we get functions

$$b_\alpha = b_\alpha(q,p). \tag{6.22}$$

Equations (6.21) and (6.22) are equivalent to Eqs. (6.20). This shows that R is the graph of a smooth map $\rho: C \to T^*A$. By a formal derivation of Eqs. (6.20) we get equations

[9] Definition 6.6.

[10] This is a simplified version of the *global Jacobi theorem* treated in (Tulczyjew 1975) (Benenti and Tulczyjew 1980, 1982a, 1982b), and (Benenti 1983a, 1988). See (Libermann and Marle 1987) for a general review.

$$\begin{cases} \dot{p}_i = \dfrac{\partial^2 W}{\partial q^i\, \partial q^j}\, \dot{q}^j + \dfrac{\partial^2 W}{\partial q^i\, \partial a^\alpha}\, \dot{a}^\alpha \\ \dot{b}_\alpha = -\dfrac{\partial^2 W}{\partial q^j\, \partial a^\alpha}\, \dot{q}^j - \dfrac{\partial^2 W}{\partial a^\alpha\, \partial a^\beta}\, \dot{a}^\beta, \end{cases} \tag{6.23}$$

which represent the tangent map $T\rho$. If we assign any arbitrary value to $(\dot{a}^\alpha, \dot{b}_\alpha)$, due to the completeness condition (6.18) the second equation (6.23) admits a solution $\dot{q}^i$ and, due to the first equation (6.23), we get values for $\dot{p}_i$. This shows that ρ is a submersion, thus, that R is a reduction. By applying Theorem 1.2 to this case we can see that R is a symplectic reduction onto an open subset of T^*A, and the first item is proved. The second and third items are a consequence of Theorems 1.3 and 3.22, respectively. □

As a complement of this theorem we observe that, when only the first item is considered, it is reversible.

Theorem 6.7. *If a function $W\colon Q \times A \to \mathbb{R}$ generates a symplectic coreduction $R^\top : T^*Q \leftarrow B$ such that $R^\top \circ B = C$, where B is an open submanifold of T^*A, then it is a complete solution of the Hamilton–Jacobi equation C.*

Proof. Assume that W is the generating function of a symplectic coreduction $R^\top \subset T^*Q \times B$, with $B \subseteq T^*A$ open and $C = R^\top \circ B$. Then the first Eqs. (6.20) describe Lagrangian submanifolds $\Lambda_a = R^\top \circ (T^*A \cap B)$ that are contained in C. Since the fibers $T^*_a A \cap B$ form a Lagrangian foliation, the Lagrangian submanifolds Λ_a form a foliation. The canonical projection π is a submersion, because it is the composition of two submersions, $\pi = \pi_A \circ \rho$, where ρ is the submersion associated with the reduction R. □

Remark 6.18. Regarding the third requirement in Definition 6.6 of complete solution we observe that there are cases in which a smooth function $W\colon Q \times A \to \mathbb{R}$ generates a Lagrangian foliation of C such that the canonical projection π is not differentiable. An example is the following: $Q = \mathbb{R}$, $C = T^*Q = \mathbb{R}^2$, $A = \mathbb{R}$, $W(q, a) = a^3 q$. The Lagrangian submanifold Λ_a is described by equation $p = a^3$ and the map π is described by $a = p^{1/3}$. This map is not differentiable for $p = 0$ $(a = 0)$. Hence, W is not a complete integral. Instead, the function $W(q, a) = aq$ is a complete integral and defines the same foliation. ◇

6.7 From a complete integral to a Hamilton principal function

Theorem 6.8. *If $W(q, a)$ is a complete integral of the Hamilton–Jacobi equation $C \subset T^*Q$, then the generating family $S(Q \times Q; A)$ defined by*

$$\boxed{S(q,q';a) = W(q,a) - W(q',a)} \tag{6.24}$$

is a Hamilton principal function.

Proof. The characteristic relation D_C of a coisotropic submanifold is given by the composition $D_C = R_C^\top \circ R_C$; see Eq. (3.17). Assume that we know a complete integral $W(q,a)$ of C. Then, according to the third item of Theorem 6.6, it generates a coreduction $R^\top$ of a reduction R that is isomorphic to the reduction R_C. This means that

$$\Phi = R_C \circ R^\top$$

is the graph of a symplectomorphism; see Theorem 3.22. Thus, instead of $D_C = R_C^\top \circ R_C$ we can write

$$D_C = R^\top \circ R.$$

Let us recall Theorem 4.10: if $W(q,a)$ generates the coreduction $R^\top$, then R is generated by the function $W^\top(a,q)$ defined by

$$W^\top(a,q) = \ - W(q,a).$$

and by composing $W^\top$ with W, according to Theorem 4.12, we get the generating family (6.24) of D_C. □

Remark 6.19. Note that in the generating family S the manifold A plays the role of supplementary manifold. Note that S is skew-symmetric, in accordance with the symmetry of D_C and Theorem 4.11. The Hamilton principal function S defined by (6.24) is a Morse family. Indeed, due to completeness condition (6.18), the matrix

$$\left[\frac{\partial^2 S}{\partial q^i\,\partial a^\alpha} \;\middle|\; \frac{\partial^2 S}{\partial q^{i'}\,\partial a^\alpha} \;\middle|\; \frac{\partial^2 S}{\partial a^\beta\,\partial a^\alpha}\right]$$

$$= \left[\frac{\partial^2 W(q,a)}{\partial q^i\,\partial a^\alpha} \;\middle|\; -\frac{\partial^2 W(q',a)}{\partial q^{i'}\,\partial a^\alpha} \;\middle|\; \frac{\partial^2 W(q,a)}{\partial a^\beta\,\partial a^\alpha} - \frac{\partial^2 W(q',a)}{\partial a^\beta\,\partial a^\alpha}\right]$$

has maximal rank m everywhere. ◇

Remark 6.20. From the third item of Theorem 6.6 it follows that a necessary condition for the existence of a global complete integral is that the reduced symplectic manifold T^*Q/C be symplectomorphic to a cotangent bundle (or at least to an open subset of a cotangent bundle). If $W(q,a)$ is a local representative of a complete solution, then Eqs. (6.20) generate an open subrelation of R. In general, by integrating the Hamilton–Jacobi equation we can find only *local complete solutions*, which generate Lagrangian foliations on open subsets of C. Then the composition formula (6.24) generates local principal functions. ◇

Remark 6.21. There are cases in which a global principal function S exists, whereas a global complete solution does not. An example is the eikonal equation on the sphere $\mathbb{S}_2 \subset \mathbb{R}^3$, for which the reduced manifold is $\mathbb{S}_2$ (indeed, any oriented geodesic on the unit sphere $\mathbb{S}_2 \subset \mathbb{R}^3$ is represented by a unit vector orthogonal to the plane on which it lies). In these cases a global principal function may be determined by other methods (for $\mathbb{S}_2$ see Chap. 8). ♢

Remark 6.22. All the above definitions and results concerning a complete solution can be extended to the case of a generating family: $W(Q \times A; V)$, where V is a supplementary manifold, with coordinates (v^a). Then W and S depend on these extra variables and equations

$$0 = \frac{\partial S}{\partial v^a}, \qquad 0 = \frac{\partial W}{\partial v^a},$$

must be added to systems (6.15) and (6.20), respectively. ♢

Example 6.11. Let us consider, for example, the eikonal equation of the Euclidean plane $Q = \mathbb{R}^2$,

$$C(x, y, p_x, p_y) \doteq p_x^2 + p_y^2 - 1 = 0.$$

In Sect. 7.3 it is proved that:

- *The reduced symplectic manifold $(T^*Q)/C$ is symplectomorphic to $T^*\mathbb{S}_1$.*
- *A global complete integral is*

$$W\colon \mathbb{R}^2 \times \mathbb{S}_1 \to \mathbb{R}, \qquad W(\boldsymbol{x}, \boldsymbol{a}) = \boldsymbol{a} \cdot \boldsymbol{x}. \tag{6.25}$$

- *The characteristic relation D_C is generated by the Hamilton principal function*

$$S(\mathbb{R}^2 \times \mathbb{R}^2; \mathbb{S}_1), \qquad S(\boldsymbol{x}, \boldsymbol{x}'; \boldsymbol{a}) \doteq (\boldsymbol{x} - \boldsymbol{x}') \cdot \boldsymbol{a}. \tag{6.26}$$

Note that R_C is a regular Lagrangian submanifold that admits a global ordinary generating function (without supplementary variables), and D_C is singular over the diagonal $\Delta_Q \subset Q \times Q$. Out of the diagonal, D_C is made of two branches, which are regular Lagrangian submanifolds generated by the functions

$$S(\boldsymbol{x}, \boldsymbol{x}') = \pm|\boldsymbol{x} - \boldsymbol{x}'|.$$

Note that these functions are not differentiable for $\boldsymbol{x} = \boldsymbol{x}'$, that is, over the diagonal. All these results have a natural extension to the space $\mathbb{R}^n$. The unit circle $\mathbb{S}_1$ is replaced by the unit sphere $\mathbb{S}_{n-1}$ (see Sect. 7.3). Note that for $n = 2$, instead of the generating families W and S defined in (6.25) and (6.26), one can use the equivalent generating families $W(\mathbb{R}^2; \mathbb{R})$ and $S(\mathbb{R}^2 \times \mathbb{R}^2; \mathbb{R})$ defined by

$$W(\boldsymbol{x}; \theta) = x \cos\theta + y \sin\theta$$

and

$$S(\boldsymbol{x}, \boldsymbol{x}'; \theta) = (x - x')\ \cos\theta + (y - y')\ \sin\theta,$$

respectively. ◇

6.8 Sources

In this and in the next sections we deal with the generation and transformation of geometrical solutions of a Hamilton–Jacobi equation. Recall that:

- A Hamilton–Jacobi equation is represented by a coisotropic submanifold of a cotangent bundle $C \subset T^*Q$.
- A geometrical solution is a Lagrangian set $\Lambda \subset C$.
- A smooth geometrical solution represents a system of rays (Definition 6.4).

However, for a better understanding of what is discussed, hereafter it is more effective to identify the latter two concepts and use the term "system of rays" for any geometrical solution $\Lambda \subset C$.

The topics touched on here are recalled in Sect. 7.4, where some basic applications to geometrical optics are discussed.

Let $\Sigma \subset Q$ be a submanifold. The composition of the canonical lift $\widehat{\Sigma} \subset T^*Q$ with the characteristic relation D_C,

$$\boxed{\Lambda = D_C \circ \widehat{\Sigma}} \tag{6.27}$$

gives rise to a geometrical solution $\Lambda \subset C$ which is the union of the (maximal) characteristics of C intersecting $\widehat{\Sigma}$. More generally, we can consider a submanifold Σ with a function $F \colon \Sigma \to \mathbb{R}$, the canonical lift a function $\widehat{(\Sigma, F)}$, and the set

$$\boxed{\Lambda = D_C \circ \widehat{(\Sigma, F)}} \tag{6.28}$$

which is still a geometrical solution of C. Then the pair (Σ, F) has the role of *source* of the system of rays represented by Λ, because it yields the *initial data* of the Cauchy problem[11] In terms of generating families we can then affirm the following.

Theorem 6.9. *If S is a Hamilton principal function of C and G is a generating family of $\widehat{(\Sigma, F)}$,[12] then the composed family $S \oplus G$ generates the geometrical solution $\Lambda = D_C \circ \widehat{\Sigma}$ of the Hamilton–Jacobi equation C.*

---

[11] Sect. 3.9, see also (Cardin 1989, 2002).

[12] See (5.4)

Remark 6.23. $S \oplus G$ is in any case a smooth solution of the Hamilton–Jacobi equation, whereas Λ may not be a smooth geometrical solution. If the intersection $C \cap \widehat{(\Sigma, F)}$ is clean then $S \oplus G$ is a Morse family and Λ is an immersed Lagrangian submanifold. ◇

Remark 6.24. Let $q = (q^\alpha, q^a)$ be coordinates adapted to Σ, so that Σ is locally described by equations $q^\alpha = 0$. It follows that $\widehat{(\Sigma, F)}$ is described by equations $q^\alpha = 0$ and $p_a - \partial_a F = 0$. Assume that C is defined by independent equations $C^A(q,p) = 0$. Then the intersection $C \cap \widehat{(\Sigma, F)}$ is clean if the matrix (the symbol $\Box$ denotes a square submatrix)

$$\begin{bmatrix} \boxed{\delta^\alpha_\beta} & -\partial_\beta \partial_a F & \partial_\beta C^A \\ 0 & \boxed{-\partial_b \partial_a F} & \partial_b C^A \\ \boxed{0} & 0 & \partial^\alpha C^A \\ 0 & \boxed{\delta^a_b} & \partial^b C^A \end{bmatrix}$$

has constant rank in a neighborhood of $C \cap \widehat{(\Sigma, F)}$, Theorem 1.7. Only the restriction of F to Σ is relevant, therefore the coordinates q^α can be chosen such that F does not depend on them, so that the above matrix becomes

$$\begin{bmatrix} \boxed{\delta^\alpha_\beta} & 0 & \partial_\beta C^A \\ 0 & \boxed{-\partial_b \partial_a F} & \partial_b C^A \\ \boxed{0} & 0 & \partial^\beta C^A \\ 0 & \boxed{\delta^a_b} & \partial^b C^A \end{bmatrix}.$$

This matrix must be computed for $C^A = 0$ and $p_i = \partial_i F$. In the case of the canonical lift of Σ (with $F = 0$ or constant) we have

$$\begin{bmatrix} \boxed{\delta^\alpha_\beta} & 0 & \partial_\beta C^A \\ 0 & \boxed{0} & \partial_b C^A \\ \boxed{0} & 0 & \partial^\beta C^A \\ 0 & \boxed{\delta^a_b} & \partial^b C^A \end{bmatrix}.$$

In the case of the eikonal equation we have $C = g^{ij} p_i p_j - n^2$, and the last columns in the above matrices reduce to a single column,

$$\begin{bmatrix} \boxed{\delta^\alpha_\beta} & 0 & \partial_\beta C \\ 0 & \boxed{-\partial_b \partial_a F} & \partial_b C \\ \boxed{0} & 0 & 2g^{\beta i} p_i \\ 0 & \boxed{\delta^a_b} & 2g^{bi} p_i \end{bmatrix}.$$

For $F = 0$ on $\widehat{\Sigma}$ we have $p_a = 0$ and the matrix becomes

$$\begin{bmatrix} \boxed{\delta^\alpha_\beta} & 0 & \partial_\beta C \\ 0 & \boxed{0} & \partial_b C \\ \boxed{0} & 0 & 2g^{\beta\gamma} p_\gamma \\ 0 & \boxed{\delta^a_b} & 2g^{b\gamma} p_\gamma \end{bmatrix}.$$

Because at least one $v^\beta = g^{\beta\gamma} p_\gamma \neq 0$ (due to the eikonal equation, otherwise all $p_i = 0$), this last matrix has maximal rank. This shows that for the eikonal equation, C and $\widehat{\Sigma}$ have a clean intersection. Consequently, $\Lambda = D_C \circ \widehat{\Sigma}$ is a Lagrangian submanifold. ♢

6.9 Mirror-relations

The first transformation that a system of rays may undergo is the reflection by a mirror. For this elementary optical phenomenon, and also for other transformations through lenses and refractions, we can build up a mathematical model based on the composition of symplectic relations of a special kind. In the next sections these relations are introduced within a quite general setting. Later on (Sect. 7.4) we show some applications to geometrical optics in Euclidean spaces.

Let $\Sigma \subset Q$ be a submanifold representing a mirror. The incident system of rays (input) and the reflected system (output) are represented by two Lagrangian sets, $\Lambda_I \subset C$ and $\Lambda_O \subset C$ (I = input, O = output). Both are geometrical solutions of the Hamilton–Jacobi equation.

We assume that they are related by equation

$$\boxed{\Lambda_O = D_C \circ \widehat{\Delta}_\Sigma \circ \Lambda_I} \tag{6.29}$$

where $\widehat{\Delta}_\Sigma \subset T^*Q \times T^*Q$ is the canonical lift of the diagonal $\Delta_\Sigma \subset Q \times Q$, defined by

$$\boxed{(p_2, p_1) \in \widehat{\Delta}_\Sigma \iff \begin{cases} p_1, p_2 \in T_q^*Q, \quad q \in \Sigma, \\ \langle v, p_2 - p_1 \rangle = 0 \ \text{ for all } v \in T_q\Sigma. \end{cases}} \tag{6.30}$$

Definition 6.7. We call a *mirror-relation* the composition

$$\boxed{M_\Sigma = D_C \circ \widehat{\Delta}_\Sigma} \qquad \heartsuit \tag{6.31}$$

Then formula (6.29) becomes

$$\boxed{\Lambda_O = M_\Sigma \circ \Lambda_I} \tag{6.32}$$

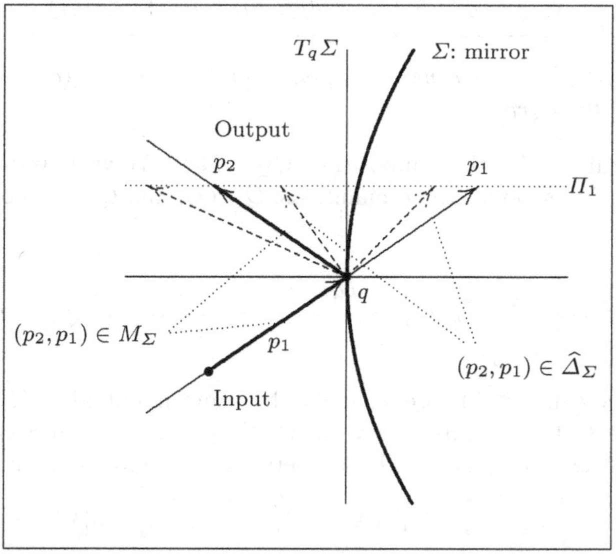

Fig. 6.1 The relation $\widehat{\Delta}_\Sigma$ and the mirror-relation M_Σ

Definition 6.7 is justified by its application to the geometrical optics in the n-dimensional Euclidean space. If $\Sigma \subset Q = \mathbb{R}^n$ is a regular r-dimensional surface, and covectors are interpreted as vectors, then a pair (p_2, p_1) based on a point $q \in \Sigma$ belongs to the relation $\widehat{\Delta}_\Sigma$ if and only if the vector $p_2 - p_1$ is orthogonal to the tangent plane $T_q\Sigma$. It follows that any p_2 in relation to a fixed p_1 belongs to the $n - r$-dimensional plane Π_1 orthogonal to $T_q\Sigma$ determined by the *endpoint* of p_1. In other words, all pairs (p_2, p_1) whose endpoints are on this plane belong to the relation. The role of the characteristic relation D_C is that of picking up among all the p_2 related to p_1 that one which belongs to C and propagating it along the characteristic to which it belongs. In other words, *the Lagrangian set* $\Delta_\Sigma \circ \Lambda_I$ *plays the role of source.*

Remark 6.25. It must be observed, especially in view of the applications, that the action of the characteristic relation D_C gives rise to two Lagrangian sets: one is the output Λ_O, the other is the input Λ_I itself. Indeed, among all the p_2 related to p_1 belonging to C there is also p_1. *This remark should be taken into account also in the following sections.* ♢

6.10 Generating family of a mirror-relation

Theorem 6.10. *The generating family of a mirror-relation is*

$$\boxed{G_{M_\Sigma}(q_O, q_I; a^\kappa, \lambda^\alpha) = S(q_O, q_I; a^\kappa) + \lambda^\alpha\, \Sigma_\alpha(q_I)} \tag{6.33}$$

where $S(q_O, q_I; a^\kappa)$ *is a generating family of* D_C*, and* $\Sigma_\alpha(q) = 0$ *are the equations of the mirror* Σ.[13]

Proof. In dealing with the composition $M_\Sigma = D_C \circ \widehat{\Delta}_\Sigma$ we have to consider, from a formal viewpoint, three manifolds, Q_1, Q_2, and Q_3, as shown by the diagram:

$$D_C \circ \widehat{\Delta}_\Sigma \colon T^*Q_3 \xleftarrow{\;D_C\;} T^*Q_2 \xleftarrow{\;\widehat{\Delta}_\Sigma\;} T^*Q_1.$$

However, these manifolds are actually the same manifold Q. Taking into account this fact, in accordance with what we have seen in Sect. 5.3, the canonical lift $\widehat{\Delta}_\Sigma \colon T^*Q_2 \leftarrow T^*Q_1$ is generated by the Morse family

$$G_1^\Sigma(q_2, q_1; \lambda^\alpha, \mu_i) = \lambda^\alpha\, \Sigma_\alpha(q_1) + \mu_i\,(q_2^i - q_1^i), \tag{6.34}$$

being $\Sigma_\alpha(q) = 0$ $(\alpha = 1, \dots m)$ the equations of the mirror Σ. If $S(q_2, q_1; a^\kappa)$ is a generating family of the characteristic relation $D_C \colon T^*Q_3 \leftarrow T^*Q_2 = T^*Q \leftarrow T^*Q$, then by the composition rule we obtain the generating family G_{M_Σ} of the mirror-relation M_Σ by the ordinary sum of functions

$$G_{M_\Sigma} = S(q_3, q_2; a^\kappa) + G_1^\Sigma(q_2, q_1; \lambda^\alpha, \mu_i)$$

and, as a second step, by considering the coordinates q_2 of the intermediate manifold $Q_2 = Q$ as supplementary variables. Hence, due to (6.34), we can

[13] In this section we denote by q_1, q_2, $\bar{q}$, ... generic coordinate systems (q_1^i), (q_2^i), $(\bar{q}^i)$, ... on the manifold Q, with $i = 1, \dots, n = \dim Q$.

write

$$
\begin{aligned}
G_{M_\Sigma}(q_3, q_1; q_2, a^\kappa, \lambda^\alpha, \mu_i) &= S(q_3, q_2; a^\kappa) + G_1^\Sigma(q_2, q_1; \lambda^\alpha, \mu_i) \\
&= S(q_3, q_2; a^\kappa) + \lambda^\alpha\, \Sigma_\alpha(q_1) + \mu_i\,(q_2^i - q_1^i).
\end{aligned}
\tag{6.35}
$$

Since Eq. (6.32) shows that the mirror-relation transforms the incoming system of rays represented by the Lagrangian set Λ_I, into the the reflected system of rays represented by the Lagrangian set Λ_O, it turns out to be convenient to apply the change of notation

$$
q_3 \mapsto q_O \quad (O = \text{output}), \quad q_1 \mapsto q_I \quad (I = \text{input}), \quad q_2 \mapsto \bar{q}.
$$

Then we can write (6.35) in the form

$$
\begin{aligned}
&G_{M_\Sigma}(q_O, q_I; \bar{q}, a^\kappa, \lambda^\alpha, \mu_i) = \\
&\qquad = S(q_O, \bar{q}; a^\kappa) + \lambda^\alpha\, \Sigma_\alpha(q_I) + \mu_i\,(\bar{q}^i - q_I^i).
\end{aligned}
\tag{6.36}
$$

However, this generating family can be reduced (Sect. 4.8). The critical set of the generating family G_{M_Σ} (6.36) is described by equations

$$
\begin{aligned}
0 &= \frac{\partial G_{M_\Sigma}}{\partial \bar{q}^i} = \frac{\partial S}{\partial \bar{q}^i} + \mu_i, & 0 &= \frac{\partial G_{M_\Sigma}}{\partial \lambda^\alpha} = \Sigma_\alpha(q_I^i), \\
0 &= \frac{\partial G_{M_\Sigma}}{\partial a^\kappa} = \frac{\partial S}{\partial a^\kappa}, & 0 &= \frac{\partial G_{M_\Sigma}}{\partial \mu_i} = \bar{q}^i - q_I^i.
\end{aligned}
$$

The last equation, $\bar{q}^i = q_I^i$, allows the removal of the parameters $\bar{q}^i$ and μ_i from the generating family, which reduces to (6.33). □

Theorem 6.11. *If the generating family $S(q_O, q_I; a^k)$ of D_C is a Morse family, then the generating family (6.33) of the mirror-relation is a Morse family.*

Proof. We write G instead of G_{M_Σ} for simplicity. The first derivatives of G are

$$
\begin{cases}
\dfrac{\partial G}{\partial q_O} = \dfrac{\partial S}{\partial q_O}, \\[2ex]
\dfrac{\partial G}{\partial q_I} = \dfrac{\partial S}{\partial q_I} + \lambda^\alpha \dfrac{\partial \Sigma_\alpha}{\partial q_I},
\end{cases}
\qquad
\begin{cases}
\dfrac{\partial G}{\partial a^k} = \dfrac{\partial S}{\partial a^k}, \\[2ex]
\dfrac{\partial G}{\partial \lambda^\alpha} = \Sigma_\alpha(q_I).
\end{cases}
$$

Hence,

$$\begin{bmatrix} \dfrac{\partial^2 G}{\partial a^k\,\partial q_O} & \dfrac{\partial^2 G}{\partial a^k\,\partial q_I} & \dfrac{\partial^2 G}{\partial a^k\,\partial a^\ell} & \dfrac{\partial^2 G}{\partial a^k\,\partial \lambda^\beta} \\ \dfrac{\partial^2 G}{\partial \lambda^\alpha\,\partial q_O} & \dfrac{\partial^2 G}{\partial \lambda^\alpha\,\partial q_I} & \dfrac{\partial^2 G}{\partial \lambda^\alpha\,\partial a^\ell} & \dfrac{\partial^2 G}{\partial \lambda^\alpha\,\partial \lambda^\beta} \end{bmatrix}$$

$$= \begin{bmatrix} \dfrac{\partial^2 S}{\partial a^k\,\partial q_O} & \dfrac{\partial^2 S}{\partial a^k\,\partial q_I} & \dfrac{\partial^2 S}{\partial a^k\,\partial a^\ell} & 0 \\ 0 & \dfrac{\partial \Sigma_\alpha(q_I)}{\partial q_I} & 0 & 0 \end{bmatrix}.$$

The generating family G_{M_Σ} is a Morse family if and only if the matrix

$$\begin{bmatrix} \dfrac{\partial^2 S}{\partial a^k\,\partial q_O} & \dfrac{\partial^2 S}{\partial a^k\,\partial q_I} & \dfrac{\partial^2 S}{\partial a^k\,\partial a^\ell} \\ 0 & \dfrac{\partial \Sigma_\alpha(q_I)}{\partial q_I} & 0 \end{bmatrix} \tag{6.37}$$

has maximal rank on the critical set Ξ of G_{M_Σ} itself. This is described by equations

$$\frac{\partial G}{\partial a^k} = \frac{\partial S}{\partial a^k} = 0, \quad \frac{\partial G}{\partial \lambda^\alpha} = \Sigma_\alpha(q_I) = 0.$$

The first equation shows that *the critical set of G_{M_Σ} is also the critical set of S*. Equations $\Sigma_\alpha(q) = 0$, $\alpha = 1, \ldots, s$, are independent, thus the matrix

$$\left[\frac{\partial \Sigma_\alpha(q_I)}{\partial q_I}\right]_{n\times s}$$

has maximal rank, equal to s. If S is a Morse family, the matrix

$$\begin{bmatrix} \dfrac{\partial^2 S}{\partial a^k\,\partial q_O} & \dfrac{\partial^2 S}{\partial a^k\,\partial q_I} & \dfrac{\partial^2 S}{\partial a^k\,\partial a^\ell} \end{bmatrix}_\Xi \tag{6.38}$$

also has maximal rank on the critical set Ξ of S. Since D_C is a symmetric relation, in accordance with Theorem 4.11 and Remark 4.10 the Morse family $S(q_O, q_I; a^k)$ is skew-symmetric on the critical set Ξ. This implies that the maximal rank of the matrix (6.38) is equivalent to the maximal rank of the matrix

$$\begin{bmatrix} \dfrac{\partial^2 S}{\partial a^k\,\partial q_O} & 0 & \dfrac{\partial^2 S}{\partial a^k\,\partial a^\ell} \end{bmatrix}_\Xi.$$

It follows that the matrix (6.37) has maximal rank. □

Remark 6.26. Theorem 6.10 shows that, under the nonrestrictive assumption that S is a Morse family, the mirror-relation $M_\Sigma \subset T^*Q \times T^*Q$ is a Lagrangian submanifold. ◇

Example 6.12. Mirror-relation in the Euclidean plane. It is opportune to give now an example of application of the overall results so far found. As we show in Sect. 7.3, the global Hamilton principal function of the eikonal equation in the Euclidean space $\mathbb{R}^2$ is

$$S(\boldsymbol{x}_O, \boldsymbol{x}_I; \boldsymbol{a}) = (\boldsymbol{x}_O - \boldsymbol{x}_I) \cdot \boldsymbol{a},$$

where $\boldsymbol{a}^2 = 1$. The generating family (6.33) of the mirror-relation is

$$G_{M_\Sigma}(\boldsymbol{x}_O, \boldsymbol{x}_I; \boldsymbol{a}) = (\boldsymbol{x}_O - \boldsymbol{x}_I) \cdot \boldsymbol{a} + \lambda^\alpha \, \Sigma_\alpha(\boldsymbol{x}_I). \tag{6.39}$$

In the Euclidean plane ($n = 2$) the unit vector $\boldsymbol{a}$ can be expressed as a function of an angle θ: $a_x = \cos\theta$, $a_y = \sin\theta$, and the mirror is given by a single equation $\Sigma(\boldsymbol{x}) = 0$. Then the generating family (6.39) of the mirror-relation assumes the expression

$$\boxed{\begin{aligned} &G_{M_\Sigma}(\boldsymbol{x}_O, \boldsymbol{x}_I; \theta) \\ &= (x_O - x_I)\,\cos\theta + (y_O - y_I)\,\sin\theta + \lambda\,\Sigma(\boldsymbol{x}_I) \end{aligned}} \tag{6.40}$$

If $G_I(\boldsymbol{x}; \boldsymbol{\zeta})$ is the generating family of the incoming Lagrangian set Λ_I, then the generating family of the Lagrangian set $\Lambda_O = M_\Sigma \circ \Lambda_I$ is

$$\boxed{\begin{aligned} G(\boldsymbol{x}; \boldsymbol{x}_I, \theta, \boldsymbol{\zeta}) &= (x - x_I)\,\cos\theta + (y - y_I)\,\sin\theta \\ &\quad + \lambda\,\Sigma(\boldsymbol{x}_I) + G_I(\boldsymbol{x}_I; \boldsymbol{\zeta}) \end{aligned}} \tag{6.41}$$

Note that in the passage from (6.40) to (6.41) the label O in $\boldsymbol{x}_O$ has been omitted.

Now, let us consider an incoming ray parallel to the x-axis, and a mirror-line Σ defined by equation $y - m\,x = 0$. In this case G_I is

$$G_I(\boldsymbol{x}) = x,$$

and the generating family of the Lagrangian set $\Lambda_O = M_\Sigma \circ \Lambda_I$ is

$$\boxed{G(\boldsymbol{x}; \boldsymbol{x}_I, \theta) = (x - x_I)\,\cos\theta + (y - y_I)\,\sin\theta + \lambda\,(y_I - m\,x_I) + x_I} \tag{6.42}$$

The Lagrangian set Λ_O representing the reflected rays is then described by the equations

$$p_x = \frac{\partial G}{\partial x} = \cos\theta, \quad p_y = \frac{\partial G}{\partial y} = \sin\theta, \tag{6.43}$$

together with the equations of the critical set:

$$\begin{cases} 0 = \dfrac{\partial G}{\partial x_I} = -\cos\theta - m\,\lambda + 1, \\ 0 = \dfrac{\partial G}{\partial y_I} = -\sin\theta + \lambda, \\ 0 = \dfrac{\partial G}{\partial \theta} = (x_I - x)\sin\theta + (y - y_I)\cos\theta, \\ 0 = \dfrac{\partial G}{\partial \lambda} = y_I - m\,x_I. \end{cases}$$

The first two equations imply

$$\begin{cases} \lambda = \sin\theta \\ \lambda = \dfrac{1}{m}(1 - \cos\theta) \end{cases} \Longrightarrow m\,\sin\theta = 1 - \cos\theta.$$

This last equation has two solutions:[14]

$$\begin{cases} \sin\theta = 0, \\ \cos\theta = 1, \end{cases} \quad \text{or} \quad \begin{cases} \sin\theta = \dfrac{2\,m}{1+m^2}, \\ \cos\theta = \dfrac{1-m^2}{1+m^2}. \end{cases}$$

Consequently, due to equations (6.43), we find two reflected system of rays:

$$\begin{cases} p_x = 1, \\ p_y = 0, \end{cases} \quad \text{and} \quad \begin{cases} p_x = \dfrac{1-m^2}{1+m^2}, \\ p_y = \dfrac{2\,m}{1+m^2}. \end{cases}$$

But the first solution represents nothing but the incoming rays, and we have to disregard it in accordance with Remark 6.25. Then the valid solution is the second one. For $m = 1$ it represents rays parallel to the y-axis, with the same orientation, as expected. ♢

[14] For $m = 0$ we have the double solution $\sin\theta = 0$, $\cos\theta = 1$.

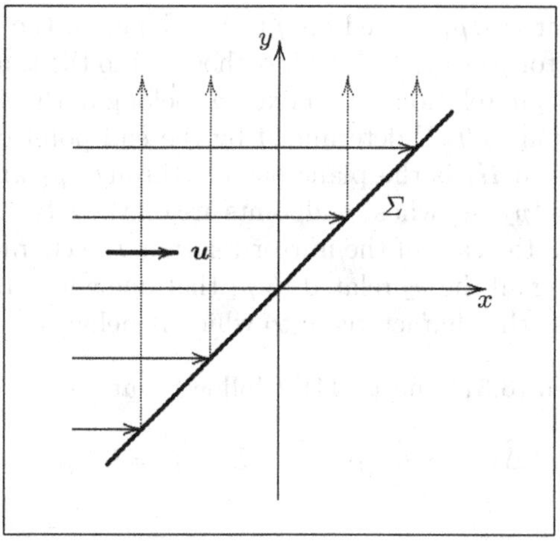

Fig. 6.2 Case $m = 1$ in Example 6.12

6.11 Lens-relations

By a method similar to that followed for a mirror we can define the lens-relation. In this case the submanifold Σ is endowed with a function F, and $\widehat{\Delta}_\Sigma$ is replaced by the canonical lift $\widehat{\Delta}_{\Sigma,F}$ defined in Eq. (5.14), Definition 5.7,

$$(p_2, p_1) \in \widehat{\Delta}_{\Sigma,F} \iff \begin{cases} p_2, p_1 \in T_q^*Q,\ q \in \Sigma, \\ \langle v, p_2 - p_1 - d_qF \rangle = 0, \quad \text{for all } v \in T_q\Sigma. \end{cases}$$

Definition 6.8. We call a *lens-relation* the composition

$$\boxed{L_{\Sigma,F} = D_C \circ \widehat{\Delta}_{\Sigma,F}} \tag{6.44}$$

for which we have

$$\boxed{\Lambda_O = L_{\Sigma,F} \circ \Lambda_I} \tag{6.45}$$

We call F the *characteristic function* of the lens. ♡

Remark 6.27. Actually, formula (6.45) represents the transition from an input Λ_I to the output Λ_O through an *ideal lens*. A *real lens* is in fact represented by a sequence (may be continuous) of compositions of refraction-relations (see below). ◇

Remark 6.28. By adapting to the present case what we have seen in Sect. 6.9), if Σ is a regular r-dimensional surface, and covectors are interpreted as vectors, then a pair (p_2, p_1) based on $q \in \Sigma$ belongs to the relation $\widehat{\Delta}_{\Sigma,F}$ if and only if the vector $p_2 - (p_1 + \nabla_q F)$ is orthogonal to the tangent plane $T_q\Sigma$. It follows that all p_2 in relation with a fixed p_1 belong to the $n-r$-dimensional plane Π_2 orthogonal to $T_q\Sigma$ determined by the end point of $p_1 + \nabla_q F$; see Fig. 6.3. Moreover, if Π_1 is the plane passing through p_1 and orthogonal to $T_q\Sigma$, then all pairs (p_2, p_1) whose endpoints are on these two planes belong to the relation. As for the case of the mirror-relation, the characteristic relation D_C picks up among all the p_2 related to p_1 that one which belongs to C and propagates it along the characteristic to which it belongs. ♢

Remark 6.29. From (6.31) and (6.44) it follows that

$$L_{\Sigma,c} = M_\Sigma, \quad \widehat{\Delta}_{\Sigma,F} = L_{\Sigma,F} \circ \widehat{\Sigma}, \quad L_{\Sigma,-F} = L^{\top}_{\Sigma,F}. \quad \diamondsuit \tag{6.46}$$

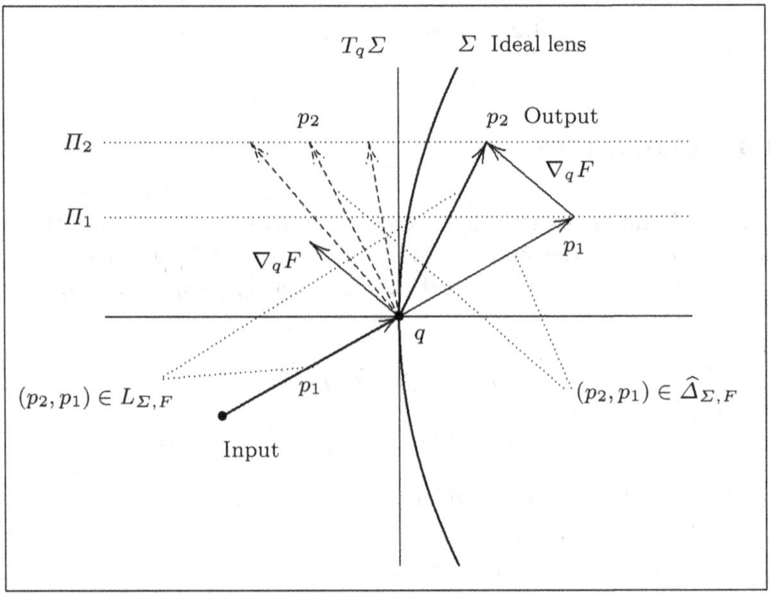

Fig. 6.3 The relation $\widehat{\Delta}_{(\Sigma,F)}$ and the lens-relation $L_{(\Sigma,F)}$

6.12 Generating family of a lens-relation

Theorem 6.12. *The generating family of a lens-relation is*

$$\boxed{G_{L_{\Sigma,F}}(q_O, q_I; a^\kappa, \lambda^\alpha) = S(q_O, q_I; a^\kappa) + F(q_I) + \lambda^\alpha\, \Sigma_\alpha(q_I)} \tag{6.47}$$

where $S(q_O, q_I; a^\kappa)$ *is a generating family of* D_C, *and* $\Sigma_\alpha(q) = 0$ *are the equations of the mirror* Σ.[15]

Proof. Due to the general Theorem 5.4 and formula (5.16), the generating family (6.36) of the mirror-relation

$$G_{M_\Sigma}(q_O, q_I; \bar{q}, a^\kappa, \lambda^\alpha, \mu_i) = S(q_O, \bar{q}; a^\kappa) + \lambda^\alpha\, \Sigma_\alpha(q_I) + \mu_i\, (\bar{q}^i - q_I^i),$$

can be adapted to the lens-relation as follows.

$$\begin{aligned} &G_{L_{\Sigma,F}}(q_O, q_I; \bar{q}, a^\kappa, \lambda^\alpha, \mu_i) \\ &= S(q_O, \bar{q}; a^\kappa) + F(q_I) + \lambda^\alpha\, \Sigma_\alpha(q_I) + \mu_i\, (\bar{q}^i - q_I^i). \end{aligned} \tag{6.48}$$

However, this generating family can be reduced. The critical set of the generating family $G_{M_\Sigma}{}'$ (6.36) is described by equations (we write G instead of $G_{L_{\Sigma,F}}$ for simplicity):

$$0 = \frac{\partial G}{\partial \bar{q}} = \frac{\partial S}{\partial \bar{q}} + \mu_i, \quad 0 = \frac{\partial G}{\partial \lambda^\alpha} = \Sigma_\alpha(q_I),$$
$$0 = \frac{\partial G}{\partial a^\kappa} = \frac{\partial S}{\partial a^\kappa}, \qquad 0 = \frac{\partial G}{\partial \mu_i} = \bar{q}^i - q_I^i.$$

Due to the last equation, $\bar{q} = q_I$, the parameter $\bar{q}$ can be removed from (6.48), as well as μ_i. □

Remark 6.30. The comparison between (6.33) and (6.47) shows a surprising fact: the generating families of the mirror-relation and of the lens-relation differ by the term $F(q_I)$, namely:

$$G_{L_{\Sigma,F}}(q_O, q_I; a^\kappa, \lambda^\alpha) = G_{M_\Sigma}(q_O, q_I; a^\kappa, \lambda^\alpha) + F(q_I). \quad \diamondsuit$$

As a consequence, a proof similar to that of Theorem 6.11 shows the following.

Theorem 6.13. *If the generating family* $S(q_O, q_I; a^k)$ *of* D_C *is a Morse family, then the generating family* (6.47) *of the lens-relation is a Morse family.*

6.13 Reflection and refraction

If, due to an eikonal equation, p_1 and p_2 have a prescribed constant length $n = v/c$, equal to the *refraction index*, then only the vectors belonging to the sphere of radius n are involved in the mirror-relation and in the lens-relation. Consequently, in the case of hypersurface of codimension 1, the mirror-relation reproduces the well-known reflection law (Fig. 6.4).

[15] In this section we use for the coordinates the same notation of Sect. 6.10.

Consider now covectors having a different length in the two half-spaces separated by a hypersurface Σ. This is the case of two different eikonal equations C_1 and C_2 in T^*Q, corresponding to two different media, with refraction index n_1 and n_2 and separated by a surface Σ. If we consider these two media coexisting in the whole space, then we have to deal with two characteristic relations D_{C_1} and D_{C_2} and with a surface Σ.

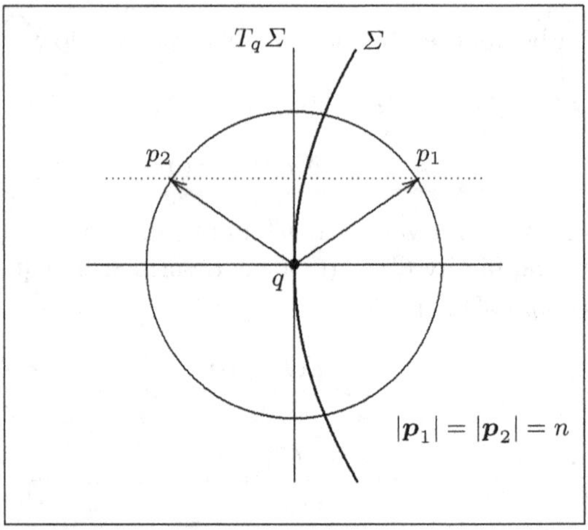

Fig. 6.4 Reflection law

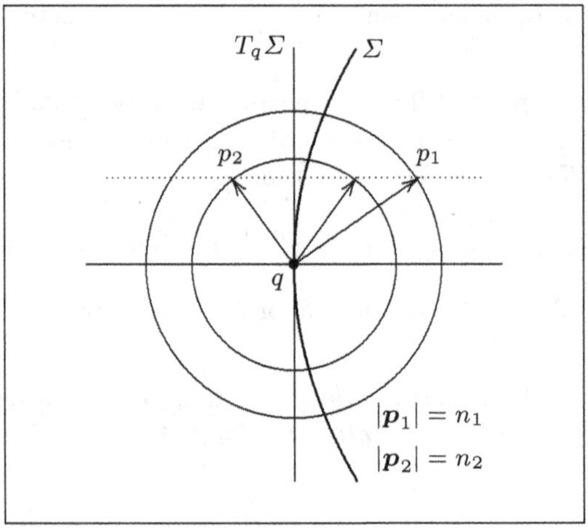

Fig. 6.5 Refraction law

If Λ_1 is a Lagrangian subset of C_1 (i.e., a system of rays in the first medium), then the composition

$$\boxed{\Lambda_2 = D_{C_2} \circ M_\Sigma \circ \Lambda_1} \tag{6.49}$$

gives a solution of C_2 representing the refracted system of rays, in the second medium.

Actually, with an *incident vector* $\boldsymbol{p}_1$ this relation associates two *refracted vectors* $\boldsymbol{p}_2$. Only one of them (and only a half-line determined by it) has a physical meaning.

Remark 6.31. Malus theorem. The possible applications to geometrical optics of the matter exposed in this chapter are quite evident. They are treated in the next chapter. The *philosophy* is the following: a system of rays emitted by a source is represented by a Lagrangian submanifold (or a Lagrangian set) Λ_0. Along the way this system can be modified by a sequence of mirrors, lenses and refractions, which in turn are represented by a sequence of compositions with symplectic relations:

$$\Lambda_0 \quad \longrightarrow \quad \Lambda_1 = R_1 \circ \Lambda_0 \quad \longrightarrow \quad \Lambda_2 = R_2 \circ \Lambda_1 \quad \longrightarrow \quad \cdots$$

Consequently, the sequence $\Lambda_0, \Lambda_1, \Lambda_2, \ldots$ is made of Lagrangian submanifolds (or Lagrangian sets) representing systems of rays. These compositions of relations are expressed in terms of compositions of generating families. This fact is nothing but the contents and the (simple) proof of the ancient and celebrated *theorem of Malus* (1807). This theorem, according to the translation of V. V. Koslov (Koslov 2003) states that *if a system of rays is orthogonal to a regular surface, then it is a Hamilton system and remains a Hamilton system after an arbitrary number of reflections and refractions.*[16] ◇

[16] According to our approach, *and lenses* should be added. A *real* lens is nothing but a sequence of refractions.

Chapter 7
Hamiltonian Optics in Euclidean Spaces

Abstract According to Hamilton (Hamilton 1828), a "system of rays" is a congruence of straight lines in the Euclidean three-space, orthogonal to a family of surfaces. This orthogonal integrability of the rays fails in the presence of a caustic. Moreover, a system of rays can be modified through the use of optical devices, as mirrors and lenses, or by passing through surfaces that delimit two media with different refraction index. In our approach, a system of rays without caustic is represented by a regular Lagrangian submanifold of the cotangent bundle of the Euclidean space, whereas all the optical devices are represented by symplectic relations. This chapter discusses some of the most important elementary examples.

7.1 The distance function

Let $Q = \mathbb{R}^n = \{\boldsymbol{x}\} = \{(x^i)\}$ be the Euclidean n-space. We can identify the tangent bundle $TQ = \{(\boldsymbol{x}, \boldsymbol{p})\}$ with the cotangent bundle $T^*Q = \{(x^i, p_i)\}$. Notation: $\boldsymbol{u} \cdot \boldsymbol{v} = \sum_i u^i v^i$ is the scalar product of two vectors and $|\boldsymbol{u}| = \sqrt{\boldsymbol{u} \cdot \boldsymbol{u}}$; for $n = 3$, $\boldsymbol{u} \times \boldsymbol{v}$ is the cross-product of two vectors.

Let $U \subset Q$ be a regular and orientable r-dimensional surface (locally) described by parametric equations $\boldsymbol{x} = \boldsymbol{u}(u^\alpha)$, $\alpha = 1, \ldots, r$. Let us consider the *distance function* $\Phi\colon Q \times U \to \mathbb{R}$ defined by

$$\boxed{\Phi(\boldsymbol{x}; u^\alpha) = |\boldsymbol{x} - \boldsymbol{u}(u^\alpha)|} \tag{7.1}$$

as a generating family on Q with supplementary manifold U, and supplementary coordinates (u^α).

Theorem 7.1. *The distance function is a smooth Morse family for $\boldsymbol{x} \neq \boldsymbol{u}$. It generates the Lagrangian submanifold Λ_U defined by*

$$(\boldsymbol{x},\boldsymbol{p}) \in \Lambda_U \iff \begin{cases} \boldsymbol{p} = \dfrac{\boldsymbol{x}-\boldsymbol{u}}{|\boldsymbol{x}-\boldsymbol{u}|}, & \boldsymbol{u}\in U, \;\; \boldsymbol{u}\neq \boldsymbol{x}, \\ \boldsymbol{p} \perp U, \end{cases} \tag{7.2}$$

and contained in the one-codimensional coisotropic submanifold $C \subset T^*Q$ *defined by equation*

$$|\boldsymbol{p}|^2 = \sum_i p_i^2 = 1. \tag{7.3}$$

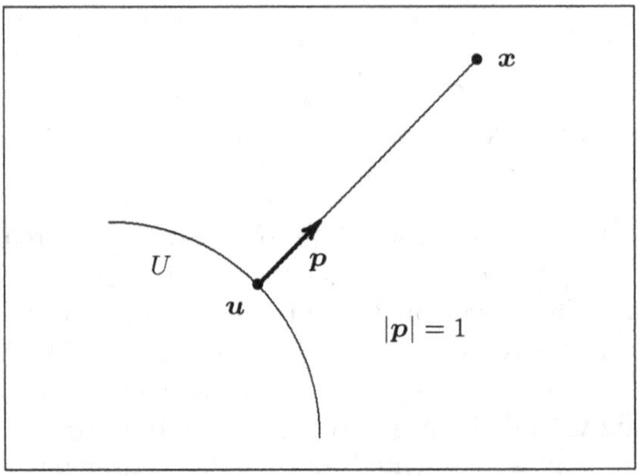

Fig. 7.1 Theorem 7.1

Proof. The equations of the Lagrangian set Λ_U generated by Φ are

$$\begin{cases} \boldsymbol{p} = \dfrac{\partial \Phi}{\partial \boldsymbol{x}} = \dfrac{\boldsymbol{x}-\boldsymbol{u}}{|\boldsymbol{x}-\boldsymbol{u}|}, \\ 0 = \dfrac{\partial \Phi}{\partial u^\alpha} = \dfrac{1}{|\boldsymbol{x}-\boldsymbol{u}|}\,\partial_\alpha(\boldsymbol{x}-\boldsymbol{u})\cdot(\boldsymbol{x}-\boldsymbol{u}) = -\,\boldsymbol{e}_\alpha\cdot\boldsymbol{p}, \end{cases} \tag{7.4}$$

where

$$\boldsymbol{e}_\alpha = \partial_\alpha \boldsymbol{u} \qquad \left(\partial_\alpha = \frac{\partial}{\partial u^\alpha}\right)$$

are the coordinate vectors tangent to U. Equations (7.4) are equivalent to (7.2). From the first equation (7.4) it follows that $|\boldsymbol{p}| = 1$. From the second Eq. (7.4) it follows that the critical set Ξ is made of pairs of vectors $(\boldsymbol{x},\boldsymbol{u})$ such that $\boldsymbol{x}-\boldsymbol{u}$ is perpendicular to U at the point $\boldsymbol{u}$. Let us set

$$A_{\alpha\beta} = \boldsymbol{e}_\alpha\cdot\boldsymbol{e}_\beta, \qquad \partial_\alpha \boldsymbol{e}_\beta = \Gamma^\gamma_{\alpha\beta}\,\boldsymbol{e}_\gamma + \boldsymbol{B}_{\alpha\beta}.$$

Then $A_{\alpha\beta}$ are the components of the first fundamental form of the surface U, $\Gamma^\gamma_{\alpha\beta}$ are the Christoffel symbols, and $\boldsymbol{B}_{\alpha\beta}$ are vector fields orthogonal to U, representing the external curvature of the surface. Note that if U is of codimension 1, then $\boldsymbol{B}_{\alpha\beta} = B_{\alpha\beta}\,\boldsymbol{n}$, where $\boldsymbol{n} =$ is a unit vector field orthogonal to U. Thus, $B_{\alpha\beta}$ are the components of the second fundamental form of the surface U. It follows that

$$\frac{\partial \boldsymbol{p}}{\partial u^\alpha} = \frac{\partial^2 \Phi}{\partial \boldsymbol{x}\, \partial u^\alpha} = -\frac{\partial_\alpha \boldsymbol{u}}{|\boldsymbol{x}-\boldsymbol{u}|} - \frac{1}{2}\frac{\boldsymbol{x}-\boldsymbol{u}}{|\boldsymbol{x}-\boldsymbol{u}|^3}\, 2\, (\boldsymbol{x}-\boldsymbol{u}) \cdot \partial_\alpha(\boldsymbol{x}-\boldsymbol{u})$$

$$= \frac{(\boldsymbol{p}\cdot\boldsymbol{e}_\alpha)\,\boldsymbol{p} - \boldsymbol{e}_\alpha}{|\boldsymbol{x}-\boldsymbol{u}|},$$

and

$$\frac{\partial^2 \Phi}{\partial u^\alpha\, \partial u^\beta} = -\,\partial_\alpha(\boldsymbol{e}_\beta \cdot \boldsymbol{p}) = -\,\partial_\alpha \boldsymbol{p}\cdot \boldsymbol{e}_\beta - \boldsymbol{p}\cdot(\Gamma^\gamma_{\alpha\beta}\boldsymbol{e}_\gamma + B_{\alpha\beta}\,\boldsymbol{n})$$

$$= \frac{A_{\alpha\beta} - (\boldsymbol{p}\cdot\boldsymbol{e}_\alpha)(\boldsymbol{p}\cdot\boldsymbol{e}_\beta)}{|\boldsymbol{x}-\boldsymbol{u}|} - \boldsymbol{p}\cdot(\Gamma^\gamma_{\alpha\beta}\boldsymbol{e}_\gamma + B_{\alpha\beta}\,\boldsymbol{n}).$$

On the critical set Ξ we have $\boldsymbol{p}\cdot\boldsymbol{e}_\alpha = 0$, so that

$$\begin{cases} \dfrac{\partial^2 \Phi}{\partial \boldsymbol{x}\, \partial u^\alpha} = \dfrac{\partial \boldsymbol{p}}{\partial u^\alpha} = -\dfrac{\boldsymbol{e}_\alpha}{|\boldsymbol{x}-\boldsymbol{u}|}, \\[2ex] \dfrac{\partial^2 \Phi}{\partial u^\alpha \partial u^\beta} = \dfrac{A_{\alpha\beta}}{|\boldsymbol{x}-\boldsymbol{u}|} - \boldsymbol{B}_{\alpha\beta}\cdot\boldsymbol{p}. \end{cases} \tag{7.5}$$

The vectors $\boldsymbol{e}_\alpha$ are independent, thus the first submatrix of the matrix

$$\left[\frac{\partial^2 \Phi}{\partial u^\alpha\, \partial x^i} \;\middle|\; \frac{\partial \Phi}{\partial u^\alpha\, \partial u^\beta} \right]_\Xi$$

has maximal rank. This proves that the distance function is a Morse family. □

Remark 7.1. The Lagrangian submanifold Λ_U is contained in the coisotropic submanifold C defined by Eq. (7.3). This is the eikonal equation of the Euclidean plane. The rays are oriented straight lines, Theorem 6.2. The system of rays corresponding to Λ_U is the set of outgoing straight lines perpendicular to U. Here, the submanifold U behaves as a source of a system of rays, according with the theory developed in Sect. 6.8 (where a source was denoted by Σ). ◇

Remark 7.2. The caustic Γ_U of Λ_U is described by Eqs. (4.34), Theorem 4.8, which in the present case become

$$\det[\partial_{\alpha\beta}\Phi] = 0, \quad \partial_\alpha \Phi = 0.$$

Due to $(7.4)_2$ and to the second Eq. (7.5), these equations are equivalent to

$$\begin{cases} \det\left[\boldsymbol{p}\cdot\boldsymbol{B}_{\alpha\beta} - \dfrac{1}{|\boldsymbol{x}-\boldsymbol{u}|}\, A_{\alpha\beta}\right] = 0, \\ \boldsymbol{p}\cdot\boldsymbol{e}_\alpha = 0. \end{cases} \quad \diamond \tag{7.6}$$

Let us consider the special case of an oriented surface U in the three-space: $n = 3$, $r = 2$. The first Eq. (7.6) is equivalent to

$$\det\left[B_{\alpha\beta} - \frac{1}{(\boldsymbol{x}-\boldsymbol{u})\cdot\boldsymbol{n}}\,A_{\alpha\beta}\right] = 0, \tag{7.7}$$

and the second one to $\boldsymbol{x} - \boldsymbol{u} \perp U$. Because the characteristic equation of the main curvatures of a surface is

$$\det\left[B_{\alpha\beta} - \lambda\, A_{\alpha\beta}\right] = 0,$$

we find

$$\lambda = \frac{1}{(\boldsymbol{x}-\boldsymbol{u})\cdot\boldsymbol{n}} = \frac{1}{|\boldsymbol{x}-\boldsymbol{u}|}.$$

This proves the following.

Theorem 7.2. *The caustic Γ_U of the Lagrangian manifold Λ_U generated by the distance function $\Phi(\boldsymbol{x},\boldsymbol{u}) = |\boldsymbol{x}-\boldsymbol{u}|$ is the set of the centers of curvature of the surface U.*

Corollary 7.1. *The only sources U that generate systems of rays without caustics are the plane surfaces.*

Remark 7.3. The caustic of a curve U in the plane (i.e., the set of the centers of curvature of U according to Theorem 7.2) is tangent to all lines orthogonal to U. ♢

Remark 7.4. Let us consider the following parametric representation of Λ_U in the three parameters (u^α, μ):

$$\begin{cases} \boldsymbol{p} = \boldsymbol{n}(u^\alpha) = \dfrac{\boldsymbol{e}_1 \times \boldsymbol{e}_2}{|\boldsymbol{e}_1 \times \boldsymbol{e}_2|}, \\ \boldsymbol{x} = \boldsymbol{u}(u^\alpha) + \mu\,\boldsymbol{p}. \end{cases} \tag{7.8}$$

The determinant of the second-order derivatives of $\boldsymbol{x}$ is equal to

$$\begin{aligned} &\boldsymbol{p}\cdot(\boldsymbol{e}_1 + \mu\,\partial_1\boldsymbol{p}) \times (\boldsymbol{e}_2 + \mu\,\partial_2\boldsymbol{p}) = \mu^2\,\boldsymbol{p}\cdot\partial_1\boldsymbol{p}\times\partial_2\boldsymbol{p} + \boldsymbol{p}\cdot\boldsymbol{e}_1\times\boldsymbol{e}_2 \\ &= \mu^2\,\boldsymbol{p}\cdot\frac{\boldsymbol{e}_1\times\boldsymbol{e}_2}{|\boldsymbol{x}-\boldsymbol{u}|^2} + \boldsymbol{p}\cdot\boldsymbol{e}_1\times\boldsymbol{e}_2 = \left(1 + \frac{\mu^2}{|\boldsymbol{x}-\boldsymbol{u}|^2}\right)\boldsymbol{n}\cdot\boldsymbol{e}_1\times\boldsymbol{e}_2 \neq 0. \end{aligned}$$

Hence, the representation (7.8) is an immersion. ♢

The previous results can be adapted to the case of a curve U in the Euclidean plane $\mathbb{R}^2 = (x, y)$.

Theorem 7.3. *If a curve $U \in \mathbb{R}^2 = (x, y)$ is described by parametric equations $x = x(t)$, $y = y(t)$, then the caustic Γ_U is described by parametric equations*

$$\begin{cases} x = x(t) - \dfrac{\dot{x}^2 + \dot{y}^2}{\dot{x}\,\ddot{y} - \dot{y}\,\ddot{x}}\;\dot{y}, \\[2ex] y = y(t) + \dfrac{\dot{x}^2 + \dot{y}^2}{\dot{x}\,\ddot{y} - \dot{y}\,\ddot{x}}\;\dot{x}. \end{cases} \tag{7.9}$$

Proof. Let us consider the Morse family

$$G(\boldsymbol{x}; \boldsymbol{a}, t) = \boldsymbol{a} \cdot (\boldsymbol{x} - \boldsymbol{x}(t))$$

with supplementary variables $\boldsymbol{a} \in \mathbb{S}_1$ and $t \in \mathbb{R}$. By setting

$$\boldsymbol{a} = \begin{bmatrix} \cos\theta \\ \sin\theta \end{bmatrix}$$

we observe that this Morse family is equivalent to

$$G(x, y; \theta, t) = (x - x(t))\,\cos\theta - (y - y(t))\,\sin\theta,$$

with supplementary variables $\theta, t \in \mathbb{R}$. The corresponding Lagrangian submanifold is then described by equations

$$\begin{cases} \dfrac{\partial G}{\partial \theta} = (y - y(t))\,\cos\theta - (x - x(t))\,\sin\theta = 0, \\[2ex] \dfrac{\partial G}{\partial t} = \dot{x}\,\cos\theta + \dot{y}\,\sin\theta = 0, \end{cases} \tag{7.10}$$

and

$$\begin{cases} p_x = \dfrac{\partial G}{\partial x} = \cos\theta, \\[2ex] p_y = \dfrac{\partial G}{\partial y} = \sin\theta. \end{cases} \tag{7.11}$$

Equations (7.10) describe the critical set. The vectorial expressions of Eqs. (7.10) and (7.11) are

$$\boldsymbol{p} = \boldsymbol{a}, \quad [\boldsymbol{x} - \boldsymbol{u}(t)] \times \boldsymbol{p} = 0, \quad \dot{\boldsymbol{u}}(t) \cdot \boldsymbol{p} = 0.$$

The last equation means that $\boldsymbol{p} \perp U$. The second equation means that $\boldsymbol{x} - \boldsymbol{u}(t)$ is parallel to $\boldsymbol{p}$. Because $|\boldsymbol{p}|^2 = 1$, this equation becomes equivalent to

$$\boldsymbol{p} = \pm \frac{\boldsymbol{x} - \boldsymbol{u}(t)}{|\boldsymbol{x} - \boldsymbol{u}(t)|}.$$

Thus, the Lagrangian submanifold generated by this function has two connected components. By choosing the + sign we find Eqs. (7.2) of the Lagrangian submanifold Λ_U generated by the distance function. The caustic of this Lagrangian submanifold is described by equation

$$\det \begin{bmatrix} \dfrac{\partial^2 G}{\partial \theta\, \partial \theta} & \dfrac{\partial^2 G}{\partial \theta\, \partial t} \\ \dfrac{\partial^2 G}{\partial t\, \partial \theta} & \dfrac{\partial^2 G}{\partial t\, \partial t} \end{bmatrix} = 0 \tag{7.12}$$

together with Eqs. (7.10). From (7.12) we get

$$\begin{aligned} &[(x - x(t))\, \cos\theta + (y - y(t))\, \sin\theta]\, (\dot{x}\, \cos\theta + \dot{y}\, \sin\theta) \\ &\qquad - (\dot{x}\, \sin\theta - \dot{y}\, \cos\theta)^2 = 0. \end{aligned}$$

By combining this equation with the first Eq. (7.10) we obtain the linear system

$$\begin{cases} \xi \cos\theta + \eta \sin\theta = X, \\ \xi \sin\theta - \eta \cos\theta = 0, \end{cases}$$

where

$$X = \frac{(\dot{x} \sin\theta - \dot{y} \cos\theta)^2}{\dot{x} \sin\theta + \dot{y} \sin\theta}, \quad \xi = x - x(t), \quad \eta = y - y(t).$$

It follows that

$$\xi = -\begin{vmatrix} X & \sin\theta \\ 0 & -\cos\theta \end{vmatrix}, \quad \eta = -\begin{vmatrix} \cos\theta & X \\ \sin\theta & 0 \end{vmatrix}.$$

From the second Eq. (7.10) we get $\cos\theta = \rho\, \dot{y}$ and $\sin\theta = -\,\rho\, \dot{x}$ with $1 = \rho^2\, (\dot{x}^2 + \dot{y}^2)$. Then,

$$\xi = -\begin{vmatrix} \rho \dfrac{(\dot{x}^2 + \dot{y}^2)^2}{\dot{x}\dot{y} - \dot{y}\dot{x}} & -\rho\dot{x} \\ 0 & -\rho\dot{y} \end{vmatrix} = \rho^2\, \frac{(\dot{x}^2 + \dot{y}^2)^2}{\dot{x}\dot{y} - \dot{y}\dot{x}}\, \dot{y} = \frac{\dot{x}^2 + \dot{y}^2}{\dot{x}\dot{y} - \dot{y}\dot{x}}\, \dot{y}.$$

This proves the first Eq. (7.9). The second equation is proved in a similar way. □

Example 7.1. Caustic of a parabola, F. 7.2. For the parabola $y = \frac{1}{2}\, x^2$, by setting $x = t$, Eqs. (7.9) give the following parametric equations of the caustic:

$$x = -\, t^3, \qquad y = 1 + \tfrac{3}{2}\, t^2. \qquad \diamond$$

Remark 7.5. Instead of the distance function we can consider the function

$$\Phi'(\boldsymbol{x}; \boldsymbol{u}) = \tfrac{1}{2}\, |\boldsymbol{x} - \boldsymbol{u}|^2.$$

This is the Euclidean version of the *world function* introduced by Synge for a generic Riemannian manifold (Synge 1960). By considering this function as a generating family we write equations

$$\boldsymbol{p} = \frac{\partial \Phi'}{\partial \boldsymbol{x}} = \boldsymbol{x} - \boldsymbol{u}, \quad 0 = \frac{\partial \Phi'}{\partial u^\alpha} = -\boldsymbol{p} \cdot \boldsymbol{e}_\alpha.$$

From the first equation it follows that

$$\frac{\partial \boldsymbol{p}}{\partial u^\alpha} = -\boldsymbol{e}_\alpha.$$

This shows that Φ' is a Morse family. This Morse family is now everywhere differentiable. The corresponding Lagrangian submanifold is defined by

$$(\boldsymbol{x}, \boldsymbol{p}) \in \Lambda'_U \iff \begin{cases} \boldsymbol{p} = \boldsymbol{x} - \boldsymbol{u}, \ \boldsymbol{u} \in U, \\ \boldsymbol{p} \perp U. \end{cases}$$

Moreover,

$$\frac{\partial^2 \Phi'}{\partial u^\beta \partial u^\alpha} = -\partial_\beta \boldsymbol{p} \cdot \boldsymbol{e}_\alpha - \boldsymbol{p} \cdot \partial_\beta \cdot \boldsymbol{e}_\alpha = A_{\alpha\beta} - \boldsymbol{p} \cdot \left(\Gamma^\gamma_{\beta\alpha}\, \boldsymbol{e}_\gamma + B_{\beta\alpha}\, \boldsymbol{n} \right).$$

Hence, under the condition $\partial_\alpha \Phi' = 0$, which is also in this case equivalent to $\boldsymbol{p} \cdot \boldsymbol{e}_\alpha = 0$, we find

$$\partial_\beta \partial_\alpha \Phi = A_{\beta\alpha} - \boldsymbol{n} \cdot (\boldsymbol{x} - \boldsymbol{u})\, B_{\beta\alpha},$$

and the equation of the caustic is identical to Eq. (7.7). Thus, $\Gamma'_U = \Gamma_U$. Note that the Lagrangian submanifold Λ'_U is not contained in the submanifold C of equation $\boldsymbol{p}^2 = 1$. ◇

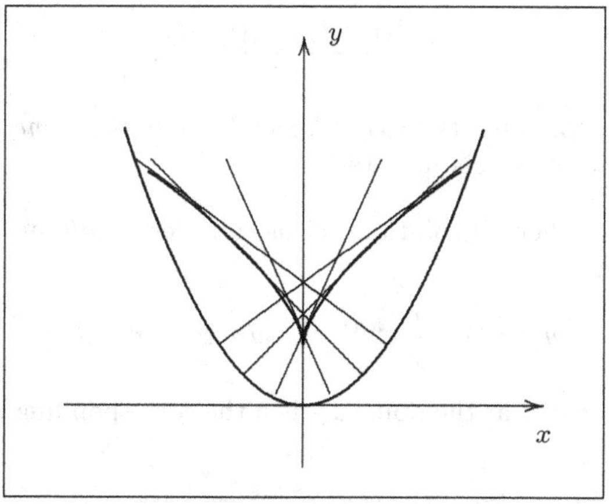

Fig. 7.2 Caustic of a parabola as a source of rays

7.2 From wave optics to geometrical optics

Any component $u(\boldsymbol{x}, t)$ of an electromagnetic potential is a solution of the *wave equation*

$$\frac{\partial^2 u}{\partial t^2} - \Delta U = 0, \tag{7.13}$$

where Δ is the Laplace–Beltrami operator. In Cartesian coordinates

$$\Delta = \frac{\partial^2}{\partial x^2} + \frac{\partial^2}{\partial y^2} + \frac{\partial^2}{\partial z^2}.$$

1 – Let us consider the special class of *spherical solutions*,

$$u = u(r, t), \quad r = |\boldsymbol{x}|.$$

Since $\Delta u = \operatorname{div}(\operatorname{grad}(u))$, for such a function we have

$$\Delta u = \frac{1}{r} \frac{\partial^2}{\partial r^2}(ru),$$

and from (7.13) it follows that

$$\frac{\partial^2 v}{\partial t^2} - \frac{\partial^2 v}{\partial r^2} = 0, \quad v = r\,u.$$

The general solution of this last equation is $v = f(r+t) + g(r-t)$, where f and g are arbitrary (smooth) functions. Solutions of the kind $f(r+t)$ and $g(r-t)$ are called *incoming waves* and *outgoing waves*, respectively. Thus, the general spherical solution of the wave equation is

$$u = \frac{f(r+t)}{r} + \frac{g(r-t)}{r}.$$

Such a function represents the *electromagnetic radiation generated by a point source* (the origin of the coordinates).

2 – Among the spherical solutions let us consider *oscillatory outgoing solutions* of the kind

$$u_\omega(r, t) = \frac{c}{r}\, e^{i\omega(r-t)}, \quad \omega \in \mathbb{R}_+, \quad c \in \mathbb{R}.$$

If the source is located at the point $\boldsymbol{u}$, then the corresponding solution of this kind is

$$u_{(\omega, \boldsymbol{u})} = c(\boldsymbol{u})\, e^{-i\omega t}\, \frac{e^{i\omega|\boldsymbol{x}-\boldsymbol{u}|}}{|\boldsymbol{x}-\boldsymbol{u}|}.$$

The factor

$$I = c(\boldsymbol{u})\, \frac{e^{i\omega|\boldsymbol{x}-\boldsymbol{u}|}}{|\boldsymbol{x}-\boldsymbol{u}|}$$

is called the *intensity* of the radiation.

3 – Let us consider the radiation generated by a surface U made of pointwise sources. The resulting intensity at any point $\boldsymbol{x}$ is given by the integral

$$\boxed{I_\omega(\boldsymbol{x}) = \int_U c(\boldsymbol{u})\, \frac{e^{i\omega|\boldsymbol{x}-\boldsymbol{u}|}}{|\boldsymbol{x}-\boldsymbol{u}|}\, d\boldsymbol{u}} \tag{7.14}$$

This is a surface integral of the kind

$$I_\omega = \int_U a(\boldsymbol{u})\, e^{i\omega\,\Phi(\boldsymbol{u})}\, d\boldsymbol{u},$$

called an *oscillatory integral.* The function $\Phi(\boldsymbol{u})$ is called the *phase function.* In the present case

$$\Phi(\boldsymbol{u}) = |\boldsymbol{x}-\boldsymbol{u}|, \quad a(\boldsymbol{u}) = \frac{c(\boldsymbol{u})}{|\boldsymbol{x}-\boldsymbol{u}|}.$$

Regarding this integral we have two fundamental *theorems of the stationary phase* (see, e.g., (Guillemin and Sternberg 1977, 1984) for proofs and references).

Theorem 7.4. *If $\boldsymbol{x}$ is such that $d_{\boldsymbol{u}}\Phi \neq 0$ at all points of U, then for all $m \in \mathbb{N}$,*

$$I_\omega(\boldsymbol{x}) = O(\omega^{-m}).$$

The meaning of this theorem is that for $\omega \to \infty$, the radiations of all sources interfere in such a way that the total intensity is negligible at any point $\boldsymbol{x}$.

Theorem 7.5. *If $\boldsymbol{x}$ is such that $d_{\boldsymbol{u}}\Phi = 0$ at a finite number of points $\boldsymbol{u}_* \in U$, then the following asymptotic formula holds*

$$\boxed{I_\omega(\boldsymbol{x}) = \left(\frac{2\pi}{\omega}\right)^{n/2} \sum_* a(\boldsymbol{u}_*)\, e^{i\omega\Phi(\boldsymbol{u}_*)}\, \frac{e^{i\pi/4}\mathrm{sign}\boldsymbol{H}_*}{\sqrt{\det \boldsymbol{H}_*}}\, (1 + O(\omega^{-1}))} \tag{7.15}$$

where $n = \dim U$, $\boldsymbol{H}$ is the Hessian matrix of Φ, and

$$\mathrm{sign}(\boldsymbol{H}) = \#(\text{positive eigenvalues}) - \#(\text{negative eigenvalues}).$$

A first consequence of this theorem is that the non negligible contribution to the intensity I_ω comes only from those points $\boldsymbol{u}_*$ where $d_{\boldsymbol{u}}\Phi = 0$; that is,

$$\frac{\partial\Phi}{\partial\boldsymbol{u}} = 0. \tag{7.16}$$

In our case $\Phi(\boldsymbol{u}) = |\boldsymbol{x}-\boldsymbol{u}|$, for any chosen $\boldsymbol{x}$, therefore the points $\boldsymbol{u}_*$ satisfying this condition are the points where the line from $\boldsymbol{u}_*$ to $\boldsymbol{x}$ (i.e., the vector $\boldsymbol{x}-\boldsymbol{u}_*$) is orthogonal to U, and this holds for any other point $\boldsymbol{x}'$ on this line. This means that for an observer located at any point $\boldsymbol{x}$ on the line perpendicular to U at the point $\boldsymbol{u}_*$ only the radiation emitted by the source $\boldsymbol{u}_*$ is detected. This line is called the *ray* issued from $\boldsymbol{u}_*$. It is parallel to the vector

$$\boldsymbol{p} = \frac{\partial \Phi}{\partial \boldsymbol{x}}. \tag{7.17}$$

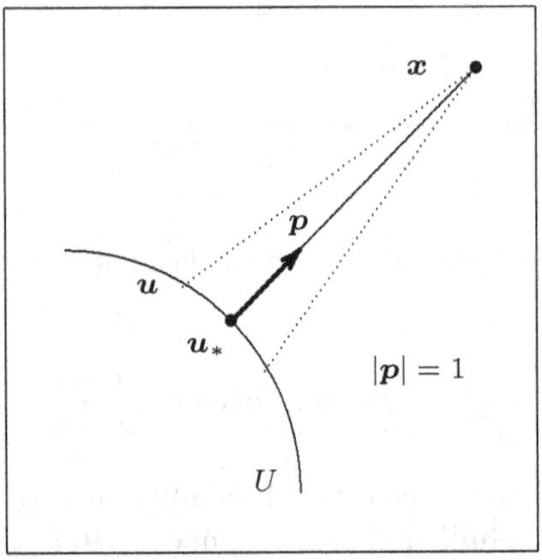

Fig. 7.3 The ray issued from a point of a surface

Furthermore, from formula (7.15) we observe that the intensity $I_\omega(\boldsymbol{x})$ is unbounded at the points $\boldsymbol{x}$ where $\det \boldsymbol{H}_* = 0$; that is,

$$\det \left[\frac{\partial^2 \Phi}{\partial u^\alpha \partial u^\beta} \right]_* = 0, \tag{7.18}$$

where (u^α) are local parameters of U and $*$ means the evaluation at the point $\boldsymbol{u}_*$. These points form the *caustic*. Equations (7.16), (7.17) and (7.18) are just the equations of the Lagrangian submanifold Λ_U and of the corresponding caustic generated by the distance function Φ, as shown in the previous section.

7.3 The global Hamilton principal function for the eikonal equation

Let $C \subset T^*\mathbb{R}^n$ be the coisotropic submanifold of codimension 1 defined by equation

$$\boxed{\textstyle\sum_i p_i^2 = |\boldsymbol{p}|^2 = 1} \tag{7.19}$$

This is the *eikonal equation* for the homogeneous (empty) Euclidean n-space. The characteristics of C can be found by integrating the Hamilton equations generated by the Hamiltonian $H = \sum_i p_i^2$,

$$\begin{cases} \dot{x}^i = \lambda\, p_i, \\ \dot{p}_i = 0, \end{cases}$$

where λ is any function. By choosing $\lambda = 1$, we find that: (i) the characteristics are the straight lines described by parametric equations

$$\begin{cases} \boldsymbol{x} = \boldsymbol{a}\, t - \boldsymbol{b}, \\ \boldsymbol{p} = \boldsymbol{a}, \end{cases} \qquad \boldsymbol{a} \in \mathbb{S}_{n-1}, \quad \boldsymbol{b} \in \mathbb{R}^n, \tag{7.20}$$

because $\boldsymbol{p} \in \mathbb{S}_{n-1}$ (the unit sphere) is equivalent to $\boldsymbol{p} \in C$; (ii) the rays are oriented straight lines in $\mathbb{R}^n$.

It is convenient to choose $\boldsymbol{b}$ orthogonal to $\boldsymbol{a}$, that is, $\boldsymbol{b}$ tangent to the sphere $\mathbb{S}_{n-1}$ at the "point" $\boldsymbol{a}$. In this way, through Eqs. (7.20), any characteristic of C is determined by a pair of vectors $(\boldsymbol{a}, \boldsymbol{b})$, where $\boldsymbol{a} \in \mathbb{S}_{n-1}$ and $\boldsymbol{b}$ is a vector tangent to the sphere and orthogonal to $\boldsymbol{a}$. This defines a one-to-one map from the set M_C of the characteristics to the tangent bundle $T\mathbb{S}_{n-1}$, which is identified with the cotangent bundle $T^*\mathbb{S}_{n-1}$. The minus sign in front of $\boldsymbol{b}$ is chosen in order to get a symplectomorphism between the reduced symplectic manifold M_C and the cotangent bundle $T^*A = T^*\mathbb{S}_{n-1}$ (see below).

Fig. 7.4 Equations (7.20)

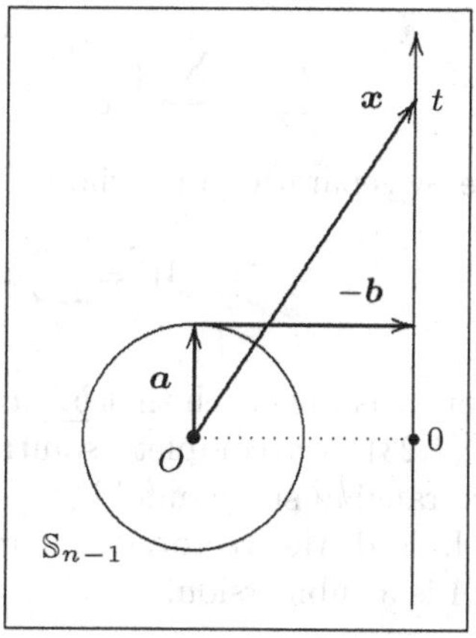

It follows from (7.20) that two pairs $(\boldsymbol{x}, \boldsymbol{p})$ and $(\boldsymbol{x}', \boldsymbol{p}')$, representing two points of T^*Q, belong to a same characteristic if and only if

$$\begin{cases} \boldsymbol{p} = \boldsymbol{p}' = \boldsymbol{a} \in \mathbb{S}_{n-1}, \\ \boldsymbol{x} - \boldsymbol{x}' \parallel \boldsymbol{p}. \end{cases} \qquad (\parallel = \text{parallel to}). \tag{7.21}$$

Thus, the characteristic relation D_C is defined by these conditions.

Remark 7.6. With each pair $(\boldsymbol{a}, \boldsymbol{x}) \in A \times Q = \mathbb{S}_{n-1} \times \mathbb{R}^n$ we associate a unique element $((\boldsymbol{a}, \boldsymbol{b}), (\boldsymbol{x}, \boldsymbol{p})) \in T^*A \times T^*Q$ belonging to the reduction relation R_C. Indeed, as proved below, R_C is a regular Lagrangian submanifold of $T^*(A \times Q)$. ♢

Remark 7.7. With each pair $(\boldsymbol{x}, \boldsymbol{x}') \in \mathbb{R}^n \times \mathbb{R}^n = Q \times Q$ such that $\boldsymbol{x} \neq \boldsymbol{x}'$ we can associate two elements of D_C differing by the sign, $((\boldsymbol{x}, \boldsymbol{p}), (\boldsymbol{x}', \boldsymbol{p}'))$ and $((\boldsymbol{x}, -\boldsymbol{p}), (\boldsymbol{x}', -\boldsymbol{p}'))$. This means that D_C is two-folded over the points of $Q \times Q$ out of the diagonal, and over the diagonal (i.e., for $\boldsymbol{x}' = \boldsymbol{x}$) it is made of pairs $((\boldsymbol{x}, -\boldsymbol{p}), (\boldsymbol{x}, -\boldsymbol{p}'))$ such that $\boldsymbol{p} = \boldsymbol{p}' \in \mathbb{S}_{n-1}$. Indeed, as proved below, D_C is a regular (and two-folded) Lagrangian submanifold of $T^*(Q \times Q)$ out of the diagonal, and it is singular over the diagonal (Sect. 6.5). ♢

For the construction and the analysis of the characteristic relation D_C we can follow another way: to look for a complete solution of C and use it for constructing a Hamilton principal function.

Theorem 7.6. *A global complete solution of the eikonal equation* (7.19) $|\boldsymbol{p}|^2 = 1$ *is the function on* $Q \times \mathbb{S}_{n-1}$ *defined by*

$$W(\boldsymbol{x}, \boldsymbol{a}) = \boldsymbol{a} \cdot \boldsymbol{x}, \quad \boldsymbol{a} \in \mathbb{S}_{n-1}. \tag{7.22}$$

Proof. The partial differential equation associated with the eikonal equation (7.19) is

$$\sum_i \left(\frac{\partial W}{\partial x^i} \right)^2 = 1.$$

It is integrable by separation of variables. A solution is

$$W = \sum_i a_i \, x^i \tag{7.23}$$

with integration constants such that $\sum_i a_i^2 = 1$. This means $\boldsymbol{a} = (a_i) \in \mathbb{S}_{n-1}$. The function (7.23) is a complete solution because for each $\boldsymbol{p} \in C$ there is a unique Lagrangian submanifold $\Lambda_{\boldsymbol{a}}$, generated by the functions $W_{\boldsymbol{a}}(\boldsymbol{x})$ containing $\boldsymbol{p}$. Indeed, the vectorial equation of Λ_a is $\boldsymbol{p} = \boldsymbol{a}$. Moreover, the map $\pi \colon C \to A$ is a submersion. □

Remark 7.8. Each Lagrangian submanifold $\Lambda_{\boldsymbol{a}}$ corresponds to a system of *parallel rays* or *plane waves* (see Remark 6.9). ♢

Remark 7.9. The function W generates the transpose $R^\top$ of a symplectic reduction $R \subset T^*A \times T^*Q$, $A = \mathbb{S}_{n-1}$, whose inverse image is C. As a consequence, (i) the reduced set M_C is symplectomorphic to the cotangent bundle $T^*A = T^*\mathbb{S}_{n-1}$; and (ii) the equations of $R^\top \subset T^*Q \times T^*A$ are

$$\begin{cases} \boldsymbol{b} = -\dfrac{\partial W}{\partial \boldsymbol{a}} = -(\boldsymbol{x} - \boldsymbol{x} \cdot \boldsymbol{a}\, \boldsymbol{a}) = -P_{\boldsymbol{a}}(\boldsymbol{x}), \\ \boldsymbol{p} = \dfrac{\partial W}{\partial \boldsymbol{x}} = \boldsymbol{a}. \end{cases} \qquad \diamondsuit \tag{7.24}$$

Remark 7.10. In the first Eq. (7.24) $P_{\boldsymbol{a}}$ denotes the projection operator onto the plane orthogonal to $\boldsymbol{a}$,

$$P_{\boldsymbol{a}}(\boldsymbol{x}) = (\boldsymbol{I} - \boldsymbol{a} \otimes \boldsymbol{a})(\boldsymbol{x}) \quad (\boldsymbol{I} = \text{identity}).$$

Here, we have used the following general property: assume that a hypersurface A (of codimension 1) in $\mathbb{R}^n$ is (locally) described by a vector function $\boldsymbol{a}(a^\alpha)$ depending on $n-1$ parameters (surface coordinates) in such a way that the vectors tangent to A,

$$\boldsymbol{e}_\alpha = \frac{\partial \boldsymbol{a}}{\partial a^\alpha} = \partial_\alpha \boldsymbol{a},$$

are pointwise independent. Let $f(\boldsymbol{a})$ be any function on A. This function is locally represented by a function $f(a^\alpha)$ of the surface coordinates. The partial derivatives

$$b_\alpha = \frac{\partial f}{\partial a^\alpha} \tag{7.25}$$

are the covariant components of a vector $\boldsymbol{b} = b^\alpha\, \boldsymbol{e}_\alpha$ tangent to the surface, being $b_\alpha = A_{\alpha\beta} b^\beta$ and $A_{\alpha\beta} = \boldsymbol{e}_\alpha \cdot \boldsymbol{e}_\beta$ the components of the first fundamental form of the surface (see Sect. 7.1). Since $b_\alpha = \boldsymbol{b} \cdot \boldsymbol{e}_\alpha$, we can write (7.25) in the vectorial form

$$\boldsymbol{b} = \frac{\partial f}{\partial \boldsymbol{a}}. \tag{7.26}$$

Let $f(\boldsymbol{x})$ be any (local) extension of $f(\boldsymbol{a})$ in a neighborhood of the surface. Its gradient,

$$\nabla f = \frac{\partial f}{\partial \boldsymbol{x}},$$

is not (in general) a vector tangent to the surface. However, its tangent component coincides with the vector (7.26). It follows that

$$\frac{\partial f}{\partial \boldsymbol{a}} = P_{\boldsymbol{n}}\left(\nabla f(\boldsymbol{a})\right),$$

where $\boldsymbol{n}$ is a unit vector orthogonal to the surface at the point $\boldsymbol{a}$, and

$$P_{\boldsymbol{n}} = \boldsymbol{I} - \boldsymbol{n} \otimes \boldsymbol{n}$$

is the projection operator onto the $(n-1)$-dimensional plane orthogonal to $\boldsymbol{n}$ (and tangent to the surface). Note that for $A = \mathbb{S}_{n-1}$ we have $\boldsymbol{n} = \boldsymbol{a}$. Note that Eqs. (7.24) coincide with Eqs. (7.20) for $t = \boldsymbol{x} \cdot \boldsymbol{a}$. $\diamondsuit$

Theorem 7.7. *The generating family* $S\colon (Q \times Q; \mathbb{S}_{n-1}) \to \mathbb{R}$, *with supplementary manifold* $A = \mathbb{S}_{n-1}$, *defined by*

$$\boxed{S(\boldsymbol{x}, \boldsymbol{x}'; \boldsymbol{a}) = (\boldsymbol{x} - \boldsymbol{x}') \cdot \boldsymbol{a}, \quad \boldsymbol{a} \in \mathbb{S}_{n-1}} \tag{7.27}$$

is a global Hamilton principal function of the eikonal equation on the Euclidean space $Q = \mathbb{R}^n$.

Proof. The co-reduction relation $R^\top$ is generated by $W(\boldsymbol{x}, \boldsymbol{a})$ (Remark 7.9). The reduction relation R is generated by $W^\top(\boldsymbol{a}, \boldsymbol{x}) = -W(\boldsymbol{x}, \boldsymbol{a})$. According to formula (7.22), by composing these generating families we get the generating family (7.27) of the characteristic relation $D_C = R^\top \circ R$. □

Remark 7.11. The equations of D_C generated by S are

$$\begin{aligned} \boldsymbol{p}' &= -\frac{\partial S}{\partial \boldsymbol{x}'} = \boldsymbol{a}, \\ \boldsymbol{p} &= \frac{\partial S}{\partial \boldsymbol{x}} = \boldsymbol{a}, \end{aligned} \qquad 0 = \frac{\partial S}{\partial \boldsymbol{a}} = P_{\boldsymbol{a}}(\boldsymbol{x} - \boldsymbol{x}').$$

These equations are equivalent (7.21). ◇

Remark 7.12. The reduction relation R_C is a regular Lagrangian submanifold, because it is generated by an ordinary generating function W (without extra variables). On the contrary, the characteristic relation is singular over the diagonal so that, in the neighborhood of the diagonal, it is generated by a generating family. ◇

Theorem 7.8. *The generating family $S(\boldsymbol{x}, \boldsymbol{x}'; \boldsymbol{a}) = (\boldsymbol{x} - \boldsymbol{x}') \cdot \boldsymbol{a}$ is a Morse family and the caustic of D_C is the diagonal of $Q \times Q$.*

Proof. Let us consider a parametric representation $\boldsymbol{a}(u^\alpha)$ of the sphere in the $n-1$ parameters (u^α). The vectors $\boldsymbol{e}_\alpha = \partial_\alpha \boldsymbol{a}$ are independent and tangent to the sphere. Since

$$\partial_\alpha S = (\boldsymbol{x} - \boldsymbol{x}') \cdot \boldsymbol{e}_\alpha, \quad \partial_\alpha = \partial / \partial u^\alpha,$$

the critical set Ξ is given by the pair of vectors such that $\boldsymbol{x} - \boldsymbol{x}' \perp \mathbb{S}_{n-1}$. Moreover,

$$\partial_\alpha \partial_\beta S = (\boldsymbol{x} - \boldsymbol{x}') \cdot \partial_\beta \boldsymbol{e}_\alpha = (\boldsymbol{x} - \boldsymbol{x}') \cdot (\Gamma^\gamma_{\beta\alpha}\, \boldsymbol{e}_\gamma + B_{\beta\alpha}\, \boldsymbol{a})$$

and

$$\begin{aligned} \partial_i \partial_\alpha S &= e^i_\alpha, \quad \partial_i = \partial / \partial x^i, \\ \partial_{i'} \partial_\alpha S &= e^i_\alpha, \quad \partial_{i'} = \partial / \partial x'^i, \end{aligned}$$

where e^i_α are the Cartesian components of the vector $\boldsymbol{e}_\alpha$. Then the matrix

$$\left[\partial_\alpha \partial_i S \,\middle|\, \partial_\alpha \partial_{i'} S \,\middle|\, \partial_\alpha \partial_\beta S \right]$$

has maximal rank everywhere, inasmuch as the submatrix $[\partial_\alpha \partial_i S] = [e^i_\alpha]$ has maximal rank, with the vectors $(\boldsymbol{e}_\alpha)$ independent. Hence, S is a Morse family. On the critical set,

$$\partial_\alpha \partial_\beta S = (\boldsymbol{x} - \boldsymbol{x}') \cdot \boldsymbol{a}\; B_{\beta\alpha}.$$

The vector $\boldsymbol{x} - \boldsymbol{x}'$ is parallel to $\boldsymbol{a}$ and on the sphere $\det[B_{\beta\alpha}] \neq 0$, thus we have $\det\left[\partial_\alpha \partial_\beta S\right] = 0$ if and only if $\boldsymbol{x} - \boldsymbol{x}' = 0$. □

Theorem 7.9. *Outside the diagonal of $Q \times Q$ the characteristic relation D_C is the union of two disjoint regular symplectic relations generated by the functions*

$$\boxed{S_\pm(\boldsymbol{x}, \boldsymbol{x}') = \pm\, |\boldsymbol{x} - \boldsymbol{x}'|} \tag{7.28}$$

Proof. The symplectic relation generated by S_+ is represented by equations

$$\boldsymbol{p}' = -\frac{\partial S_+}{\partial \boldsymbol{x}'} = \frac{\boldsymbol{x} - \boldsymbol{x}'}{|\boldsymbol{x} - \boldsymbol{x}'|}, \quad \boldsymbol{p} = \frac{\partial S_+}{\partial \boldsymbol{x}} = \frac{\boldsymbol{x}' - \boldsymbol{x}}{|\boldsymbol{x} - \boldsymbol{x}'|}.$$

The requirements (7.21) are fulfilled. With S_- we get the opposite pair $(-\boldsymbol{p}, -\boldsymbol{p}')$. □

Remark 7.13. According to Theorem 4.11 (see also Remark 6.19), the global Hamilton principal function (7.27) is skew-symmetric in $(\boldsymbol{x}, \boldsymbol{x}')$. Instead, the generating functions (7.28) $S_\pm$ are symmetric. This is not a contradiction, because these functions are "nonglobal" Hamilton principal functions: each generates only a branch of the symplectic relation D_C. ◇

Up to now, it seems that in the literature the concept of a *global* Hamilton principal function for the eikonal equation does not exists, except in the form given by Theorem 7.9 which, however, excludes the case of two coincident points: $\boldsymbol{x} = \boldsymbol{x}'$. It should therefore be emphasized that with the concept of generating family, as done through Theorems 7.7 and 7.8, it is possible to give a comprehensive definition of this function, which plays a basic role in analytical mechanics, geometrical optics, and other branches of mathematical physics.

7.4 Generating families of systems of rays

The Hamilton principal function (7.27) is the basis for an algorithm that allows the analysis of the formation and evolution of systems of rays in the Euclidean spaces. This algorithm is explained, by elementary examples, in the following subsections.

The majority of these examples involve the Euclidean plane, where things are easier to handle, but the reader can extend them to three or more dimensions with a little effort. The advantage of the dimension 2 is basically this: the Hamilton principal function (7.27) is reducible to the Morse family

$$S(\boldsymbol{x}_0, \boldsymbol{x}_I; \theta) = (x_O - x_I)\cos\theta + (y_O - y_I)\sin\theta,$$

with a single parameter $\theta \in \mathbb{R}$.

Let us recall Eqs. (6.31) and (6.44) defining the mirror-relation and the lens-relation:

$$M_\Sigma = D_C \circ \widehat{\Delta}_\Sigma, \qquad L_{\Sigma,F} = D_C \circ \widehat{\Delta}_{\Sigma,F}. \tag{7.29}$$

For a Euclidean space, the respective generating families can be obtained, by a simple change of notation, from the general formulae (6.33) of Theorem 6.10 and (6.47) of Theorem 6.12. They are:

$$G_{M_\Sigma}(\boldsymbol{x}_O, \boldsymbol{x}_I; \boldsymbol{a}) = (\boldsymbol{x}_O - \boldsymbol{x}_I) \cdot \boldsymbol{a} + \lambda^\alpha\, \Sigma_\alpha(\boldsymbol{x}_I), \tag{7.30}$$

and

$$G_{M_\Sigma}(\boldsymbol{x}_O, \boldsymbol{x}_I; \boldsymbol{a}) = (\boldsymbol{x}_O - \boldsymbol{x}_I) \cdot \boldsymbol{a} + \lambda^\alpha\, \Sigma_\alpha(\boldsymbol{x}_I) + F(\boldsymbol{x}_I). \tag{7.31}$$

In these formulae the submanifold $\Sigma \subset \mathbb{R}^n$, defined by equations $\Sigma_\alpha(\boldsymbol{x}) = 0$, represents the mirror or the lens. The function $F(\boldsymbol{x})$ over Σ represents the optical characteristics of the lens.

Furthermore, because $S(\boldsymbol{x}_O, \boldsymbol{x}_I; \boldsymbol{a}) = (\boldsymbol{x}_O - \boldsymbol{x}_I) \cdot \boldsymbol{a}$ is a Morse family (Theorem 7.8) we can apply Theorems 6.11 and 6.13 and state the following.

Theorem 7.10. *Both the generating families* (7.30) *and* (7.31) *are Morse families.*

Consequently, we have the following.

Theorem 7.11. *In the Euclidean space* $\mathbb{R}^n$ *the mirror-relations and the lens-relations are Lagrangian submanifolds.*

7.4.1 System of rays generated by a hypersurface

Let $\Sigma \subset \mathbb{R}^n$ be an r-dimensional regular surface described by $n - s$ independent equations $\Sigma_\alpha(\boldsymbol{x}) = 0$. The canonical lift $\widehat{\Sigma}$ is generated by the Morse family, see Eq. (5.4),

$$G_\Sigma(\boldsymbol{x}; \boldsymbol{\lambda}) = \lambda^a\, \Sigma_a(\boldsymbol{x}), \;\; \boldsymbol{\lambda} = (\lambda^a) \in \mathbb{R}^{n-s}. \tag{7.32}$$

The system of rays outcoming from Σ is represented (see Sect. 7.1) by the Morse family

$$G_1(\boldsymbol{x};\boldsymbol{u}) = |\boldsymbol{x}-\boldsymbol{u}|, \quad \boldsymbol{u} \in \Sigma. \tag{7.33}$$

Due to Theorem 6.9, the system of outgoing and incoming rays is described by the generating family

$$G_2(\boldsymbol{x};\boldsymbol{x}',\boldsymbol{a},\boldsymbol{\lambda}) = (\boldsymbol{x}-\boldsymbol{x}')\cdot\boldsymbol{a} + \lambda^\alpha\,\Sigma_\alpha(\boldsymbol{x}'), \tag{7.34}$$

with supplementary variables $\boldsymbol{x}' \in \mathbb{R}^n$, $\boldsymbol{a} \in \mathbb{S}_{n-1}$, $\boldsymbol{\lambda} = (\lambda^a) \in \mathbb{R}^{n-s}$. This follows from the composition of G_U (7.32) with the Hamilton principal function (7.27). It is remarkable that this is always a Morse family, whatever Σ (see Remark 6.24).

7.4.2 System of rays generated by a point

If the surface Σ reduces to a point $\boldsymbol{x}_0$, then $\widehat{\Sigma} = \widehat{\boldsymbol{x}_0}$ is the fiber over this point and the generating families (7.32), (7.33), and (7.34) become

$$\begin{aligned} &G_{\boldsymbol{x}_0}(\boldsymbol{x};\boldsymbol{\lambda}) = \boldsymbol{\lambda}\cdot(\boldsymbol{x}-\boldsymbol{x}_0),\\ &G_1(\boldsymbol{x}) = |\boldsymbol{x}-\boldsymbol{x}_0|, \\ &G_2(\boldsymbol{x};\boldsymbol{x}',\boldsymbol{a},\boldsymbol{\lambda}) = (\boldsymbol{x}-\boldsymbol{x}')\cdot\boldsymbol{a} + \boldsymbol{\lambda}\cdot(\boldsymbol{x}'-\boldsymbol{x}_0), \quad \boldsymbol{a}\in\mathbb{S}_{n-1}, \quad \boldsymbol{\lambda}\in\mathbb{R}^n, \end{aligned} \tag{7.35}$$

respectively. However, one of the equations of the critical set of G_2 is

$$0 = \partial G_2/\partial \boldsymbol{x}' = \boldsymbol{\lambda} - \boldsymbol{a}.$$

Thus, the generating family G_2 reduces to (we use the same symbol)

$$\boxed{G_2(\boldsymbol{x};\boldsymbol{a}) = (\boldsymbol{x}-\boldsymbol{x}_0)\cdot\boldsymbol{a}, \quad \boldsymbol{a}\in\mathbb{S}_{n-1}} \tag{7.36}$$

Note that *the third generating family in* (7.35) *describes only outgoing rays, whereas the family* (7.36) *describes both incoming and outgoing rays.*

7.4.3 Mirrors

Let us recall from Example 6.12 that in the plane the mirror-relation M_Σ and the output Lagrangian set $\Lambda_O = M_\Sigma \circ \Lambda_I$ are generated by

$$G_{M_\Sigma}(\boldsymbol{x}_O,\boldsymbol{x}_I;\theta) = (x_O - x_I)\,\cos\theta + (y_O - y_I)\,\sin\theta + \lambda\,\Sigma(\boldsymbol{x}_I). \tag{7.37}$$

and

$$G(\boldsymbol{x}; \boldsymbol{x}_I, \theta, \boldsymbol{\zeta}) = (x - x_I)\cos\theta + (y - y_I)\sin\theta + \lambda\,\Sigma(\boldsymbol{x}_I) + G_I(\boldsymbol{x}_I; \boldsymbol{\zeta}), \quad (7.38)$$

respectively. The equations of Λ_O are

$$\begin{cases} 0 = \dfrac{\partial G}{\partial x_I} = -\cos\theta + \lambda\dfrac{\partial \Sigma}{\partial x_I} + \dfrac{\partial G_I}{\partial x_I}, \\ 0 = \dfrac{\partial G}{\partial y_I} = -\sin\theta + \lambda\dfrac{\partial \Sigma}{\partial y_I} + \dfrac{\partial G_I}{\partial y_I}, \\ 0 = \dfrac{\partial G}{\partial \theta} = -(x - x_I)\sin\theta + (y - y_I)\cos\theta, \\ 0 = \dfrac{\partial G}{\partial \lambda} = \Sigma(\boldsymbol{x}_I), \\ 0 = \dfrac{\partial G}{\partial \zeta^\kappa} = \dfrac{\partial G_I}{\partial \zeta^\kappa} = f^\kappa(\boldsymbol{x}_I, \boldsymbol{\zeta}), \end{cases} \qquad \begin{cases} p_x = \dfrac{\partial G}{\partial x} = \cos\theta, \\ p_y = \dfrac{\partial G}{\partial y} = \sin\theta, \end{cases} \quad (7.39)$$

where the first group represents the critical set.

7.4.4 *The coffee cup*

If you hold in your hand a cup of Italian "espresso" and in the room there is a sufficiently concentrated source of light, you can observe on the surface of your coffee, of a nice brown color, a strange double curve drawn by the light, like this:

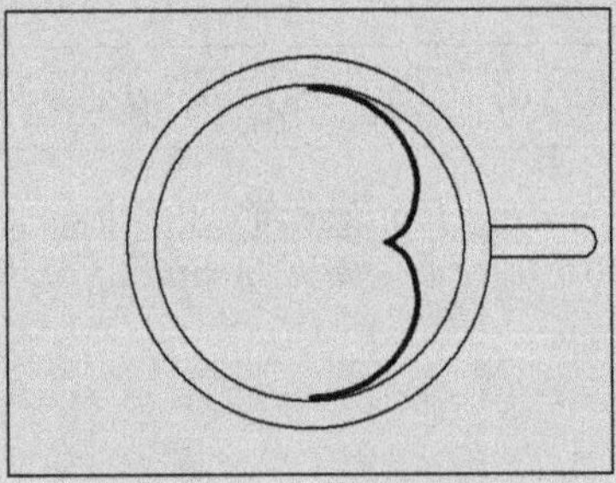

Let a mirror be a semicircle of radius R, centered at the origin, in the positive half-plane $x > 0$. Let the incoming rays be parallel to the x-axis and with the same orientation, represented by the unit vector $\boldsymbol{p}_I$. In this case,

$$\Sigma(\boldsymbol{x}_I) = \tfrac{1}{2}\,(x_I^2 + y_I^2 - R^2), \quad G_I = x_I,$$

and (7.39) become

$$\begin{cases} \cos\theta = \lambda\, x_I + 1, \\ \sin\theta = \lambda\, y_I, \\ x_I^2 + y_I^2 = R^2, \\ 0 = -(x - x_I)\, \sin\theta + (y - y_I)\, \cos\theta, \end{cases} \qquad \begin{cases} p_x = \cos\theta, \\ p_y = \sin\theta, \end{cases} \tag{7.40}$$

In order to find the reflected Lagrangian set Λ_O[1] we have to solve these equations with respect to (x, y, p_x, p_y), in the sense that $(x, y, p_x, p_y) \in \Lambda_O$ if and only if there are values of $(x_I, y_I, \lambda, \theta)$ such that $(x, y, p_x, p_y; x_I, y_I, \lambda, \theta)$ is a solution of (7.40).

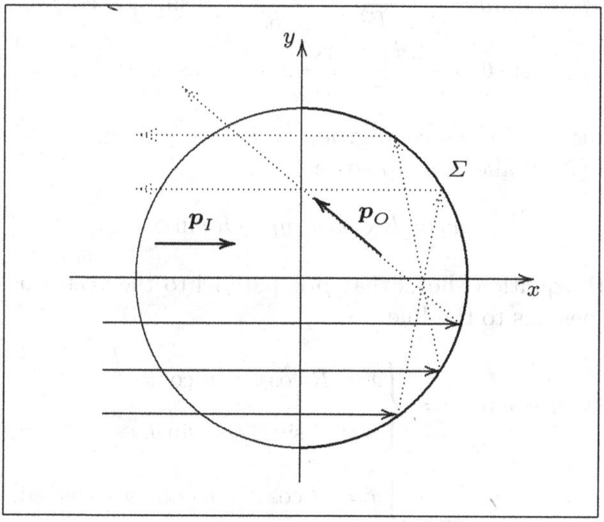

Fig. 7.5 Parallel rays reflected by a circular mirror

Theorem 7.12. *The reflected Lagrangian set Λ_O is a Lagrangian submanifold described by the parametric equations*

$$\begin{cases} x = R\, \cos\phi + u\, (\sin^2\phi - \cos^2\phi), \\ y = R\, \sin\phi - 2\, u\, \sin\phi\, \cos\phi, \end{cases} \qquad \begin{cases} p_x = \sin^2\phi - \cos^2\phi, \\ p_y = -2\, \sin\phi\, \cos\phi, \end{cases} \tag{7.41}$$

with parameters u and ϕ.

Proof. 1. The first two (7.40) imply $1 = (\lambda\, x_I + 1)^2 + \lambda^2\, y_I^2$, that is, $\lambda^2\, R^2 + 2\, \lambda\, x_I = 0$. The two roots are

[1] We disregard here the fact that a ray can be multireflected by the mirror. See the figure.

$$\lambda = 0, \;\; \lambda = -\frac{2\,x_I}{R^2}.$$

For $\lambda = 0$:

$$\begin{cases} \cos\theta = 1, \\ \sin\theta = 0, \end{cases} \qquad \begin{cases} p_x = 1, \\ p_y = 0. \end{cases}$$

This solution gives $\Lambda_O = \Lambda_I$ in accordance with Remark 6.25. The relevant root is then

$$\lambda = -\frac{2\,x_I}{R^2},$$

for which the first two (7.40) give

$$\begin{aligned} p_x &= \cos\theta = 1 - \frac{2\,x_I^2}{R^2} = \frac{y_I^2 - x_I^2}{R^2} = \sin^2\phi - \cos^2\phi, \\ p_y &= \sin\theta = -\frac{2\,x_I\,y_I}{R^2} = -2\,\sin\phi\,\cos\phi, \end{aligned}$$

and the equations of the second group (7.41) are proved.
2. The third (7.40) allows us to write

$$x_I = R\,\cos\phi, \;\; y_I = R\,\sin\phi. \tag{7.42}$$

3. The fourth equation shows that $\boldsymbol{p}$ is parallel to the vector $\boldsymbol{x} - \boldsymbol{x}_I$. Then the point $\boldsymbol{x}$ belongs to the line

$$\boldsymbol{x} = \boldsymbol{x}_I + u\,\boldsymbol{p} \iff \begin{cases} x = R\,\cos\phi + u\,\cos\theta, \\ y = R\,\sin\phi + u\,\sin\theta. \end{cases}$$

$$\iff \begin{cases} x = R\,\cos\phi + u\,(\sin^2\phi - \cos^2\phi), \\ y = R\,\sin\phi - 2\,u\,\sin\phi\,\cos\phi. \end{cases}$$

and the equations of the first group (7.41) are proved.
4. In Sect. 4.5.1 we have seen that the parametric equations represent an immersion if and only if the matrix (4.15)

$$\left[\frac{\partial q^i}{\partial u^k} \;\middle|\; \frac{\partial p_i}{\partial u^k}\right]$$

has maximal rank. In the present case this matrix is

$$\begin{bmatrix} \dfrac{\partial x}{\partial u} & \dfrac{\partial y}{\partial u} & \dfrac{\partial p_x}{\partial u} & \dfrac{\partial p_y}{\partial u} \\[2ex] \dfrac{\partial x}{\partial \phi} & \dfrac{\partial y}{\partial \phi} & \dfrac{\partial p_x}{\partial \phi} & \dfrac{\partial p_y}{\partial \phi} \end{bmatrix}, \tag{7.43}$$

where, due to Eqs. (7.41),

$$\begin{cases} \dfrac{\partial x}{\partial u} = \sin^2\phi - \cos^2\phi = p_x, \\ \dfrac{\partial x}{\partial \phi} = -R\,\sin\phi + 4\,u\,\sin\phi\,\cos\phi = -R\,\sin\phi - 2\,u\,p_y, \\ \dfrac{\partial y}{\partial u} = -\,2\,\sin\phi\,\cos\phi = p_y, \\ \dfrac{\partial y}{\partial \phi} = R\,\cos\phi + 2\,u\,(\sin^2\phi - \cos^2\phi) = R\,\cos\phi + 2\,u\,p_x, \end{cases} \tag{7.44}$$

$$\begin{cases} \dfrac{\partial p_x}{\partial u} = 0, \\ \dfrac{\partial p_x}{\partial \phi} = 4\,\sin\phi\,\cos\phi = -\,2\,p_y, \end{cases} \qquad \begin{cases} \dfrac{\partial p_y}{\partial u} = 0, \\ \dfrac{\partial p_y}{\partial \phi} = 2\,(\sin^2\phi - \cos^2\phi) = 2\,p_x. \end{cases}$$

Hence, the matrix (7.43) becomes

$$\begin{bmatrix} p_x & p_y & 0 & 0 \\ -R\,\sin\phi - 2\,u\,p_y & R\,\cos\phi + 2\,u\,p_x & -2\,p_y & 2\,p_x \end{bmatrix}. \tag{7.45}$$

Since

$$\det \begin{bmatrix} p_x & 0 \\ -R\,\sin\phi - 2\,u\,p_y & 2\,p_x \end{bmatrix} = 2\,p_x^2,$$

$$\det \begin{bmatrix} p_y & 0 \\ R\,\cos\phi + 2\,u\,p_x & -2\,p_y \end{bmatrix} = -2\,p_y^2,$$

and $p_x^2 + p_y^2 = 1$, the matrix (7.45) has maximal rank everywhere on Λ_O. □

Theorem 7.13. *The caustic of Λ_O is represented by the parametric equations*

$$\boxed{x = R\,\cos\phi\,(\tfrac{1}{2} + \sin^2\phi), \quad y = R\,\sin^3\phi} \tag{7.46}$$

with $-\pi/2 < \phi < \pi/2$.

Proof. We apply the general Theorem 4.2 to this case: the equation of the caustic is

$$\det \left[\frac{\partial q^i}{\partial u^k}\right]_{\Lambda_O} = \det \begin{bmatrix} \dfrac{\partial x}{\partial u} & \dfrac{\partial y}{\partial u} \\ \dfrac{\partial x}{\partial \phi} & \dfrac{\partial y}{\partial \phi} \end{bmatrix}_{\Lambda_O} = 0. \tag{7.47}$$

We have already seen in (7.44) that on Λ_O,

$$\frac{\partial x}{\partial u} = p_x, \qquad \frac{\partial y}{\partial u} = p_y,$$
$$\frac{\partial x}{\partial \phi} = -R\,\sin\phi - 2\,u\,p_y, \quad \frac{\partial y}{\partial \phi} = R\,\cos\phi + 2\,u\,p_x$$

Then (7.47) becomes

$$R\,(\cos\phi\,p_x + \sin\phi\,p_y) + 2\,u = 0. \tag{7.48}$$

But on Λ_O we have

$$\cos\phi\,p_x + \sin\phi\,p_y = \cos\phi\,(\sin^2\phi - \cos^2\phi) - 2\,\sin^2\phi\,\cos\phi = \;-\cos\phi.$$

and (7.48) gives $2\,u = R\,\cos\phi$. By inserting this result into the first Eqs. (7.41) we obtain the parametric equations (7.46). About the range of the parameter ϕ we observe that, due to (7.42), the limitation $\pi/2 < \phi < 3\pi/2$ implies $x_I < 0$, whereas the points of reflection (x_I, y_I) lie in the half-plane $X > 0$. □

If we plot Eqs. (7.46) we see this:

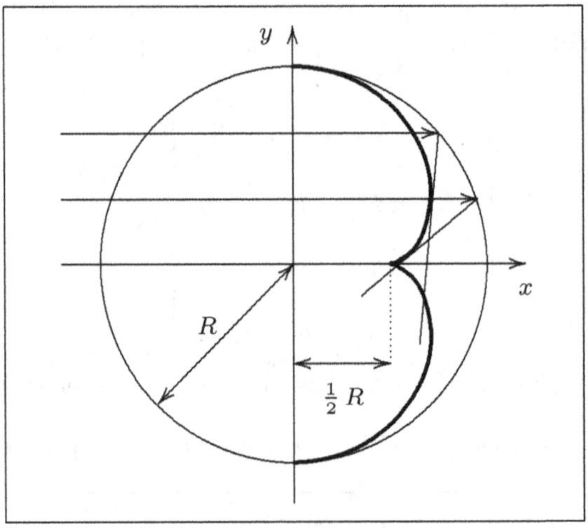

Fig. 7.6 Caustic of the coffee cup

7.4.5 Concave and convex ideal lenses

A generating family of the lens-relation has been given in (6.47), Theorem 6.12. In the Euclidean space $\mathbb{R}^n$ it can be written in the form

$$G_{L(\Sigma,F)}(\boldsymbol{x}_O, \boldsymbol{x}_I; \boldsymbol{a}, \boldsymbol{\lambda}) = \boldsymbol{a}\cdot(\boldsymbol{x}_O - \boldsymbol{x}_I) + F(\boldsymbol{x}_I) + \lambda^a\,\Sigma_a(\boldsymbol{x}_I), \tag{7.49}$$

with $\boldsymbol{a}^2 = 1$. In the Euclidean plane it becomes

$$\boxed{\begin{aligned} & G_{L(\Sigma,F)}(\boldsymbol{x}_O, \boldsymbol{x}_I; \theta, \lambda) \\ & = (x_O - x_I)\,\cos\theta + (y_O - y_I)\,\sin\theta + F(\boldsymbol{x}_I) + \lambda\,\Sigma(\boldsymbol{x}_I) \end{aligned}} \tag{7.50}$$

where the lens-curve is given by the implicit equation $\Sigma(\boldsymbol{x}) = 0$.

Let us examine an application of this generating family.

ASSUME that:

1. The lens is the y-axis: $\Sigma(\boldsymbol{x}) = x$.
2. The incoming rays are parallel to the x-axis. The incoming Lagrangian set Λ_I is then described by the Morse family $G_I(\boldsymbol{x}) = x$.

PROBLEM: find the characteristic function $F(y)$ on Σ such that the lens-relation transforms the incoming parallel rays to a system of rays focused at a fixed point $(c, 0)$ on the x-axis.

The generating family of the lens-relation (7.50) is now

$$\begin{aligned} & G_{L(\Sigma,F)}(\boldsymbol{x}_O, \boldsymbol{x}_I; \theta, \lambda) = \\ & = (x_O - x_I)\,\cos\theta + (y_O - y_I)\,\sin\theta + F(y_I) + \lambda\,x_I, \end{aligned} \tag{7.51}$$

and its composition with $G_I(\boldsymbol{x}) = x$ gives the generating family of the output Lagrangian set Λ_O:

$$\begin{aligned} & G_O(\boldsymbol{x}; \boldsymbol{x}_I, \theta, \lambda) = \\ & = (x - x_I)\,\cos\theta + (y - y_I)\,\sin\theta + F(y_I) + (\lambda + 1)\,x_I. \end{aligned} \tag{7.52}$$

The equations of its critical set are:

$$0 = \frac{\partial G_O}{\partial x_I} = \lambda + 1 - \cos\theta, \quad 0 = \frac{\partial G_O}{\partial \theta} = (y - y_I)\cos\theta - (x - x_I)\sin\theta,$$

$$0 = \frac{\partial G_O}{\partial y_I} = F'(y_I) - \sin\theta, \quad 0 = \frac{\partial G_O}{\partial \lambda} = x_I.$$

They allow us to reduce the generating family G_O to the family (for simplicity we put $y_I = u$)

$$\boxed{G(x, y; \theta, u) = x\,\cos\theta + (y - u)\,\sin\theta + F(u)} \tag{7.53}$$

This is a Morse family, Λ_O is a Lagrangian submanifold for $p_x = \cos\theta \neq 0$ or for $p_x = \cos\theta = 0$ and $F''(v) \neq 0$. Indeed, the matrix

$$\begin{bmatrix} \dfrac{\partial^2 G}{\partial\theta\,\partial\theta} & \dfrac{\partial^2 G}{\partial\theta\,\partial u} & \dfrac{\partial^2 G}{\partial\theta\,\partial x} & \dfrac{\partial^2 G}{\partial\theta\,\partial y} \\ \dfrac{\partial^2 G}{\partial u\,\partial\theta} & \dfrac{\partial^2 G}{\partial u\,\partial u} & \dfrac{\partial^2 G}{\partial u\,\partial x} & \dfrac{\partial^2 G}{\partial u\,\partial y} \end{bmatrix} \tag{7.54}$$

$$= \begin{bmatrix} -x\cos\theta-(y-u)\sin\theta & -\cos\theta & -\sin\theta\cos\theta \\ -\cos\theta & F''(u) & 0 & 0 \end{bmatrix}$$

has maximal rank for $p_x = \cos\theta \neq 0$. For $p_x = \cos\theta = 0$ it becomes

$$\begin{bmatrix} \pm(y-u) & 0 & \pm 1 & 0 \\ 0 & F''(u) & 0 & 0 \end{bmatrix},$$

and the rank is maximal only for $F''(u) \neq 0$. Since (7.53) is a Morse family, we can apply Theorem 4.8 for computing the caustic of Λ_O. The first equation is obtained by putting to zero the determinant of the first square submatrix (7.54),

$$\det \begin{bmatrix} \dfrac{\partial^2 G}{\partial\theta\,\partial\theta} & \dfrac{\partial^2 G}{\partial\theta\,\partial u} \\ \dfrac{\partial^2 G}{\partial u\,\partial\theta} & \dfrac{\partial^2 G}{\partial u\,\partial u} \end{bmatrix} = -F''(u)\,[x\cos\theta+(y-u)\sin\theta]-\cos^2\theta = 0.$$

The remaining equations are the equations of the critical set of G:

$$F'(u) = \sin\theta, \qquad x\sin\theta-(y-u)\cos\theta = 0.$$

For $F''(u) \neq 0$ we get a linear system in $(x, y-u)$,

$$\begin{cases} x\cos\theta+(y-u)\sin\theta = -\dfrac{\cos^2\theta}{F''(u)}, \\ x\sin\theta-(y-u)\cos\theta = 0, \end{cases}$$

whose solution is

$$x = -\frac{\cos^3\theta}{F''(u)}, \quad y-u = -\frac{\sin\theta\cos^2\theta}{F''(u)}.$$

From $F'(u) = \sin\theta$ and $p_x = \cos\theta \geq 0$ it follows that $\cos\theta = 1 - F'^2(u)$. Then we have proved the following.

Theorem 7.14. *If the ideal lens Σ coincides with the y-axis and the incoming rays are parallel to the x-axis, then the caustic of the outgoing system of rays is given by the parametric equations*

$$x = -\frac{\left[1 - F'^2(u)\right]^{3/2}}{F''(u)}, \quad y = u - \frac{F'(u)\left[1 - F'^2(u)\right]}{F''(u)} \tag{7.55}$$

where $F(u)$ is the characteristic function of the lens.

The caustic reduces the single point $(c, 0)$ when

$$c = -\frac{\left[1 - F'^2(u)\right]^{3/2}}{F''(u)}, \quad 0 = u - \frac{F'(u)\left[1 - F'^2(u)\right]}{F''(u)}. \tag{7.56}$$

Due to the first equation, the second equation gives $-u\left[1 - F'^2(u)\right]^{1/2} = c\,F'(u)$, and by taking the square of both sides we find

$$F'^2(u) = \frac{u^2}{c^2 + u^2}, \quad \text{that is,} \quad F'(u) = \pm\frac{u}{\sqrt{c^2 + u^2}}.$$

Hence, up to an inessential additive constant, we find two solutions of our problem (recall that $u = y$):

$$F(y) = \pm\sqrt{c^2 + y^2} \tag{7.57}$$

Since

$$F''(u) = \pm\frac{c^2}{(c^2 + u^2)^{3/2}},$$

due to the first Eq. (7.56), we conclude that:

- $F(y) = \sqrt{c^2 + y^2} \implies c < 0$. *This is the case of a concave lens:*

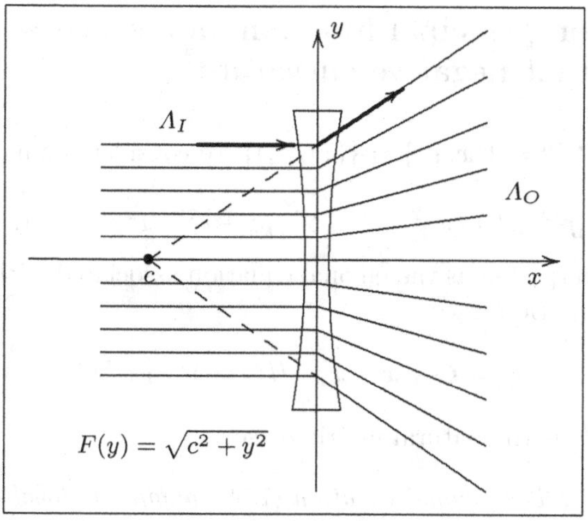

Fig. 7.7 Concave ideal lens

- $F(y) = -\sqrt{c^2 + y^2} \implies c > 0$. *This is the case of a convex lens*:

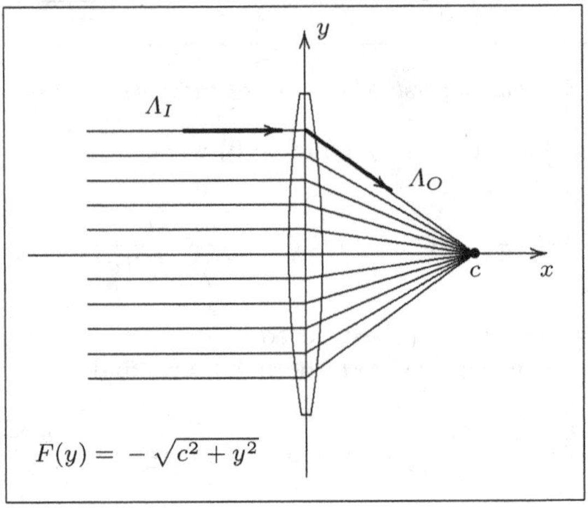

Fig. 7.8 Convex ideal lens

The corresponding lens-relations are generated by (see Eq. (7.53)):

$$\boxed{G(x, y; \theta, u) = x \cos\theta + (y - u) \sin\theta \pm \sqrt{c^2 + u^2}} \tag{7.58}$$

Remark 7.14. In this case the characteristic function $F(y)$ has an interesting meaning: it is, up to the sign, the distance from the focus $(c, 0)$ to the point of intersection of an incoming ray with the lens Σ. ♢

7.5 Hamilton principal function on a space of constant negative curvature

In the space $T^*\mathbb{R}^n = \{(\boldsymbol{x}, \boldsymbol{p})\} = \{(x^a, p_a)\}$ we consider the modified eikonal equation

$$|\boldsymbol{p}|^2 + (\boldsymbol{p} \cdot \boldsymbol{x})^2 = 1, \quad \sum_a p_a^2 + \left(\sum_a x^a p_a\right)^2 = 1. \tag{7.59}$$

This can be interpreted as the eikonal equation associated with the modified contravariant metric tensor

$$\boldsymbol{H} = \boldsymbol{G} + \boldsymbol{x} \otimes \boldsymbol{x}, \quad H^{ab} = \delta^{ab} + x^a x^b, \tag{7.60}$$

where $\boldsymbol{G} = [\delta^{ab}]$ is the natural metric tensor.

Theorem 7.15. *The eikonal equation* (7.59) *admits a global complete solution* $W\colon Q \times A = \mathbb{R}^n \times \mathbb{S}_{n-1} \to \mathbb{R}$ *defined by*

$$W(\boldsymbol{x}, \boldsymbol{a}) = \log\left(\boldsymbol{a} \cdot \boldsymbol{x} + \sqrt{1 + (\boldsymbol{a} \cdot \boldsymbol{x})^2}\right) = \operatorname{arcsinh}(\boldsymbol{a} \cdot \boldsymbol{x}), \tag{7.61}$$

with $\boldsymbol{a} \in \mathbb{S}_{n-1}$.

Proof. The vector

$$\boldsymbol{p} = \begin{bmatrix} p_1 \\ \vdots \\ p_n \end{bmatrix} = \frac{\partial W}{\partial \boldsymbol{x}} = \frac{\boldsymbol{a}}{\sqrt{1 + (\boldsymbol{a} \cdot \boldsymbol{x})^2}} \tag{7.62}$$

satisfies Eq. (7.59) for any $\boldsymbol{a}$. From this equation we derive

$$\begin{cases} \boldsymbol{p} \cdot \boldsymbol{x} = \dfrac{\boldsymbol{x} \cdot \boldsymbol{a}}{\sqrt{1 + (\boldsymbol{x} \cdot \boldsymbol{a})^2}}, \\ (\boldsymbol{p} \cdot \boldsymbol{x})^2 = \dfrac{(\boldsymbol{x} \cdot \boldsymbol{a})^2}{1 + (\boldsymbol{x} \cdot \boldsymbol{a})^2} < 1, \end{cases} \qquad \begin{cases} (\boldsymbol{x} \cdot \boldsymbol{a})^2 = \dfrac{(\boldsymbol{p} \cdot \boldsymbol{x})^2}{1 - (\boldsymbol{p} \cdot \boldsymbol{x})^2}, \\ 1 + (\boldsymbol{x} \cdot \boldsymbol{a})^2 = \dfrac{1}{1 - (\boldsymbol{p} \cdot \boldsymbol{x})^2}. \end{cases}$$

Then the map $\pi \colon C \to A \colon \boldsymbol{p} \mapsto \boldsymbol{a}$ is given by

$$\boldsymbol{a} = \frac{\boldsymbol{p}}{\sqrt{1 - (\boldsymbol{p} \cdot \boldsymbol{x})^2}}.$$

This is a submersion. □

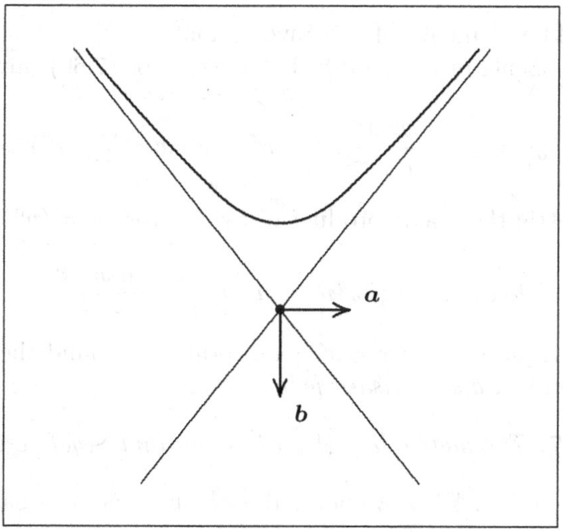

Fig. 7.9 Remark 7.15

Remark 7.15. The symplectic reduction $R \subset T^*\mathbb{S}_{n-1} \times T^*\mathbb{R}^n$ corresponding to this complete solution is described by Eq. (7.62) together with equation

$$\boldsymbol{b} = -\frac{\partial W}{\partial \boldsymbol{a}} = -\frac{1}{\sqrt{1+(\boldsymbol{a}\cdot\boldsymbol{x})^2}}\,P_{\boldsymbol{a}}(\boldsymbol{x}) = \frac{\boldsymbol{x}\cdot\boldsymbol{a}\,\boldsymbol{a}-\boldsymbol{x}}{\sqrt{1+(\boldsymbol{a}\cdot\boldsymbol{x})^2}}. \tag{7.63}$$

Note that $\boldsymbol{b}\cdot\boldsymbol{a}=0$. From this formula it follows that the ray determined by the pair of orthogonal vectors $(\boldsymbol{a},\boldsymbol{b})$ is the hyperbola on the plane $(\boldsymbol{a},\boldsymbol{b})$ with center at the origin $\boldsymbol{x}=0$, asymptotes determined by the vectors $\boldsymbol{b}\pm\boldsymbol{a}$ and vertex at the point $\boldsymbol{x}=-\boldsymbol{b}$.

The vector $\boldsymbol{v}=(v^a)$ defined by $v^a = H^{ab}\,p_a$ is tangent to the ray. It follows from (7.60) and (7.62) that

$$\boldsymbol{v} = \boldsymbol{p}+\boldsymbol{x}\cdot\boldsymbol{p}\,\boldsymbol{x} = \frac{\boldsymbol{a}+\boldsymbol{a}\cdot\boldsymbol{x}\,\boldsymbol{x}}{\sqrt{1+(\boldsymbol{a}\cdot\boldsymbol{x})^2}}. \qquad \diamondsuit$$

From Theorem 7.15 (see also Remark 6.19) we derive the following.

Theorem 7.16. *The function*

$$\begin{aligned} S(\boldsymbol{x},\boldsymbol{x}';\boldsymbol{a}) &= W(\boldsymbol{x},\boldsymbol{a}) - W(\boldsymbol{x}',\boldsymbol{a}) \\ &= \log\frac{\boldsymbol{a}\cdot\boldsymbol{x}+\sqrt{1+(\boldsymbol{a}\cdot\boldsymbol{x})^2}}{\boldsymbol{a}\cdot\boldsymbol{x}'+\sqrt{1+(\boldsymbol{a}\cdot\boldsymbol{x}')^2}} \\ &= \operatorname{arcsinh}(\boldsymbol{x}\cdot\boldsymbol{a}) - \operatorname{arcsinh}(\boldsymbol{x}'\cdot\boldsymbol{a}), \end{aligned}$$

with $\boldsymbol{a}\in\mathbb{S}_{n-1}$, is a global Hamilton principal function of the eikonal equation (7.59). *It is a Morse family.*

Remark 7.16. The elements of the inverse matrix $\left[h_{ab}\right]$ of $\left[H^{ab}\right]$ (i.e., the covariant components of the modified metric tensor (7.60)) are

$$h_{ab} = \delta_{ab} - \frac{x^a\,x^b}{1+r^2}, \qquad r^2 = \boldsymbol{x}\cdot\boldsymbol{x} = \textstyle\sum_a (x^a)^2. \tag{7.64}$$

In this new metric the scalar product of two vectors $\boldsymbol{u}=(u^a)$ and $\boldsymbol{v}=(v^a)$ is given by

$$\boldsymbol{h}(\boldsymbol{u},\boldsymbol{v}) = h_{ab}\,u^a v^b = \boldsymbol{u}\cdot\boldsymbol{v} - \frac{\boldsymbol{x}\cdot\boldsymbol{u}\;\boldsymbol{x}\cdot\boldsymbol{v}}{1+r^2}.$$

This metric is invariant under Euclidean rotations around the origin $\boldsymbol{x}=0$. Thus, the origin is a *distinguished point*. $\diamondsuit$

Theorem 7.17. *The metric $\boldsymbol{h}=(h_{ab})$ has constant negative curvature.*

Proof. Let us consider $\mathbb{R}^n\times\mathbb{R}$ endowed with the canonical basis

$$\boldsymbol{c}_1 = \begin{bmatrix}1\\ \vdots\\ 0\\ 0\end{bmatrix}, \quad \dots \quad \boldsymbol{c}_n = \begin{bmatrix}0\\ \vdots\\ 1\\ 0\end{bmatrix}, \quad \boldsymbol{c}_{n+1} = \boldsymbol{t} = \begin{bmatrix}0\\ \vdots\\ 0\\ 1\end{bmatrix},$$

and the Minkowskian metric

$$m(\boldsymbol{u},\boldsymbol{v}) = \sum_{a=1}^{n} u^a v^a - u^{n+1} v^{n+1}.$$

Let us consider the set $\mathbb{H}_n$ of the unit time-like vectors $\boldsymbol{q}$, $m(\boldsymbol{q},\boldsymbol{q}) = -1$, pointing to the "future", that is, such that $m(\boldsymbol{q},\boldsymbol{t}) < 0$. It is known that $\mathbb{H}_n$ is a proper Riemannian manifold with constant negative curvature (see, for instance, (Wolf 1984). Taking $(x^a) = (x^1, \ldots, x^n)$ as parameters, this hyperboloid is described by the parametric equation

$$\boldsymbol{q} = x^a\, \boldsymbol{c}_a + \sqrt{1 + \textstyle\sum_a (x^a)^2}\; \boldsymbol{t}.$$

The corresponding tangent frame $(\boldsymbol{e}_a)$ is then defined by

$$\boldsymbol{e}_a = \partial_a \boldsymbol{q} = \boldsymbol{c}_a + \frac{x^a}{z}\, \boldsymbol{t},$$

being

$$z = \sqrt{1 + \textstyle\sum_a (x^a)^2} = \sqrt{1 + r^2}.$$

It follows that the components of the induced metric tensor (the first fundamental form) of $\mathbb{H}_n$ are

$$m(\boldsymbol{e}_a, \boldsymbol{e}_b) = m(\boldsymbol{c}_a, \boldsymbol{c}_b) - \frac{x^a\, x^b}{z^2} = \delta_{ab} - \frac{1}{1 + r^2}\, x^a x^b.$$

This is the metric (7.64). □

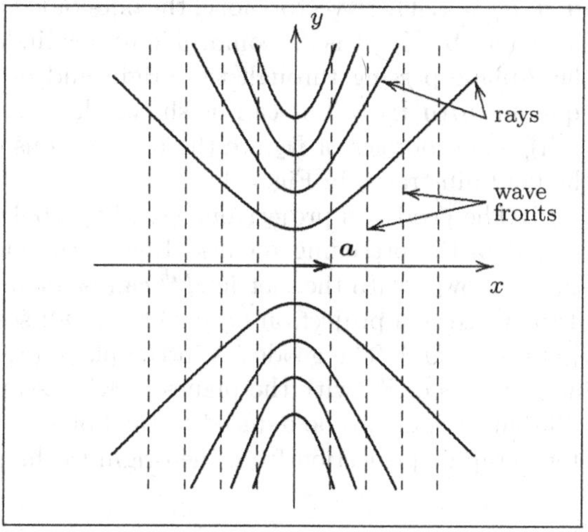

Fig. 7.10 Wave fronts and rays in $\mathbb{R}^2$ endowed with a metric of constant negative curvature

Remark 7.17. Note that $\boldsymbol{q}$ is a unit vector orthogonal to $\mathbb{H}_n$ and that $\partial_b \boldsymbol{e}_a \cdot \boldsymbol{q} = \boldsymbol{B}_{ba} \cdot \boldsymbol{q} = -g_{ab}$. For simplicity, in the following we consider the case $n = 2$. All results can be easily extended to any dimension n. ♢

Remark 7.18. Let us consider for instance $\boldsymbol{a} = \boldsymbol{c}_1$ (the first vector of the canonical basis of $\mathbb{R}^2$). In this case (7.63) becomes

$$\boldsymbol{b} = \frac{x\,\boldsymbol{c}_1 - \boldsymbol{x}}{\sqrt{1+x^2}} = -\frac{y\,\boldsymbol{c}_2}{\sqrt{1+x^2}}.$$

By setting $\boldsymbol{b} = b\,\boldsymbol{c}_2$ we find

$$b\sqrt{1+x^2} = -\,y. \tag{7.65}$$

For $b \neq 0$ it follows that

$$\frac{y^2}{b^2} - x^2 = 1.$$

This is the equation of the system of rays associated with the Lagrangian submanifold $\Lambda_{\boldsymbol{c}_1}$ generated by the function

$$G(\boldsymbol{x}) = \log\left(x + \sqrt{1+x^2}\right).$$

Equation (7.65) describes a family of hyperbolas centered at the origin of $\mathbb{R}^2$ and vertices the points $(0, \pm b)$. For $b = 0$ Eq. (7.65) reduces to $y = 0$, the x-axis is a ray. The corresponding wave fronts, described by equations $G =$ const. (i.e., $y =$ const.), are the straight lines parallel to the y-axis (see Fig. 7.10). ♢

Remark 7.19. Let $\boldsymbol{n}$ be a unit space-like vector in the Minkowski three-space, $\boldsymbol{m}(\boldsymbol{n}, \boldsymbol{n}) = 1$. These vectors form the one-folded rotational hyperboloid which we denote by $\mathbb{K}_2$; it is diffeomorphic to a cylinder; see Sect. 10.3. Let $\Pi_{\boldsymbol{n}}$ be the 2-plane passing through the origin and orthogonal to $\boldsymbol{n}$, described by equation $\boldsymbol{m}(\boldsymbol{q}, \boldsymbol{n}) = 0$. It can be shown that:

(i) The geodesics of $\mathbb{H}_2$ are the intersections of $\mathbb{H}_2$ with the planes $\Pi_{\boldsymbol{n}}$ of the kind illustrated in Fig. 7.11.

(ii) The geodesics project onto the hyperbolas of the two-plane (x, y) described in the preceding remark. Hence, the metric properties of the plane (x, y) endowed with the metric H^{ab} can be deduced by those of $\mathbb{H}_2$ by means of the Cartesian projection $(x, y, z) \mapsto (x, y)$; see Fig. 7.12 and, for instance, (Petersen 1998). The geodesics in the plane (x, y) are the projections of the intersections of $\mathbb{H}_2$ with the planes in $\mathbb{R}^3$ passing through the origin. We recall that $\mathbb{H}_2$ can also be reduced to the Lobachevskij disk $\mathbb{D}_2$ by means of the stereographic projection from the origin to the plane $z = 1$; Fig. 7.13.

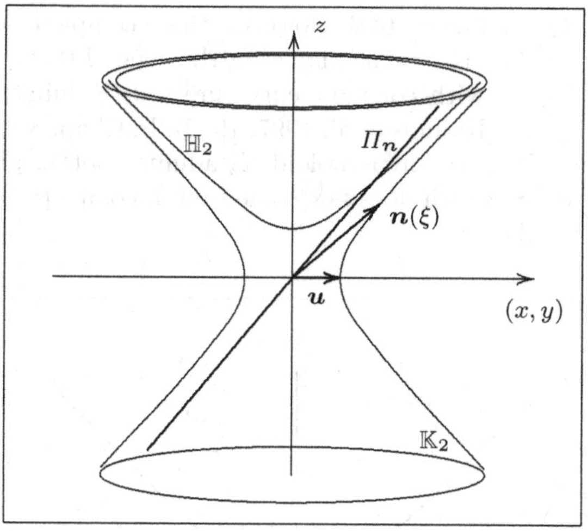

Fig. 7.11 The hyperboloids $\mathbb{H}_2$ and $\mathbb{K}_2$.

(iii) The systems of rays of the kind described in Remark 7.19 are obtained by considering the unit vectors

$$\boldsymbol{n}(\xi, \boldsymbol{u}) = \cosh \xi \, \boldsymbol{u} + \sinh \xi \, \boldsymbol{c}_3,$$

where $\xi \in \mathbb{R}$ is a parameter and $\boldsymbol{u}$ is a unit vector orthogonal to $\boldsymbol{a}$ in the plane (x, y). The space-like unit vectors $\boldsymbol{p}$ associated with this family of geodesics (parametrized by ξ) form a section $\Lambda_{\boldsymbol{u}}$ of $T^*\mathbb{H}_2$. If we take $\boldsymbol{u} = -\boldsymbol{c}_2$, then with respect to the frame (7.65) the components of these covectors are

$$p_y = 0, \qquad p_x = \frac{1}{z} \cosh \xi = \frac{1}{\sqrt{1+x^2}}.$$

Since

$$\int \frac{1}{\sqrt{1+x^2}} = \log\left(x + \sqrt{1+x^2}\right) = \operatorname{arcsinh}(x),$$

the set $\Lambda_{\boldsymbol{u}}$ is a Lagrangian submanifold generated by the function

$$G(x, y) = \log\left(x + \sqrt{1+x^2}\right) = \operatorname{arcsinh}(x).$$

Since $x = \boldsymbol{x} \cdot \boldsymbol{c}_1$, by replacing $\boldsymbol{c}_1$ by any unit vector $\boldsymbol{u}$ we get the complete solution (7.61). This is an example of a complete solution of a Hamilton–Jacobi equation obtained by means of a geometrical process and not by separation of variables. Indeed, the Hamilton–Jacobi equation (7.59) is integrable by separation of variables in the polar coordinates (ρ, θ), with θ ignorable, because the metric of $\mathbb{H}_2$ is invariant under rotations around the z-axis. Other systems of separable coordinates are known, which are associated with pairs of rotations around time-like vectors. However, these complete separated solutions are not defined on the whole plane (for the general theory of separation of variables in spaces with constant curvature see (Kalnins 1986); for the separability in $\mathbb{H}_2$ see (Kalnins et al. 1997, 1999). In Chap. 8 it is shown that the eikonal equation on the hyperboloid $\mathbb{H}_2$ admits another global principal Hamilton function S, which does not come from a complete integral W and is not a Morse family. ◊

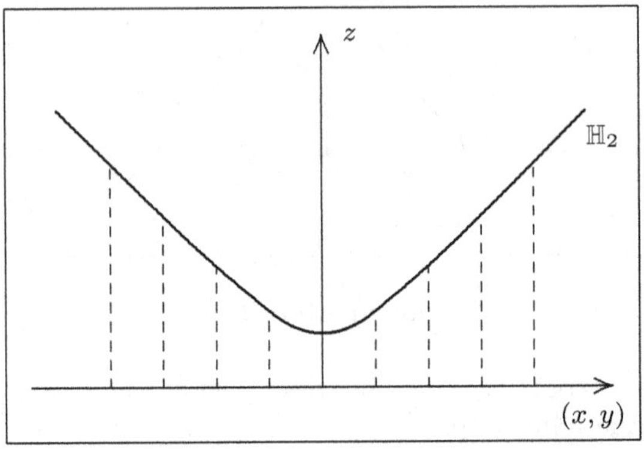

Fig. 7.12 Cartesian projection of $\mathbb{H}_2$ onto $\mathbb{R}^2$

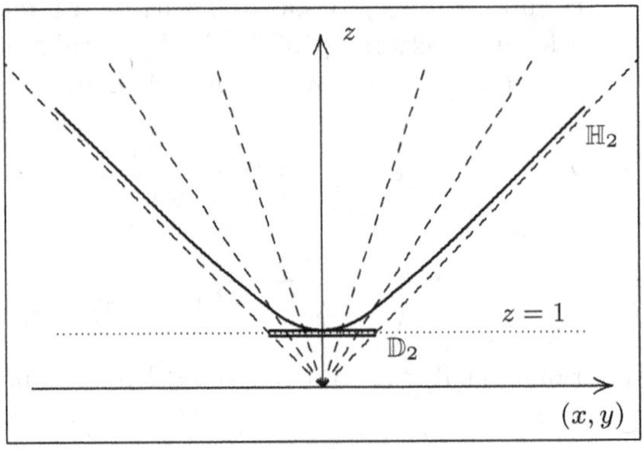

Fig. 7.13 Lobachevskij disk $\mathbb{D}_2$

Chapter 8
Control of Static Systems

Abstract What we have done until now was created as part of geometrical optics and analytical mechanics. But, surprisingly, it can be applied to other topics of mathematical physics; for instance, the study of the behavior of static systems, purely mechanical as well as thermodynamical.

8.1 Control relations

If a manifold Q represents the configuration space of a mechanical system, then any tangent vector $v \in TQ$ represents a *virtual velocity* or a *virtual displacement*, and any covector $f \in T^*Q$ represents a *force*. The evaluation $\langle v, f \rangle$ represents the *virtual power* or the *virtual work* produced by the force f in the virtual velocity (or displacement) v. If $(q^i, \delta q^i)$ are the fibered coordinates on TQ associated with coordinates (q^i) on Q, then the symbols δq^i represent the components of the tangent vectors v, so that the expression in coordinates of the virtual work is

$$\langle v, f \rangle = f_i \, \delta q^i, \tag{8.1}$$

with f_i the components of the force f.

Definition 8.1. A *control relation* is a relation $R\colon Q \leftarrow \bar{Q}$ of the form

$$\boxed{R = \mathrm{graph}(\phi) \cap (Q \times \Sigma)} \tag{8.2}$$

where $\Sigma \subset \bar{Q}$ is a submanifold and $\phi\colon \bar{Q} \to Q$ is a fibration. We call

- Q the *control manifold*,
- $\bar{Q}$ the *extended configuration manifold*,
- ϕ the *control fibration*,

- Σ the *constraint*,

of the control relation. ♡

Hence, a control relation is the graph of the restriction to a constraint Σ of a fibration $\phi\colon \bar{Q} \to Q$. Note that R may not be a smooth relation.

Definition 8.2. Two virtual displacements $v \in TQ$ and $\bar{v} \in T\bar{Q}$ are *compatible* with respect to the control relation R if $\bar{v} \in T\Sigma$ and $T\phi(\bar{v}) = v$. ♡

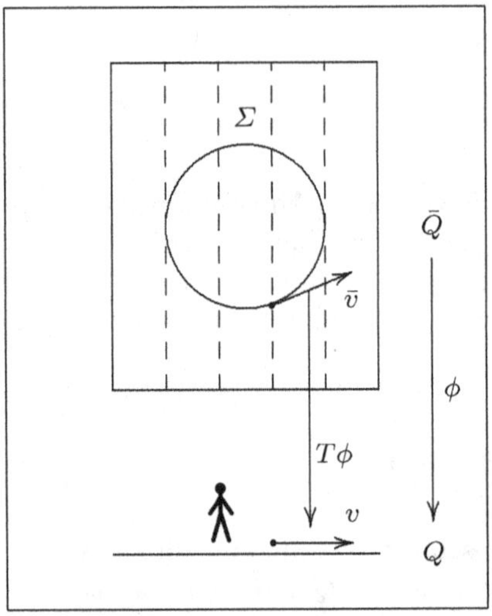

Fig. 8.1 Control relation

The fibration ϕ represents the existence of *hidden* or *internal variables* in the extended configuration manifold $\bar{Q}$. These variables are not controlled by the "little man" on Q and may assume any value belonging to the constraint Σ. Control relations arise in control problems of static mechanical systems, in catastrophe theory, and in thermodynamics.

There is a useful equivalent definition of control relation:

Theorem 8.1. *Equation* (8.2) *is equivalent to*

$$\boxed{R = \Phi \circ \Delta_\Sigma} \tag{8.3}$$

where $\Phi = \mathrm{graph}(\phi)$, *and* $\Delta_\Sigma \subset \bar{Q} \times \bar{Q}$ *is the diagonal of* $\Sigma \times \Sigma$, *the identity relation on* Σ.

Proof. Equation (8.2) is equivalent to

$$R = \left\{ (q, \bar{q}) \in Q \times \bar{Q} \mid q = \phi(\bar{q}), \ \ \bar{q} \in \Sigma \right\}.$$

Equation (8.3) is equivalent to

$$\begin{aligned} R = \Phi \circ \Delta_\Sigma &= \Big\{ (q, \bar{q}) \in Q \times \bar{Q} \ \text{ such that there exists } \bar{q}' \in \bar{Q} \\ &\qquad \text{with } q = \phi(\bar{q}') \text{ and } (\bar{q}, \bar{q}') \in \Delta_\Sigma \Big\} \\ &= \Big\{ (q, \bar{q}) \in Q \times \bar{Q} \ \text{ such that there exists } \bar{q}' \in \bar{Q} \\ &\qquad \text{with } q = \phi(\bar{q}') \text{ and } \bar{q} = \bar{q}' \in \Sigma \Big\} \\ &= \{ (q, \bar{q}) \in Q \times \bar{Q} \text{ with } q = \phi(\bar{q}') \text{ and } \bar{q} = \bar{q}' \in \Sigma \}. \qquad \square \end{aligned}$$

Definition 8.3. The *canonical lift of a control relation* $R = \Phi \circ \Delta_\Sigma$ is the composition

$$\boxed{\widehat{R} = \widehat{\Phi} \circ \widehat{\Delta}_\Sigma \subseteq T^*Q \times T^*\bar{Q}} \tag{8.4}$$

of the canonical lifts of Δ_Σ and of Φ. ♡

Theorem 8.2. *Equation* (8.4) *is equivalent to*

$$\boxed{\begin{aligned} \widehat{R} = \Big\{ &(f, \bar{f}) \in T^*Q \times T^*\bar{Q} \ \text{ such that } \ \pi_Q \times \pi_{\bar{Q}}(f, \bar{f}) = (q, \bar{q}) \in R, \\ &\text{and} \\ &\langle T\phi(\bar{v}), f \rangle = \langle \bar{v}, \bar{f} \rangle, \ \text{ for all } \bar{v} \in T_{\bar{q}}\Sigma \Big\} \end{aligned}} \tag{8.5}$$

Proof. By the definition of canonical lift of a smooth relation we have:

$$\begin{aligned} \widehat{\Phi} &= \{ (f, \bar{f}) \in T^*Q \times T^*\bar{Q} \ \text{ such that } \ (\pi_Q \times \pi_{\bar{Q}})(f, \bar{f}) = (q, \bar{q}) \in \Phi, \\ &\qquad \text{and } \ \langle v, f \rangle = \langle \bar{v}, \bar{f} \rangle, \ \text{ for all } \ (v, \bar{v}) \in T_{(q,\bar{q})}\Phi \} \\ &= \{ (f, \bar{f}) \in T^*Q \times T^*\bar{Q} \ \text{ such that } \ \pi_Q \times \pi_{\bar{Q}}(f, \bar{f}) = (q, \bar{q}) \in \Phi, \\ &\qquad \text{and } \ \langle T\phi(\bar{v}), f \rangle = \langle \bar{v}, \bar{f} \rangle, \ \text{ for all } \bar{v} \in T_{\bar{q}}\Sigma \}, \end{aligned}$$

and

$$\begin{aligned} \widehat{\Delta}_\Sigma = \{ &(\bar{f}, \bar{f}') \in T^*\bar{Q} \times T^*\bar{Q} \ \text{ such that } \ \pi_{\bar{Q}}(\bar{f}) = \pi_{\bar{Q}}(\bar{f}') = \bar{q} \in \Sigma, \\ &\text{and } \ \langle \bar{v}, \bar{f} - \bar{f}' \rangle = 0, \ \text{ for all } \ \bar{v} \in T_{\bar{q}}\Sigma \}. \end{aligned}$$

By applying the composition rule of relations we get

$$\widehat{\Phi} \circ \widehat{\Delta}_\Sigma = \Big\{ (f, \bar{f}) \in T^*Q \times T^*\bar{Q} \text{ such that there exists a } \bar{f}' \in T^*\bar{Q} \text{ with } (f, \bar{f}') \in \widehat{\Phi},\ (\bar{f}', \bar{f}) \in \widehat{\Delta}_\Sigma \Big\}.$$

That is,

$$\begin{aligned} \widehat{\Phi} \circ \widehat{\Delta}_\Sigma = \Big\{ &(f, \bar{f}) \in T^*Q \times T^*\bar{Q} \text{ such that there exists a } \bar{f}' \in T^*\bar{Q} \\ &\text{with:} \\ &f \in T^*_q Q,\ \bar{f}' \in T^*_{\bar{q}} \bar{Q},\ q = \phi(\bar{q}),\ \text{with } \bar{q} \in \Sigma, \\ &\langle T\phi(\bar{v}), f \rangle = \langle \bar{v}, \bar{f}' \rangle,\ \text{for all } \bar{v} \in T_{\bar{q}} \bar{Q},\ \text{with } \bar{q} = \pi_{\bar{Q}}(\bar{f}), \\ &\text{and } \langle \bar{v}, \bar{f} - \bar{f}' \rangle = 0,\ \text{for all } \bar{v} \in T_{\bar{q}} \Sigma \Big\}. \end{aligned} \tag{8.6}$$

From these last conditions it follows that $(q, \bar{q}) \in R$ and $\langle T\phi(\bar{v}), f \rangle = \langle \bar{v}, \bar{f} \rangle$ for all $\bar{v} \in T_{\bar{q}} \Sigma$. This shows that $\widehat{\Phi} \circ \widehat{\Delta}_\Sigma \subseteq \widehat{R}$ as defined in (8.5). Conversely, if $(f, \bar{f}) \in \widehat{R}$, then the last conditions (8.6) are satisfied for $\bar{f}' = \bar{f}$. Thus, $\widehat{\Phi} \circ \widehat{\Delta}_\Sigma \supseteq \widehat{R}$. □

Remark 8.1. Formula (8.5) shows that a pair of forces $(f, \bar{f})$ belongs to the relation $\widehat{R}$ if and only if these two covectors are based at a pair $(q, \bar{q})$ of points belonging to the relation R and such that

$$\langle T\phi(\bar{v}), f \rangle = \langle \bar{v}, \bar{f} \rangle, \tag{8.7}$$

for any virtual displacement $\bar{v}$ tangent to the constraint Σ. This means that $\langle v, f \rangle = \langle \bar{v}, \bar{f} \rangle$ for two compatible virtual displacements $(v, \bar{v})$. ◊

We use the above-given definitions and theorems for stating the following two *axioms*:

> **1** The system with configuration manifold $\bar{Q}$ remains in static equilibrium under forces $\bar{f} \in T^*\bar{Q}$ belonging to a Lagrangian submanifold $\bar{\mathscr{E}} \subset T^*\bar{Q}$ generated by a function $\bar{V} \colon \bar{Q} \to \mathbb{R}$, called the *extended potential energy*:
>
> $$\boxed{\bar{\mathscr{E}} = d\bar{V}(\bar{Q})} \tag{8.8}$$

Remark 8.2. We could consider the general case of a generating family $\bar{V} \colon \bar{Q} \times \bar{U} \to \mathbb{R}$. However, in all the examples illustrated below, the potential energy $\bar{V}$ is an ordinary generating function defined on $\bar{Q}$, without supplementary variables. ◊

2 The system with configuration manifold $\bar{Q}$ remains in static equilibrium under the action of an "external device", represented by the control relation R, only with forces f belonging to a certain *set of equilibrium states* $\mathscr{E} \subseteq T^*Q$, also called the *constitutive set* of the system, defined by

$$\boxed{\mathscr{E} = \widehat{R} \circ \overline{\mathscr{E}}} \tag{8.9}$$

Remark 8.3. Equation (8.9) means that

$$f \in \mathscr{E} \iff \text{thetere exists a } \bar{f} \in \bar{\mathscr{E}} \text{ such that } (f, \bar{f}) \in \widehat{R}. \tag{8.10}$$

Then (8.10) and (8.7) show that $f \in \mathscr{E}$ if and only if for all compatible virtual displacements $(v, \bar{v})$ we have

$$\boxed{\langle v, f \rangle = \langle \bar{v}, d\bar{V} \rangle} \qquad \diamondsuit \tag{8.11}$$

We analyze the local coordinate representations of the above concepts by using generating families.

Let ϕ be (locally) represented by equations $q^i = \phi^i(\bar{q}^\alpha)$ and Σ by independent equations $\Sigma_a(\bar{q}^\alpha) = 0$. Thus, in accordance with Eq. (8.2), the control relation R is described by equations

$$\begin{cases} q^i - \phi^i(\bar{q}^\alpha) = 0, \\ \Sigma_a(\bar{q}^\alpha) = 0. \end{cases} \tag{8.12}$$

Theorem 8.3. *If the control relation R is locally described by equations (8.12), then its canonical lift $\widehat{R}$ is locally described by the generating family*

$$\boxed{G_R(q^i, \bar{q}^\alpha; \lambda_i, \mu^a) = \lambda_i \, \left(q^i - \phi^i(\bar{q}^\alpha)\right) + \mu^a \, \Sigma_a(\bar{q}^\alpha)} \tag{8.13}$$

with Lagrangian multipliers (λ_i, μ^a).

Proof. The generating families of the canonical lifts $\widehat{\Phi}$ and $\widehat{\Delta}_\Sigma$ are, respectively,

$$\begin{cases} G_\Phi(q^i, \bar{q}^\alpha; \lambda_i) = \lambda_i \, \left(q^i - \phi^i(\bar{q}^\alpha)\right), \\ G_\Sigma(\bar{q}_0^\alpha, \bar{q}^\alpha; \mu^a, \nu_\alpha) = \mu^a \, \Sigma_a(\bar{q}^\alpha) + \nu_\alpha(\bar{q}_0^\alpha - \bar{q}^\alpha). \end{cases}$$

Here, $(\lambda_i, \mu^a, \nu_\alpha)$ are supplementary variables. By the composition rule of the generating families we get the generating family G_R of $\widehat{\Phi} \circ \widehat{\Delta}_\Sigma$,

$$G_R(q^i, \bar{q}^\alpha; \bar{q}_0^\alpha, \lambda_i, \mu^a, \nu_\alpha) = \lambda_i \, \left(q^i - \phi^i(\bar{q}_0^\alpha)\right) + \mu^a \, \Sigma_a(\bar{q}_0^\alpha) + \nu_\alpha(\bar{q}_0^\alpha - \bar{q}^\alpha),$$

with supplementary variables $(\bar{q}^\alpha_o, \lambda_i, \mu^a, \nu_\alpha)$. Since $\widehat{R}$ is then described by equation

$$f_i \, dq^i - \bar{f}_\alpha \, d\bar{q}^\alpha = dG_R, \tag{8.14}$$

the vanishing of the coefficients of $d\nu_\alpha$ implies $\bar{q}^\alpha_0 = \bar{q}^\alpha$. Thus, the generating family is reducible to (8.13). □

Remark 8.4. Equation (8.14) with G_R defined by formula (8.13) is equivalent to the equations obtained by putting to zero the coefficients of $(dq^i, d\bar{q}^\alpha, d\lambda_i, d\mu^a)$,

$$\begin{cases} f_i = \lambda_i, \\ \bar{f}_\alpha = \lambda_i \, \dfrac{\partial \phi^i}{\partial \bar{q}^\alpha} - \mu^a \, \dfrac{\partial \Sigma_a}{\partial \bar{q}^\alpha}, \\ q^i = \phi^i(\bar{q}^\alpha), \\ \Sigma_a(\bar{q}^\alpha) = 0. \end{cases}$$

By eliminating the Lagrangian multipliers λ_i we get equations

$$\begin{cases} \bar{f}_\alpha = f_i \, \dfrac{\partial \phi^i}{\partial \bar{q}^\alpha} - \mu^a \, \dfrac{\partial \Sigma_a}{\partial \bar{q}^\alpha}, \\ q^i = \phi^i(\bar{q}^\alpha), \\ \Sigma_a(\bar{q}^\alpha) = 0. \end{cases}$$

These are the equations describing $\widehat{R}$. The last two equations are the equations of R (fibration and constraint, respectively). The first equation is in accordance with (8.6). Indeed, if $\bar{v} = (\delta \bar{q}^\alpha)$ and $v = (\delta q^i)$, then

$$\begin{cases} \bar{v} \in T\Sigma \iff \dfrac{\partial \Sigma_a}{\partial \bar{q}^\alpha} \, \delta \bar{q}^\alpha = 0, \\ v = T\phi(\bar{v}) \iff \delta q^i = \dfrac{\partial \phi^i}{\partial \bar{q}^\alpha} \, \delta \bar{q}^\alpha. \end{cases} \qquad \diamond$$

By applying the composition rule of generating families, we can prove the following.

Theorem 8.4. *The constitutive set $\mathscr{E} = \widehat{R} \circ \bar{\mathscr{E}}$ is the Lagrangian set in T^*Q (possibly a Lagrangian submanifold) generated by the composite generating family*

$$\boxed{V = G_R \oplus \bar{V}} \tag{8.15}$$

It is described by equation

$$\boxed{f_i \, dq^i = d(G_R + \bar{V})} \tag{8.16}$$

equivalent to the four equations

$$\boxed{\begin{aligned} f_i &= \frac{\partial G_R}{\partial q^i}, & 0 &= \frac{\partial G_R}{\partial \lambda_i}, \text{ i.e. } q^i = \phi^i(\bar{q}^\alpha), \\ 0 &= \frac{\partial G_R}{\partial \bar{q}^\alpha} + \frac{\partial \bar{V}}{\partial \bar{q}^\alpha}, & 0 &= \frac{\partial G_R}{\partial \mu^a}, \text{ i.e. } \Sigma_a(\bar{q}^\alpha) = 0. \end{aligned}} \tag{8.17}$$

If the potential energy $\bar{V}$ depends on supplementary variables $\underline{u}$, then to this system we add equation $0 = \partial\bar{V}/\partial\underline{u}$.

Remark 8.5. With any smooth function $F\colon Q \to \mathbb{R}$ we associate a function δF on the tangent bundle TQ defined by

$$\delta F(v) = \langle v, dF \rangle.$$

The coordinate representation of this function is

$$\delta F = \frac{\partial F}{\partial q^i}\,\delta q^i.$$

This function is linear on each fiber of TQ. Thus, from the expression (8.1) of the virtual work it follows that the equilibrium states defined by (8.9) are characterized by the following *variational equation*

$$\boxed{f_i\,\delta q^i = \delta(G_R + \bar{V})} \tag{8.18}$$

Although the two symbols d and δ have different meanings, they have the same formal properties (linearity, Leibniz rule, etc.). Thus, Eq. (8.18) is formally equivalent to (8.16). However, although (8.16) has a pure mathematical character, (8.18) has a physical meaning: it states that a force $f = (f_i)$ is an equilibrium force (i.e., when applied to the system it is able to maintain the system in equilibrium) if and only if the corresponding virtual work, for any virtual displacement $v = (\delta q^i)$, is given by the value of the function $\delta(G_R + \bar{V})$. ◇

Equation (8.18) represents a generalization of the classical *virtual work principle* of *D'Alembert–Lagrange*.

We can think of more general kinds of control relations. For another general approach to this matter see (Tulczyjew 1989). The definition of control relation proposed here is suitable for dealing with the applications illustrated below. For a further approach to thermostatics see (Duboisand and Dufour 1974, 1976, 1978).

There are important special cases of control relations.

Case 1, *complete control without constraint.* In this case $\bar{Q} = \Sigma = Q$ and ϕ is the identity. It follows that $\mathscr{E} = \bar{\mathscr{E}}$ is the Lagrangian submanifold generated by a potential energy $V\colon Q \to \mathbb{R}$.

Case 2, *pure constraint:* $\Sigma \subset Q = \bar{Q}$ and ϕ is the identity. In this case we have $R = \Delta_\Sigma$ and $\mathscr{E} = (\Sigma, V)^\frown \subset T^*Q$. $\mathscr{E}$ is generated by the potential energy V over the constraint Σ. We can interpret this case in another way: $\Sigma = Q = \bar{Q}$. It follows that $\mathscr{E}$ is the Lagrangian submanifold of $T^*\Sigma$ generated by the restriction $V|\Sigma$ of the potential energy to the constraint. In other words, we look at Σ as the configuration manifold of the system.

Case 3, *pure fibration.* We have no constraint, but there are internal degrees of freedom (internal or hidden variables) of $\bar{Q}$ that are not controlled.

Case 4. The constraint Σ is such that the restriction of the fibration (or the surjective submersion) $\phi\colon \bar{Q} \to Q$ to Σ is a fibration (or a surjective submersion) $\phi\colon \Sigma \to Q$. In this case we can replace $\bar{Q}$ with Σ and the control relation reduces to the Case 3 of pure fibration.

Let us consider some basic examples.

Example 8.1. Let P be a point free to move in the plane $\mathbb{R}^2 = (x, y) = (\boldsymbol{x})$ and subjected to internal forces with potential energy V. Let us act on it by imposing its position $\boldsymbol{x}$. In this case $\bar{Q} = Q = \mathbb{R}^2$ (ϕ is the identity) and we have no constraint. In this control, we first impose the position of P_1, and then we measure the force f we have to apply for maintaining the point in that position. Then, according to (8.18), the equilibrium states are described by equation

$$f\,\delta x + g\,\delta y = \delta V,$$

which yields equations

$$f = \frac{\partial V}{\partial x}, \quad g = \frac{\partial V}{\partial y}.$$

These equations give the components (f, g) of the force $\boldsymbol{f}$ to be applied for maintaining the point at the assigned position. Thus, the set of the equilibrium states is the Lagrangian submanifold of $T^*\mathbb{R}^2$ generated by the function V. This is a case of complete control (Case 1). ◇

Example 8.2. A point P on the plane $(\bar{x}, \bar{y})$ is constrained to the unit circle $\mathbb{S}_1$, $\bar{x}^2 + \bar{y}^2 = 1$. We control only its coordinate $x = \bar{x}$, by moving a bar parallel to the y-axis along which P can slide freely. We can consider $\bar{Q} = \mathbb{R}^2 = (\bar{x}, \bar{y})$, $\Sigma = \mathbb{S}_1$, $Q = \mathbb{R} = (x)$, and the projection onto the x-axis as the fibration $\phi\colon \mathbb{R}^2 \to \mathbb{R}$ (Fig. 8.2). Inasmuch as this fibration is defined by equation $x = \bar{x}$, the generating family of the canonical lift $\widehat{R}$ is, according to (8.13),

$$G_R(x, \bar{x}, \bar{y}; \lambda, \mu) = \lambda\,(x - \bar{x}) + \mu\,(\bar{x}^2 + \bar{y}^2 - 1). \tag{8.19}$$

Now we apply the general variational equation (8.18) in the case of no active force, $\bar{V} = 0$,

$$f\,\delta x = \delta G_R, \tag{8.20}$$

with G_R given by (8.19). We derive the equations

$$\begin{cases} f = \lambda, \\ 0 = -\lambda + 2\,\mu\,\bar{x}, \\ 0 = \mu\,\bar{y}. \end{cases} \qquad \begin{cases} x - \bar{x} = 0, \\ \bar{x}^2 + \bar{y}^2 - 1 = 0, \end{cases}$$

which reduce to

$$f = 2\,\mu\,x, \quad \mu\,\bar{y} = 0, \quad x^2 + \bar{y}^2 - 1 = 0. \tag{8.21}$$

We observe that $\mu \neq 0$ implies $\bar{y} = 0$ and $x = \pm 1$. So that $x \neq \pm 1$ implies $\bar{y} \neq 0$, $\mu = 0$ and $f = 0$. Moreover, $x = \pm 1$ implies $\bar{y} = 0$, $f = \pm 2\mu$ and $\mu \in \mathbb{R}$. Then the equilibrium set $\mathscr{E}$ is represented in the plane $(x, f) = \mathbb{R}^2 = T^*\mathbb{R}$ by Fig. 8.3.

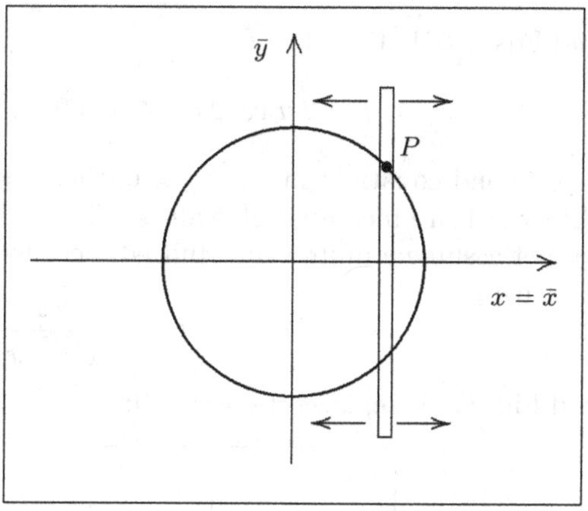

Fig. 8.2 Example 8.2

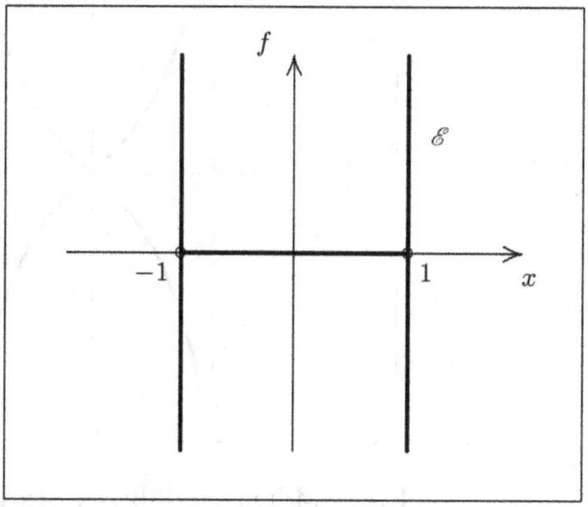

Fig. 8.3 The equilibrium states of Example 8.2

The generating family (8.19) of $\widehat{R}$ can be reduced in this case to a generating family of $\mathscr{E}$,

$$G_{\mathscr{E}}(x; \bar{y}, \mu) = \mu \, (x^2 + \bar{y}^2 - 1)$$

with extra variables $(\bar{y}, \mu)$. Indeed, the variational equation

$$f \, \delta x = \delta G_{\mathscr{E}}$$

is equivalent to Eqs. (8.21). It can be seen that this is a Morse family except at the two points $(\pm 1, 0)$, in accordance with the fact that without these two points $\mathscr{E}$ is a Lagrangian submanifold. ◇

Example 8.3. In the preceding example assume that a gravitational constant force (parallel to the $\bar{y}$-axis) acts on P. The potential energy is $\bar{V} = g\,\bar{y}$, $g > 0$. In this case Eq. (8.20) is replaced by

$$f \, \delta x = \delta G_R + \delta \bar{V} \tag{8.22}$$

and Eqs. (8.21) by

$$f = 2\,\mu\,x, \;\; 2\,\mu\,\bar{y} + g = 0, \;\; x^2 + \bar{y}^2 - 1 = 0.$$

The second equation implies $\mu\bar{y} \neq 0$, thus $\bar{y} \neq 0$, and the last equation shows that $x \pm 1$ are incompatible values. This means that for $x = \pm 1$ the force f cannot assume a finite value. Indeed, because $\mu = -g/\bar{y}$ and $\bar{y} = \pm\sqrt{1 - x^2}$, we have

$$f = \pm \frac{gx}{\sqrt{1 - x^2}}$$

and Fig. 8.3 is replaced by Fig. 8.4:

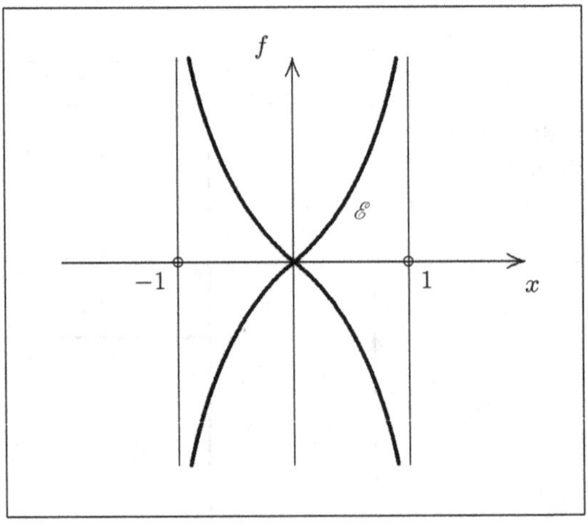

Fig. 8.4 The equilibrium states of Example 8.3

The sign of the force at a given point $x \neq \pm 1$ depends on the position of the constraint of the point P, which is not controlled. ♢

Example 8.4. On the plane $\mathbb{R}^2 = (x, y) = (\boldsymbol{x})$ we consider a point $P_1 = (x_1)$ moving on the x-axis and tied elastically to a point $P_2 = (x_2, y_2)$ free to move on the plane. Thus, $\bar{Q} = \mathbb{R}^2 \times \mathbb{R} = (x_2, y_2, x_1)$ is the configuration manifold of the holonomic system made of these two points. Let us act simultaneously on both points by imposing their positions. We are in the case of a complete control, $Q = \bar{Q}$. Then the set of the equilibrium states is the regular Lagrangian submanifold $\bar{\mathscr{E}}$ generated by the potential energy

$$\bar{V} = \tfrac{k}{2}\,(\boldsymbol{x}_1 - \boldsymbol{x}_2)^2 = \tfrac{k}{2}\,\left[(x_1 - x_2)^2 + y_2^2\right]$$

and described by equations

$$f_1 = \frac{\partial \bar{V}}{\partial x_1}, \quad f_2 = \frac{\partial \bar{V}}{\partial x_2}, \quad g_2 = \frac{\partial \bar{V}}{\partial y_2},$$

which provide the external forces $\boldsymbol{f}_1 = (f_1)$ and $\boldsymbol{f}_2 = (f_2, g_2)$ needed for maintaining the system in equilibrium. ♢

Example 8.5. Let us operate on the system of Example 8.4 by constraining the point P_2 to move on the circle $\mathbb{S}_1$ of radius 1 and centered at the origin and by controlling only the position of the point P_1 on the x-axis. This is a particular case of the so-called *Zeeman machine* (Poston and Stewart 1978) where the point P_1 is free to move in the plane (see also (Dubois and Dufour 1976).

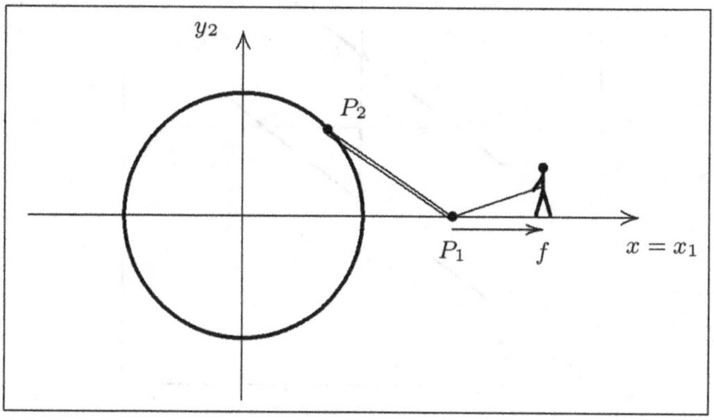

Fig. 8.5 Example 8.5: the Zeeman machine

The control configuration manifold is now $Q = \mathbb{R} = (x)$, the constraint is $\Sigma = \mathbb{S}_1 \times \mathbb{R}$, and the fibration ϕ is just the Cartesian projection onto the x-axis. Thus, the control relation R is represented by equations $x_2^2 + y_2^2 - 1 = 0$, $x - x_1 = 0$, and its canonical lift $\widehat{R}$ is generated by the family

$$G_R(x, x_1, x_2, y_2; \lambda, \mu) = \lambda(x - x_1) + \mu(x_2^2 + y_2^2 - 1).$$

Then the set of the equilibrium states $\mathscr{E} \subset T^*Q = (x, f)$ of the system under this control is described by the variational equation

$$f\,\delta x = \delta(G_R + \bar{V}) = \delta\left(\lambda(x - x_1) + \mu(x_2^2 + y_2^2 - 1) + \tfrac{k}{2}\left[(x_1 - x_2)^2 + y_2^2\right]\right)$$

which is equivalent to equations

$$\begin{cases} f = \lambda, \\ 0 = -\lambda + k(x_1 - x_2), \\ 0 = 2x_2\mu - k(x_1 - x_2), \end{cases} \qquad \begin{cases} 0 = 2y_2\mu + ky_2, \\ 0 = x - x_1, \\ 0 = x_2^2 + y_2^2 - 1. \end{cases}$$

These equations are reducible to

$$\begin{cases} f = k(x - x_2), \\ (2\mu + k)x_2 = kx, \end{cases} \qquad \begin{cases} (2\mu + k)y_2 = 0, \\ x_2^2 + y_2^2 = 1. \end{cases}$$

For $y_2 = 0$ we have $x_2 = \pm 1$ and $f = k(x \pm 1)$. For $y_2 \neq 0$ we have $2\mu + k = 0$, thus $x = 0$ and $f = -kx_2$, with $|x_2| \leq 1$. The set $\mathscr{E}$ of the equilibrium states is then represented by Fig. 8.6. ♢

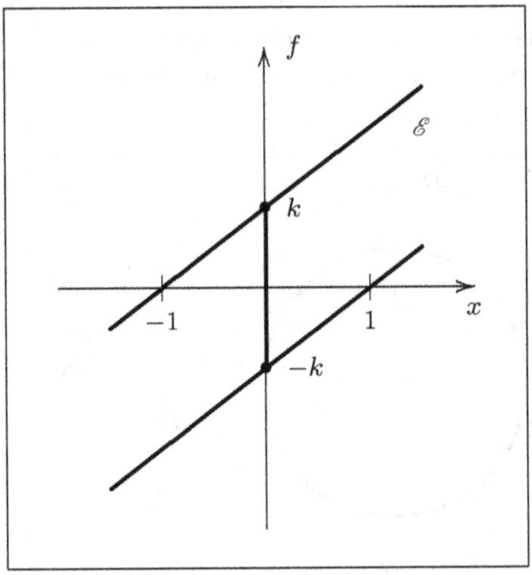

Fig. 8.6 The equilibrium states of the Zeeman machine

Example 8.6. For the same system of Example 8.5 we can think of another control relation. We can consider the configuration manifold $\bar{Q} = \mathbb{S}_1 \times \mathbb{R} = (\theta, x)$. If we control the positions of both points, then the equilibrium states are represented by the regular Lagrangian submanifold $\bar{\Lambda}$ generated by the potential energy

$$\bar{V} = \tfrac{k}{2}\,(\boldsymbol{x}_1 - \boldsymbol{x}_2)^2 = \frac{k}{2}\left[(x - \cos\theta)^2 + \sin^2\theta\right]$$

and described by equations

$$f = \frac{\partial \bar{V}}{\partial x} = k\,(x - \cos\theta), \quad \tau = \frac{\partial \bar{V}}{\partial \theta} = k\,x\,\sin\theta,$$

where θ is the angle between $\boldsymbol{x}_2$ and the x-axis, and τ is the torque applied to the point P_2. If we control only the point P_1, leaving the point P_2 free on $\mathbb{S}_1$, then the control manifold is $Q = \mathbb{R}$ (the x-axis) and the control relation is given by the trivial fibration $\phi\colon \mathbb{S}_1 \times \mathbb{R} \to \mathbb{R}$ only. The equilibrium states of the system form the set $\mathscr{E} \subset T^*Q \simeq \mathbb{R}^2 = (x, f)$ represented by equations (we put $\tau = 0$ in the equations above)

$$f = \frac{\partial \bar{V}}{\partial x_1} = k(x - \cos\theta), \quad 0 = \frac{\partial \bar{V}}{\partial \theta} = k\,x\,\sin\theta.$$

We get the same set $\mathscr{E}$ as above. Because

$$\frac{\partial^2 \bar{V}}{\partial \theta^2} = k\,x\,\cos\theta, \quad \frac{\partial^2 \bar{V}}{\partial \theta\,\partial x} = k\,\sin\theta,$$

the generating family $\bar{V}(x;\theta)$ is a Morse family except for $x = 0$, $\sin\theta = 0$, that is, over the points $(0, \pm k)$. In accordance with the theory, by excluding these two points, the set $\mathscr{E}$ is a Lagrangian submanifold. It is made of five branches (open segments and half-lines). The "vertical" segment is the set of the singular points, in accordance with the fact that the caustic is represented by equations

$$0 = \frac{\partial^2 \bar{V}}{\partial \theta^2} = k\,x\,\cos\theta, \quad 0 = \frac{\partial \bar{V}}{\partial \theta} = k\,x\,\sin\theta. \qquad \diamondsuit$$

Example 8.7. Two points P_1 and P_2 are constrained to the x-axis and the y-axis, respectively. They are linked by a rigid rod of length a. The point P_2 is tied elastically to the origin by a spring. Let b be the length at rest of the spring. We act only on the point P_1, see Fig. 8.7. An interpretation of this static system is the following. The extended configuration manifold is $\bar{Q} = \mathbb{R}^2 = (x_1, y_2)$, the constraint Σ is represented by the rod, that is, by equation $x_1^2 + y_2^2 = a^2$, the control manifold is $Q = \mathbb{R} = (x)$, and the fibration ϕ is represented by equation $x = x_1$; y_2 is considered as an internal variable. The internal potential energy is $\bar{V}(x_1, y_2) = k/2\,(b - y_2)^2$. The generating family of the control relation R is

$$G_R(x, x_1, y_2; \lambda, \mu) = \lambda(x - x_1) + \mu(x_1^2 + y_2^2 - a^2),$$

and the set $\mathscr{E} \subset T^*Q$ of the equilibrium states of the system under this control is described by equation

$$f\,\delta x = \delta(G_R + \bar{V}) = \delta\Big(\lambda(x - x_1) + \mu(x_1^2 + y_2^2 - a^2) + \tfrac{k}{2}\,(b - y_2)^2\Big)$$

which yields equations

$$\begin{cases} f = \lambda, \\ 0 = x - x_1, \\ 0 = -\lambda + 2\,\mu\,x_1, \end{cases} \qquad \begin{cases} 0 = x_1^2 + y_2^2 - a^2, \\ 0 = 2\,\mu\,y_2 - k\,(b - y_2). \end{cases}$$

These equations reduce to

$$f = 2\,\mu\,x, \;\; x^2 + y_2^2 = a^2, \;\; 2\,\mu\,y_2 = k\,(b - y_2). \tag{8.23}$$

For $y_2 \neq 0$ we find

$$\mu = \frac{k}{2}\left(\frac{b}{y_2} - 1\right)$$

thus,

$$f = k\,x\left(\frac{b}{\sqrt{a^2 - x^2}} - 1\right).$$

For $y_2 = 0$, the third equation (8.23) has a meaning only for $b = 0$ (ideal spring). For $b = 0$ Eqs. (8.23) become

$$f = 2\,\mu\,x, \;\; x^2 + y_2^2 = a^2, \;\; (2\,\mu + k)\,y_2 = 0,$$

so that, for $y_2 = 0$ (i.e., for $x = \pm a$) the extra variable μ is not determined and we find that f may assume any arbitrary value. For $y_2 = 0$, we find $\mu = -k/2$, thus $f = -kx$. The set $\mathscr{E}$ is then represented by the following picture, for all possible values of b; for $b = 0$ is not a submanifold. ◇

Example 8.8. Let us consider Example 8.2 modified as follows: (i) the point P is constrained on a curve $\bar{y} = h(\bar{x})$, and (ii) it is subjected to a force parallel to the $\bar{y}$-axis with potential energy $V(\bar{y})$. The equilibrium set $\mathscr{E}$ is then described by the variational equation

$$f\,\delta x = \delta\big[V(\bar{y}) + \mu\,(\bar{y} - h(x))\big],$$

which yields equations

$$\bar{y} = h(x), \;\; f = \,-\,\mu\,h'(x), \;\; V'(\bar{y}) + \mu = 0.$$

It follows that $\mathscr{E}$ is described by equation

$$f(x) = F\,(h(x))\;h'(x), \;\; F = V'. \tag{8.24}$$

If, for instance, $V(\bar{y}) = k/2\,\bar{y}^2$ (ideal spring), then

$$f(x) = k\,h(x)\,h'(x). \tag{8.25}$$

In this way, we can construct any kind of (smooth) force function $f(x)$ (at least in a neighborhood of a point x_0) by taking a curve $h(x)$ which is a solution of the differential equation (8.24) or (8.25). For instance, if we want a repulsive linear force

$$f(x) = -k\,x$$

in the neighborhood of $x = 0$, then (8.25) reads $x dx = -h\,dh$ and leads to solutions of the kind $h^2(x) = c^2 - x^2$. The curve that realizes such a force is then any circle centered at the origin. ◊

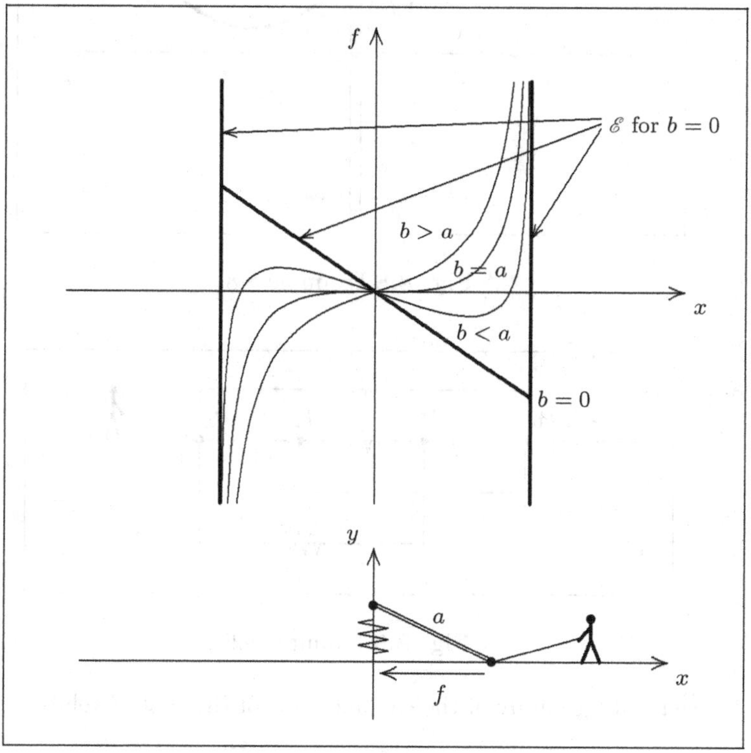

Fig. 8.7 Example 8.7

Example 8.9. The static control of n-body systems. Let us consider a static system made of four points $(P_i) = (P_0, P_1, P_2, P_3)$ on a straight line (the x-axis), with interacting forces with potentials $V_{ij}(r_{ij})$ that are even functions of the distances $r_{ij} = x_i - x_j$. We consider for simplicity the case of four points, but the following discussion can be easily extended to the generic case of n points. Assume that the point P_0 is constrained at the origin, so that $x_0 = 0$, and that we act only on the last point P_3. The total potential energy is

$$\bar{V} = V_{01}(x_1 - x_0) + V_{02}(x_2 - x_0) + V_{03}(x_3 - x_0) + V_{12}(x_2 - x_1) + V_{13}(x_3 - x_1) + V_{23}(x_3 - x_2).$$

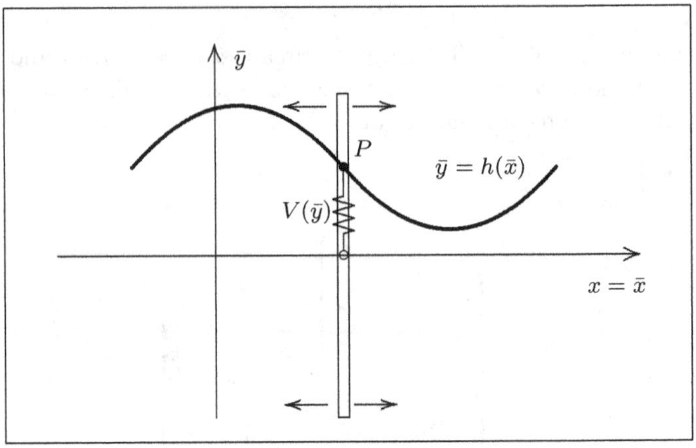

Fig. 8.8 Example 8.8

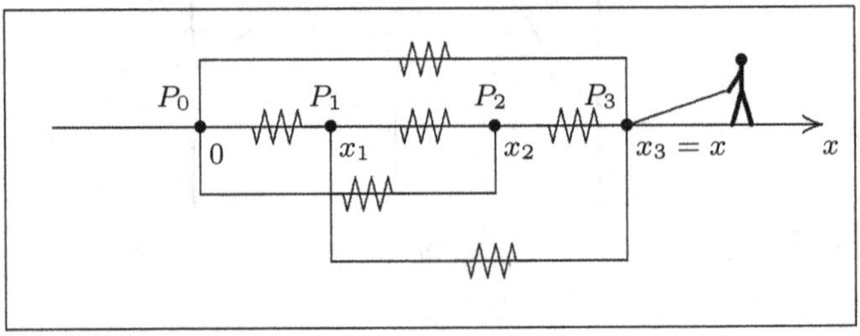

Fig. 8.9 Example 8.9

The generating family of the canonical lift of the control relation is

$$G_R = \lambda(x - x_3) + \mu x_0.$$

Equation (8.22) now reads

$$\begin{aligned} f\delta x = \delta(G_R + \bar{V}) &= \delta\lambda\,(x - x_3) + \lambda(\delta x - \delta x_3) + \mu\,\delta x_0 + x_0\,\delta\mu \\ &+ f_{01}(x_1 - x_0)\,(\delta x_1 - \delta x_0) + f_{02}(x_2 - x_0)\,(\delta x_2 - \delta x_0) \\ &+ f_{03}(x_3 - x_0)\,(\delta x_3 - \delta x_0) + f_{12}(x_2 - x_1)\,(\delta x_2 - \delta x_1) \\ &+ f_{13}(x_3 - x_1)\,(\delta x_3 - \delta x_1) + f_{23}(x_3 - x_2)\,(\delta x_3 - \delta x_2), \end{aligned}$$

where $f_{ij} = V'_{ij}$ are the odd functions representing the internal interacting forces. This is equivalent to the following equations

$$\begin{cases} f = \lambda, \\ x = x_3, \\ x_0 = 0, \end{cases}$$

$$\begin{cases} 0 = \mu - f_{01}(x_1 - x_0) - f_{02}(x_2 - x_0) - f_{03}(x_3 - x_0), \\ 0 = f_{01}(x_1 - x_0) - f_{12}(x_2 - x_1) - f_{13}(x_3 - x_1), \\ 0 = f_{02}(x_2 - x_0) + f_{12}(x_2 - x_1) - f_{23}(x_3 - x_2), \\ 0 = -\lambda + f_{03}(x_3 - x_0) + f_{13}(x_3 - x_1) + f_{23}(x_3 - x_2). \end{cases} \tag{8.26}$$

Due to the first three equations, from the last one we get the expression of the controlling force,

$$f = f_{03}(x) + f_{13}(x - x_1) + f_{23}(x - x_2), \tag{8.27}$$

which depends only on the interacting forces between the point P_3 and the points (P_0, P_1, P_2). The remaining forces are internal forces. The fourth equation (8.26) gives the expression of μ as a reaction force at the fixed point P_0,

$$\mu = f_{01}(x_1) + f_{02}(x_2) + f_{03}(x_3).$$

The remaining two equations (8.26) read

$$\begin{cases} f_{01}(x_1) = f_{12}(x_2 - x_1) + f_{13}(x - x_1), \\ f_{23}(x - x_2) = f_{02}(x_2) + f_{12}(x_2 - x_1). \end{cases} \tag{8.28}$$

For any fixed value of x, Eqs. (8.28) define a subset $D_x \subseteq \mathbb{R}^2 = (x_1, x_2)$. By replacing this subset of values of (x_1, x_2) in (8.27) we get a set $F_x \subseteq \mathbb{R}$ of forces f associated with the controlled value of x. The union $\mathscr{E} = \cup_{x \in \mathbb{R}} F_x$ of all these sets gives the equilibrium states of the system. In general, it is a very complicated subset of $\mathbb{R}^2 = (x, f)$. ◇

Remark 8.6. In the model of control of static systems we have considered, we have not introduced and discussed the notion of *stability* of an equilibrium state. Example 8.5 (the Zeeman machine) suggests the following definition. An equilibrium state of $\mathscr{E}$ is *stable* if it corresponds to stable states on the constraint manifold Σ. ◇

8.2 Simple closed thermostatic systems

Let us consider a system of particles (atoms, molecules) in a closed vessel. Let us act on it by means of an external device. The energy transfered to the system in a "quasi-static process" c, made of slow transformations of equilibrium states, is defined by

$$E_c = \int_c (\delta Q - P\,dV),$$

where P is the pressure, V is the volume, and δQ is a one-form representing the heat absorbed by the system. If we *postulate* that this one-form admits an integrating factor,

$$\delta Q = T\,dS, \tag{8.29}$$

where T is the *absolute temperature* T and S is the *entropy*, then the integral E_c can be written as the integral

$$E_c = \int_c \theta,$$

of a one-form

$$\boxed{\theta = T\,dS - P\,dV} \tag{8.30}$$

Let us call θ the *fundamental one-form of thermodynamics.*

Following (Tulczyjew 1977a), This suggests to take the four-dimensional space

$$\boxed{M = (S, V, P, T) = \mathbb{R}^4}$$

as the *space of states* (or *state manifold*). A quasi-static process c is one-dimensional path in this space.

According to their physical meaning, the observables (S, V, P, T) assume only positive values. However, for the moment, it is not necessary to consider this restriction. This simplifies our discussion.[1]

We call (S, V) *extensive observables* and (T, P) *intensive observables.*

The great advantage of considering this four-dimensional space is that the fundamental one-form θ induces a symplectic form

$$\boxed{\omega = d\theta = dT \wedge dS + dV \wedge dP} \tag{8.31}$$

on the space $\mathbb{R}^4 = (S, V, P, T)$. In this way, as a consequence of the first principle of thermodynamics and formula (8.29), the state manifold is endowed

[1] Note that the conditions $S, V, P, T > 0$ follow as a consequence of the axioms stated below.

with a canonical symplectic structure.[2] The corresponding Poisson bracket is

$$\boxed{\{F,G\} = \frac{\partial F}{\partial T}\frac{\partial G}{\partial S} + \frac{\partial F}{\partial V}\frac{\partial G}{\partial P} - \frac{\partial F}{\partial S}\frac{\partial G}{\partial T} - \frac{\partial F}{\partial P}\frac{\partial G}{\partial V}} \tag{8.32}$$

The simplest way to find this expression for the Poisson bracket is to write (8.31) as $\omega = dp_i \wedge dq^1 + dp_2 \wedge dq^2$, with $p_1 = T$, $p_2 = V$, $q^1 = S$, $q^2 = P$, and to apply formula (4.8).

Definition 8.4. The *equilibrium states* that are physically admissible form a subset

$$\mathscr{E} \subseteq M$$

called the *constitutive set.* We say that the system is *simple* if $\mathscr{E}$ is an *exact Lagrangian submanifold* ; that is, the restriction of the one-form θ to the vectors tangent to $\mathscr{E}$ (which is closed, inasmuch as $\mathscr{E}$ is Lagrangian) is an exact form:

$$\theta|\mathscr{E} = dW,$$

where $W\colon \mathscr{E} \to \mathbb{R}$ is a smooth function we call the *intrinsic potential energy* of the system. ♡

Remark 8.7. The definition of *simple thermostatic system* is equivalent to assuming that $\mathscr{E}$ is a two-dimensional submanifold and that the integral E_c is zero for all quasi-static cycles over $\mathscr{E}$. This definition is in accordance with that of (Carathéodory 1909). The notion of "intrinsic potential energy" does not appear in the texts on thermodynamics. However, as we show below, its mathematical importance is due to the fact that we can derive the four fundamental thermostatic potentials from its expression. ◇

Remark 8.8. Being a Lagrangian submanifold of a four-dimensional symplectic manifold, the constitutive set $\mathscr{E}$ is represented (we assume globally) by *two* independent *equations of state* or *constitutive equations*,

$$\boxed{E_1(S,V,P,T) = 0, \quad E_2(S,V,P,T) = 0} \tag{8.33}$$

with functions (E_1, E_2) in involution on $\mathscr{E}$:

$$\boxed{\{E_1, E_2\}|\mathscr{E} = 0} \tag{8.34}$$

◇

[2] The common geometrical setting of thermodynamics is odd-dimensional, in terms of contact manifolds; see, for instance, (Hermann 1973) and (Mrugała, 1995). However, the even-dimensional framework, in terms of symplectic manifolds and Lagrangian submanifolds, seems to be more symmetric and elegant. A remarkable example of this structural symmetry is the general setting of the Legendre transform and the definition of thermodynamic potentials illustrated in Sect. 7.5.

We can summarize the above discussion as follows.

We consider (S, V, P, T) (including the entropy S) as independent variables (i.e., as *coordinates* of $\mathbb{R}^4$). The first principle of thermodynamics endows this space with an exact symplectic structure, and the space of the smooth functions $f(S, V, P, T)$ with a Poisson structure. The equilibrium states form a subset $\mathscr{E}$ of this space. The first principle of thermodynamics implies a special structure for this subset: it is in general a Lagrangian set, but it may be a Lagrangian submanifold, as in the case of a simple thermostatic system. As a consequence, $\mathscr{E}$ is described by *two* equations of state in involution (and not only one, as in the common approach to thermodynamics).

Then,

Our expectation is to find a certain number of generating families for $\mathscr{E}$. When these generating families reduce to ordinary functions, then they coincide with ordinary *thermodynamic potentials.* We show that these generating families are of four types. The first is *internal energy.*

8.3 The internal energy

Let us consider the extensive variables (S, V) as global coordinates of a configuration manifold $Q_1 = \mathbb{R}^2$. Let (S, V, p_S, p_V) be the corresponding canonical coordinates on the cotangent bundle T^*Q_1. The Liouville form in this space is

$$\theta_{Q_1} = p_S\, dS + p_V\, dV.$$

If we compare this one-form with fundamental one-form (8.30) $\theta = T\, dS - P\, dV$, then we observe that the injective map $\alpha_1 \colon M \to T^*Q_1$ defined by

$$p_S = T, \quad p_V = -P, \tag{8.35}$$

is a symplectomorphism onto T^*Q_1, and the following diagram is commutative,

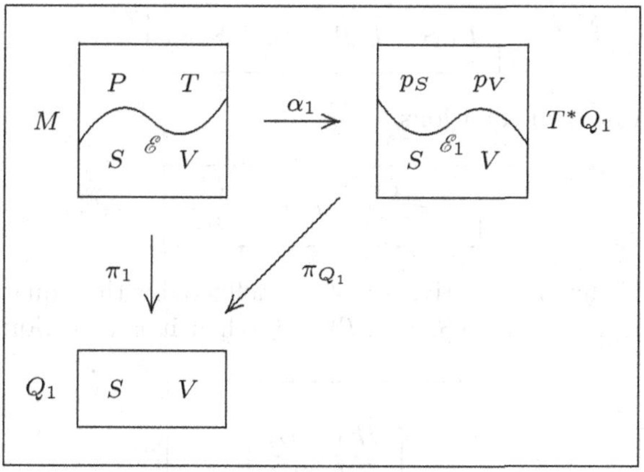

Fig. 8.10 The space of states projected onto the entropy-volume space

It follows that the image $\mathscr{E}_1 = \alpha_1(\mathscr{E})$ is a Lagrangian submanifold of the cotangent bundle T^*Q_1. Hence, we are led to look for global or local generating families $U(S, V; \underline{\lambda})$ of $\mathscr{E}_1$, which are functions of (S, V) and auxiliary parameters $\underline{\lambda}$, and to introduce the following.

Definition 8.5. A generating function $U(S, V; \underline{\lambda})$ of $\mathscr{E}_1$ is called *internal energy.* ♡

So, within a general framework, we have to consider two cases:

1. $\mathscr{E}_1 \subset T^*Q_1$ is a section of the cotangent fibration $\pi_1 \colon T^*Q_1 \to Q_1$.
2. $\mathscr{E}_1 \subset T^*Q_1$ is not a section of π_1 and/or it has singular points.

Case 1 is in fact the case fitting with the two fundamental examples, the ideal gas and the Van der Waals gas, examined below. It relies on the following general theorem (whose proof is given in Sect. 9.2):

Theorem 8.5. *Let $\alpha_1 \colon M \to T^*Q_1$ be a symplectomorphism. If a Lagrangian submanifold $\mathscr{E} \subset M$ is the image of a section of $\pi_1 = \pi_{Q_1} \circ \alpha_1 \colon M \to Q_1$, then it admits a global generating function $U \colon Q_1 \to \mathbb{R}$ if and only if it admits a global function $W \colon \mathscr{E} \to \mathbb{R}$ such that $dW = \theta_1|\mathscr{E}$, where $\theta_1 = \alpha_1^*\theta_{Q_1}$. The link between these two functions is $W = \pi_1^*U = U \circ \pi_1$.*

Due to this theorem we can affirm the following.

Theorem 8.6. *A closed simple thermostatic system admits an internal energy $U(S, V)$ if and only if the constitutive set $\mathscr{E}$ is a section of π_1.*

This means that $\mathscr{E}_1$ is completely described by equation $p_S\, dS + p_V\, dV = dU(S, V)$ so that, due to (8.35), $\mathscr{E}$ is described by equation

$$\boxed{T\,dS - P\,dV = dU(S, V)} \tag{8.36}$$

which is equivalent to equations

$$\boxed{T = \frac{\partial U}{\partial S}, \quad P = -\frac{\partial U}{\partial V}} \tag{8.37}$$

Remark 8.9. If the constitutive set $\mathscr{E}$ is defined by the equations of state $E_1(S, V, P, T) = 0$ and $E_2(S, V, P, T) = 0$, then it is a section of π_1 if and only if

$$\boxed{\det \begin{bmatrix} \dfrac{\partial E_1}{\partial T} & \dfrac{\partial E_1}{\partial P} \\ \dfrac{\partial E_2}{\partial T} & \dfrac{\partial E_2}{\partial P} \end{bmatrix} \neq 0} \tag{8.38}$$

at the points of $\mathscr{E}$. ◊

8.4 The ideal gas

Let us apply the general considerations of the preceding sections to the *ideal gas.* The basic and well-known constitutive equation, or equation of state, of an ideal gas is

$$\boxed{E_1(P, V, T) \doteq P\,V - n\,R\,T = 0} \tag{8.39}$$

where $n > 0$ is the *mole number* and $R > 0$ is a physical constant. This equation summarizes the Boyle, Gay–Lussac, and Avogadro's laws. In accordance with the assumption that the constitutive set $\mathscr{E}$ is a two-dimensional manifold, this equation is not sufficient to describe the behavior of the gas; we need a second independent constitutive equation $E_2 = 0$, in involution with $E_1 = 0$.

The second constitutive equation must involve the entropy. Let us assume that it has the form

$$\boxed{S = f(V, P, T)} \tag{8.40}$$

that is, the entropy can be expressed as a function of the remaining observables. Then $\mathscr{E}$ is Lagrangian if and only if the function

$$F(S, V, P, T) \doteq S - f(V, P, T) = 0$$

is in involution with E: $\{F, E_1\} = 0$.

As shown in the proof of the next theorem, the involution condition $\{F, E_1\} = 0$ is translated into a Hamilton–Jacobi equation which is integrable by separation of variables on the domain where $V \neq 0$ and $P \neq 0$, so that the expression of the entropy $S = f(V, P, T)$ is determined by a pure analytical method, and the physical meaningful conditions $V, P, T > 0$ appear as an analytical consequence of the first equation of state $E_1 = 0$.

Theorem 8.7. *For the ideal gas the second equation of state of the kind* (8.40) *is*

$$\boxed{S = A + (R + C)\,\log V + C\,\log P} \tag{8.41}$$

where A and C are functions of T. This equation is equivalent to

$$\boxed{E_2 \doteq P\,V^{\gamma} - K\,\exp\frac{S}{C} = 0} \tag{8.42}$$

with

$$\boxed{\gamma \doteq 1 + \frac{nR}{C}, \quad K \doteq \exp\left(-\frac{A}{C}\right)} \tag{8.43}$$

Proof. Let us compute the Poisson bracket (8.32) for $G = E_1$ and $F = S - f(P, V, T)$:

$$\begin{aligned}\{F, E_1\} &= \frac{\partial F}{\partial T}\,\frac{\partial E_1}{\partial S} + \frac{\partial F}{\partial V}\,\frac{\partial E_1}{\partial P} - \frac{\partial F}{\partial S}\,\frac{\partial E_1}{\partial T} - \frac{\partial F}{\partial P}\,\frac{\partial E_1}{\partial V} \\ &= -\frac{\partial f}{\partial V}\,\frac{\partial E_1}{\partial P} - \frac{\partial E_1}{\partial T} + \frac{\partial f}{\partial P}\,\frac{\partial E_1}{\partial V},\end{aligned}$$

bevause E_1 does not depend on S. Hence $\{F, E_1\} = 0$ is equivalent to

$$\boxed{V\,\frac{\partial f}{\partial V} - P\,\frac{\partial f}{\partial P} = nR} \tag{8.44}$$

This is a Hamilton–Jacobi equation in the cotangent bundle of the configuration manifold of variables (P, V). The variable T is not involved, therefore we can try to solve (8.44) by separation of variables, by considering a solution of the kind

$$f(V, P) = g(V) + h(P)$$

up to an additive constant. We get the equation

$$V\,\frac{dg}{dV} - P\,\frac{dh}{dP} = n\,R$$

which splits into two equations,

$$V\,\frac{dg}{dV} = C + nR, \quad P\,\frac{dh}{dP} = C, \quad C = \text{constant}.$$

It follows that a solution is

$$f = A + (nR + C)\,\log|V| + C\,\log|P|,$$

where (A, C) are constant parameters. This is a solution defined only for $V, P \neq 0$, as we said above. So, it is the time to take into account, from now on, the physical meaningful restrictions

$$\boxed{V > 0,\ \ P > 0,\ \ T > 0} \tag{8.45}$$

where $T > 0$ is a consequence of the first two because of (8.39), and write this last equation as

$$f = A + (nR + C)\,\log V + C\,\log P, \tag{8.46}$$

This is a complete solution with respect to the nonadditive constant C, because the matrix

$$\left[\frac{\partial^2 f}{\partial V \partial C} \;\middle|\; \frac{\partial^2 f}{\partial P \partial C}\right] = \left[\frac{1}{V} \;\middle|\; \frac{1}{P}\right]$$

has maximal rank everywhere. However, in (8.44) the temperature T is not involved, thus we can consider A and C as functions of T. So, with the "natural" choice of (8.46) among all the solutions of (8.44), we find the second constitutive equation (8.41) of an ideal gas. We can write this equation in the form

$$\frac{S}{C} = \frac{A}{C} + \left(\frac{nR}{C} + 1\right) \log V + \log P, \tag{8.47}$$

and by introducing the quantities (8.43) we get (8.42). □

Equation (8.42) is the standard form of the second constitutive equation of an ideal gas, involving the entropy, as it appears in books of thermodynamics with

$$\boxed{C = n\,c_V,\ \ k = K\,n^{\gamma}} \tag{8.48}$$

It remains to prove that the constitutive set $\mathscr{E}$ of the ideal gas is a section of π_1 so that, according to Theorem 8.6, it admits the internal energy as an ordinary generating function.

Theorem 8.8. *If $\mathscr{E}$ is described by the two constitutive equations* (8.39) *and* (8.42), *with $V > 0$ and $P > 0$,*[3] *then (up to an additive constant) the internal energy is given by*

$$\boxed{U(S, V) = \frac{K}{\gamma - 1}\,V^{1-\gamma}\,\exp\frac{S}{C}} \tag{8.49}$$

[3] In the following this formal assumption is understood.

Proof. This follows from the integration of the closed one-form (8.36) $\theta_1 = T\,dS - P\,dV$. By solving Eqs. (8.39) and (8.42) with respect to P and T, we get these observables as functions of (V, S):

$$P = K\,V^{-\gamma} \exp\frac{S}{C}, \quad T = \frac{K}{nR}\,V^{1-\gamma}\,\exp\frac{S}{C}.$$

Hence,

$$\theta_1 = K\,\exp\frac{S}{C}\,V^{-\gamma}\left(\frac{1}{n\,R}\,V\,dS - dV\right)$$

Let us integrate this one-form on the plane (S, V) following the path $(0, \varepsilon) \to (0, V) \to (S, V)$, with a small $\varepsilon > 0$, see Fig. 8.11.

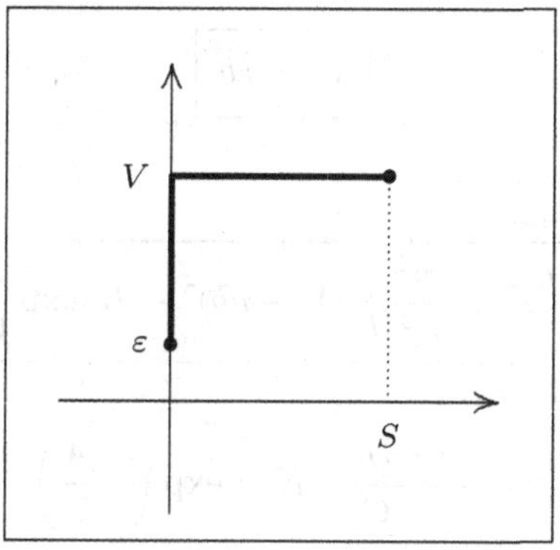

Fig. 8.11 Path of integration in the proof of Theorem 8.8

We obtain:

$$\begin{aligned}
\int_{(0,\varepsilon)}^{(S,V)} \theta_1 &= -K \int_{\varepsilon}^{V} V^{-\gamma}\,dV + K \int_0^S \exp\frac{S}{C}\,V^{-\gamma}\left(\frac{1}{n\,R}\,V\,dS\right) \\
&= \frac{K}{\gamma - 1}\left(V^{1-\gamma} - \varepsilon^{1-\gamma}\right) + \frac{K}{n\,R}\,V^{1-\gamma} \int_0^S \exp\frac{S}{C}\,dS \\
&= \frac{K}{\gamma - 1}\,V^{1-\gamma} + \frac{K}{n\,R}\,V^{1-\gamma}\,C\left(\exp\frac{S}{C} - 1\right) + \text{constant}
\end{aligned}$$

By recalling (8.43), $\gamma - 1 = nR/C$, Eq. (8.49) is found. □

8.5 The Van der Waals gas

A similar analysis can be done for the *Van der Waals gas*. The basic equation of state is assumed to be

$$\boxed{E_1 \doteq (V - n\,b)\left(P + a\,\frac{n^2}{V^2}\right) - n\,R\,T = 0} \tag{8.50}$$

with a and b non-negative constants.

Theorem 8.9. *For a Van der Waals gas the second equation of state of the kind* (8.40) *is*

$$\boxed{S = A + (nR + C)\,\log(V - nb) + C\,\log\left(P + \frac{an^2}{V^2}\right)} \tag{8.51}$$

where A and C are functions of T, and

$$\boxed{V > n\,b} \tag{8.52}$$

This equation is equivalent to

$$\boxed{E_2 \doteq \left(P + \frac{an^2}{V^2}\right)(V - nb)^{\gamma} - K\,\exp\frac{S}{C} = 0} \tag{8.53}$$

with

$$\gamma \doteq 1 + \frac{nR}{C}, \quad K \doteq \exp\left(-\frac{A}{C}\right).$$

Proof. As in the proof concerning the ideal gas we consider the Poisson bracket

$$\{F, E_1\} = -\frac{\partial f}{\partial V}\frac{\partial E_1}{\partial P} - \frac{\partial E_1}{\partial T} + \frac{\partial f}{\partial P}\frac{\partial E_1}{\partial V},$$

for $F = S - f(P, V, T)$. Since in the present case we have

$$\begin{aligned}
\frac{\partial E_1}{\partial T} &= -\,n\,R,\\
\frac{\partial E_1}{\partial P} &= V - nb,\\
\frac{\partial E_1}{\partial V} &= P + a\,\frac{n^2}{V^2} - 2a\,(V - n\,b)\,\frac{n^2}{V^3} = P - a\,\frac{n^2}{V^2} + 2ab\,\frac{n^3}{V^3},
\end{aligned}$$

we find

$$\{F, E_1\} = -(V - nb)\,\frac{\partial f}{\partial V} + n\,R + \left(P - a\,\frac{n^2}{V^2} + 2ab\,\frac{n^3}{V^3}\right)\frac{\partial f}{\partial P},$$

and we get the Hamilton–Jacobi equation

$$\boxed{(V - nb)\,\frac{\partial f}{\partial V} - n\,R - \left(P - \frac{an^2}{V^2} + \frac{2abn^3}{V^3}\right)\frac{\partial f}{\partial P} = 0} \tag{8.54}$$

Note that the temperature T is not involved. In order to integrate this equation we take inspiration from the special case of the ideal case. Let us compare the two basic equations of state,

$$E_1 = PV - nRT = 0, \qquad \text{ideal gas},$$

$$E_1 \doteq (V - n\,b)\left(P + a\,\frac{n^2}{V^2}\right) - n\,RT = 0, \quad \text{VdW},$$

and the two corresponding Hamilton–Jacobi equations,

$$V\,\frac{\partial f}{\partial V} - P\,\frac{\partial f}{\partial P} - n\,R = 0, \qquad \text{ideal gas},$$

$$(V - nb)\,\frac{\partial f}{\partial V} - \left(P - \frac{an^2}{V^2} + \frac{2abn^3}{V^3}\right)\frac{\partial f}{\partial P} - n\,R = 0, \quad \text{VdW}.$$

The first equation was solved by considering the kind of solution (8.46)

$$f = A + B\,\log V + C\,\log P,$$

then, by analogy, for the second equation we can try to find a solution of the form

$$f = A + B\,\log(V - nb) + C\,\log\left(P + \frac{an^2}{V^2}\right), \tag{8.55}$$

where A, B, and C are constant, or functions of T. Since

$$\frac{\partial f}{\partial V} = \frac{B}{V - nb} - \frac{2an^2C}{\left(P + \dfrac{an^2}{V^2}\right)V^3} = \frac{B}{V - nb} - \frac{2an^2C}{V\,(PV^2 + an^2)},$$

$$\frac{\partial f}{\partial P} = \frac{C}{P + \dfrac{an^2}{V^2}} = \frac{CV^2}{PV^2 + an^2},$$

the Hamilton–Jacobi equation reads

$$(V - nb)\left(\frac{B}{V - nb} - \frac{2an^2C}{V\,(PV^2 + an^2)}\right) - n\,R$$
$$- \left(P - \frac{an^2}{V^2} + \frac{2abn^3}{V^3}\right)\frac{CV^2}{PV^2 + an^2} = 0.$$

A straightforward calculation shows that it can be reduced to the simple form

$$(B - n\,R - C)(P\,V^2 + a\,n^2) = 0.$$

This equation is satisfied by setting $B = n\,R + C$. Going back to (8.55) we conclude that

$$f = A + (nR + C)\,\log(V - nb) + C\,\log\left(P + \frac{an^2}{V^2}\right)$$

is a complete solution of our Hamilton–Jacobi equation (8.54). This proves (8.51). Moreover, since

$$\frac{S}{C} - \frac{A}{C} = \left(1 + \frac{nR}{C}\right)\log(V - nb) + \log\left(P + \frac{an^2}{V^2}\right),$$

we find

$$\exp\left(\frac{S}{C} - \frac{A}{C}\right) = \left(P + \frac{an^2}{V^2}\right)\,(V - nb)^{\gamma},$$

with $\gamma \doteq 1 + nR/C$. This equation can be put in the form (8.53). □

Remark 8.10. In order to verify the correctness of Theorem 8.9 from a physical viewpoint, we can refer to the expression of the entropy, as a function of T and V, written in [Fermi, 1936], Chap. 4, Formula [16.7], for $n = 1$:

$$S = c_V\,\log T + R\,\log(V - b) + \text{constant}.$$

From the first state equation (8.50), $E_1 = 0$, we get

$$\log T = \log(V - b) + \log\left(P + \frac{a}{V^2}\right) - \log R,$$

so that

$$\begin{aligned} S &= c_V\,\log(V - b) + c_V\,\log\left(P + \frac{a}{V^2}\right) + R\,\log(V - b) + \text{constant} \\ &= (c_V + R)\,\log(V - b) + c_V\,\log\left(P + \frac{a}{V^2}\right) + \text{constant}. \end{aligned}$$

With S and V extensive observables, for a number n of moles we find

$$S = n\,(c_V + R)\,\log(V - nb) + n\,c_V\,\log\left(P + \frac{an^2}{V^2}\right) + \text{constant}.$$

This result is in perfect accordance with Eq. (8.51),

$$S = A + (nR + C)\,\log(V - nb) + C\,\log\left(P + \frac{an^2}{V^2}\right)$$

if we put $C = n\,c_V$, according to (8.48), and $A = \text{constant}$. ◇

It remains to prove that also for the Van der Waals gas the constitutive set $\mathscr{E}$ is a section of π_1 so that it admits the internal energy as a global ordinary generating function.

Theorem 8.10. *If $\mathscr{E}$ is described by the two constitutive equations* (8.50) *and* (8.53), *then (up to an additive constant) the internal energy is given by*

$$\boxed{U(S,V) = \frac{K}{\gamma - 1}(V - nb)^{1-\gamma} \exp\frac{S}{C} - \frac{a\,n^2}{V}} \tag{8.56}$$

Proof. The proof is analogous to that of Theorem 8.8. Equations (8.50) and (8.53) can be rewritten as

$$\left(P + \frac{an^2}{V^2}\right) = \frac{n\,R\,T}{(V - n\,b)},$$
$$\left(P + \frac{an^2}{V^2}\right) = K\,(V - nb)^{-\gamma} \exp\frac{S}{C},$$

respectively. The two right-hand sides are equal,

$$\frac{n\,R\,T}{(V - n\,b)} = K\,(V - nb)^{-\gamma} \exp\frac{S}{C}.$$

Hence, from the first equation of state above, we derive

$$P = K\,(V - nb)^{-\gamma} \exp\frac{S}{C} - \frac{an^2}{V^2}.$$

As a consequence, the one-form $\theta_1 = T\,dS - P\,dV$ becomes

$$\theta_1 = \frac{K}{nR}\,(V - nb)^{1-\gamma} \exp\frac{S}{C}\,dS - \left(K\,(V - nb)^{-\gamma} \exp\frac{S}{C} - \frac{an^2}{V^2}\right) dV.$$

We integrate this one-form following the path $(0, \varepsilon) \to (0, V) \to (S, V)$, with a small $\varepsilon > 0$:, as in the proof of Theorem 8.8:

$$\begin{aligned}
\int_{(0,\varepsilon)}^{(S,V)} \theta_1 &= - \textstyle\int_\varepsilon^V \left(K\,(V - nb)^{-\gamma} - \dfrac{an^2}{V^2}\right) dV \\
&\quad + \frac{K}{nR}\,(V - nb)^{1-\gamma} \int_0^S \exp\frac{S}{C}\,dS \\
&= \frac{K}{\gamma - 1}\,(V - nb)^{1-\gamma} - \frac{an^2}{V} + \text{constant} \\
&\quad + \frac{KC}{nR}\,(V - nb)^{1-\gamma} \left(\exp\frac{S}{C} - 1\right).
\end{aligned}$$

Since $\gamma - 1 = nR/C$, we get Eq. (8.56). □

8.6 Control modes

Definition 8.6. Let (M, ω) be a symplectic manifold. A *control mode* on M is a surjective submersion $\pi_c \colon M \to Q_c$ onto a manifold Q_c, called *control manifold*, associated with a symplectomorphism $\alpha_c \colon M \to T^*Q_c$ such that the diagram

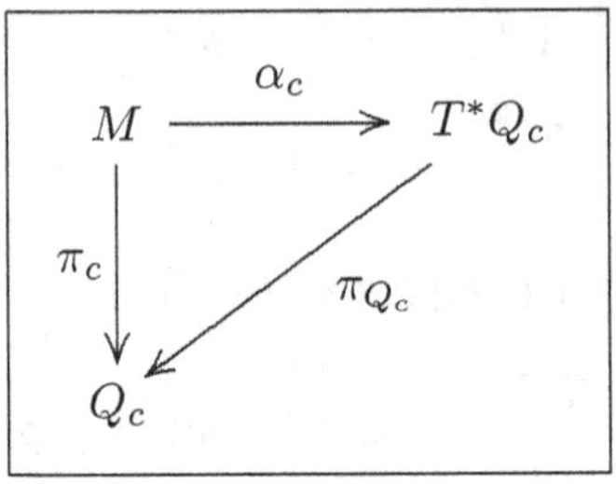

is commutative; that is, $\pi_c = \pi_{Q_c} \circ \alpha_c$. The one-form on M

$$\theta_c = \alpha_c^* \, \theta_{Q_c} \tag{8.57}$$

is called the *control form.* ♡

Remark 8.11. It follows from this definition that

$$d\theta_c = \omega.$$

This means that (M, ω) is an *exact symplectic manifold*: the symplectic form ω is exact. ◇

Definition 8.7. Let $\mathscr{E} \subset M$ be a Lagrangian submanifold. Since α_c is a symplectomorphism, the set

$$\mathscr{E}_c = \alpha_c(\mathscr{E})$$

is a Lagrangian submanifold of T^*Q_c. A generating function G_c of $\mathscr{E}_c$ is called a *generating function of $\mathscr{E}$ with respect to the control mode α_c.* ♡

Remark 8.12. In the applications to the control theory of static systems, M represents the space of the states and $\mathscr{E}$ the equilibrium states. The generating function G_c is the *potential energy* with respect to the control mode α_c. Functions on M are called *observables.* ◇

Remark 8.13. The fibers of π_c are Lagrangian submanifolds, so observables that are constant on the fibers are in involution. Conversely, if F^i are n independent global observables in involution, then they define a control mode. Indeed, equations $F^i = q^i$ (constant) define a Lagrangian foliation and the set of all the admissible constant values $(q^i) \in \mathbb{R}^n$ forms a control manifold $Q_c \subseteq \mathbb{R}^n$. It can be seen that if we choose a section of the corresponding projection π_c, then we can define a symplectomorphism into T^*Q_c. ◇

Remark 8.14. From Remark 8.13 we derive the following *control rule*:

> We cannot "control" simultaneously and independently n observables of a static system if they are not in involution.

Here, "to control" means "to force the observables to assume any value we like", at least in a suitable domain. For example, for a point P in the plane, as in Example 8.1, we cannot control simultaneously the position x and the force f along the x-axis: these two observables are not in involution.[4] ◇

8.7 The Legendre transform

The case of the simultaneous existence of two control modes on M, α_1 and α_2 is interesting. Then the transition from the generating functions G_1 to G_2 of Lagrangian submanifolds of M is called the *Legendre transform* (Tulczyjew 1977). A symplectic diffeomorphism from a symplectic manifold to a cotangent bundle has been called a "special symplectic structure" (Ławruk et al. 1975). If X is a vector space, then the direct sum $X \oplus X^*$ is endowed with a canonical symplectic form. A symplectic isomorphism from a symplectic vector space (A, α) to a direct sum $X \oplus X^*$ has been called a "frame"; see, for instance, (Leray 1981). A special important case of the Legendre transform is that connecting the Hamiltonian description and the Lagrangian description of dynamics. Other important special cases are related to the control of thermostatic systems. For another general approach to the catastrophe theory in thermodynamics see (Dubois and Dufour 1978).

The general setting of the *Legendre transform* is the following.

1 – Assume that a symplectic manifold (M, ω) is symplectomorphic to two distinct cotangent bundles,

$$T^*Q_2 \xleftarrow{\alpha_2} M \xrightarrow{\alpha_1} T^*Q_1.$$

Then,

$$\alpha_2^* d\theta_{Q_2} = \omega = \alpha_1^* d\theta_{Q_1},$$

or

$$d(\alpha_2^* \theta_{Q_2}) = \omega = d(\alpha_1^* \theta_{Q_1}).$$

If we introduce on M the one-forms

$$\theta_2 = \alpha_2^* \theta_{Q_2}, \quad \theta_1 = \alpha_1^* \theta_{Q_1},$$

[4] This seems to be an argument of quantum physics.

then we can write

$$d\theta_2 = \omega = d\theta_1,$$

and consequently

$$d(\theta_2 - \theta_1) = 0.$$

2 – The graph of the symplectomorphism $\alpha_2 \circ \alpha_1^{-1} \colon T^*Q_1 \to T^*Q_2$ is a symplectic relation

$$R_{21} \subset T^*Q_2 \times T^*Q_1.$$

Assume that it admits a global generating family L_{21} over the product $Q_2 \times Q_1$.

3 – Let $\mathscr{E}$ be a Lagrangian submanifold of (M, ω). Then $\mathscr{E}$ originates two Lagrangian submanifolds,

$$\mathscr{E}_2 = \alpha_2^{-1}(\mathscr{E}), \quad \mathscr{E}_2 = \alpha_2^{-1}(\mathscr{E})$$

of T^*Q_2 and T^*Q_1, respectively, such that

$$\mathscr{E}_2 = R_{21} \circ \mathscr{E}_1,$$

4 – Assume that $\mathscr{E}_1$ admits a generating family G_1 over Q_1. Then we obtain a generating family G_2 of $\mathscr{E}_2$ by means of the composition rule of generating families:

$$G_2 = L_{21} \oplus G_1.$$

For a better understanding of the Legendre transform let us look at the following commutative diagram.

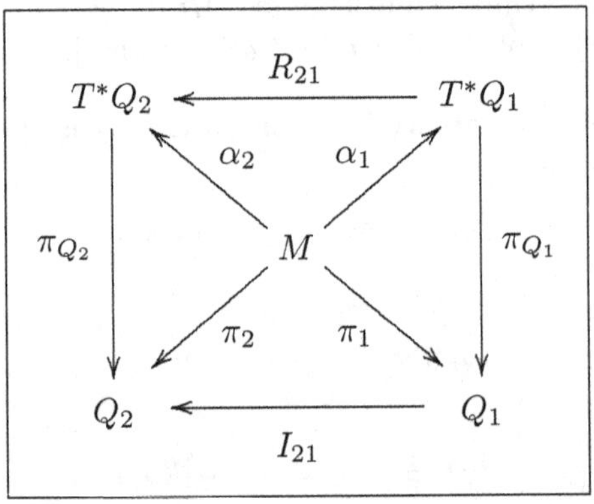

In this diagram we have introduced the surjective submersions

$$\pi_i = \pi_{Q_i} \circ \alpha_i, \quad i = 1, 2,$$

and the set

$$I_{21} = (\pi_{Q_2} \times \pi_{Q_1})(R_{21}) \subseteq Q_2 \times Q_1$$

as a relation from Q_1 to Q_2. It is convenient to introduce the maps

$$\alpha_{21} \colon M \to T^*Q_2 \times T^*Q_1, \quad \pi_{21} \colon M \to Q_2 \times Q_1,$$

defined by

$$\alpha_{21}(x) = (\alpha_2(x), \alpha_1(x)), \quad \pi_{21}(x) = (\pi_2(x), \pi_1(x)), \quad x \in M.$$

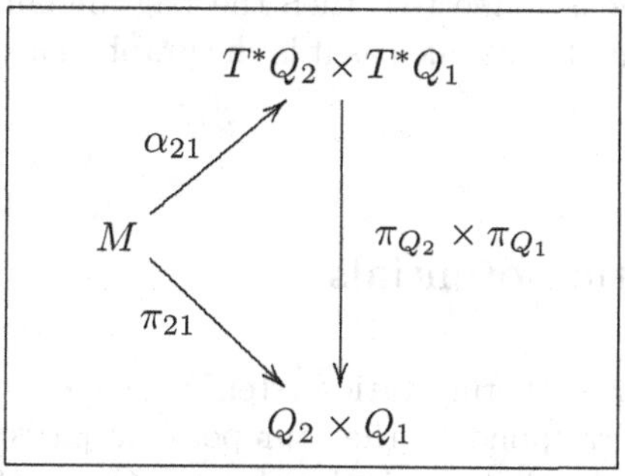

Then we have

$$R_{21} = \alpha_{21}(M), \quad I_{21} = \pi_{21}(M).$$

We observe that α_{21} is a one-to-one map: from

$$(\alpha_2(x), \alpha_1(x)) = (\alpha_2(x'), \alpha_1(x'))$$

it follows that $\alpha_i(x) = \alpha_i(x')$, $i = 1, 2$. Thus $x = x'$, because the α_i are one-to-one.

Remark 8.15. Since $d(\theta_2 - \theta_1) = \omega - \omega = 0$, the one-form $\theta_2 - \theta_1$ on M is closed. Thus, there exist local functions $W_{21} \colon M \to \mathbb{R}$ such that

$$\boxed{\theta_2 - \theta_1 = dW_{21}} \tag{8.58}$$

In the most interesting examples of the Legendre transform the set I_{21} is a submanifold of $Q_2 \times Q_1$ and the function W_{21} is globally defined on M. ◇

This is the case illustrated by the following.

Theorem 8.11. *Assume that:*

1. *M is connected and $I_{21} = \pi_{21}(M)$ is a submanifold.*
2. *The map $\pi = \pi_{21}|M \colon M \to I_{21}$ is a surjective submersion.*
3. *There exists a global function $W_{21} \colon M \to \mathbb{R}$ satisfying* (8.58).
4. *There exists a function $E_{21} \colon I_{21} \to \mathbb{R}$ such that $W_{21} = E_{21} \circ \pi = \pi^* E_{21}$.*

Then,

$$\boxed{R_{21} = (\widehat{I_{21}, E_{21}})} \tag{8.59}$$

that is, the symplectic relation R_{21} is generated by the function E_{21} on the relation I_{21}.

This theorem recalls, although in a different form, the results of (Tulczyjew 1977).[5] It is a corollary of a general theorem (Theorem 9.7) about the exact Lagrangian submanifolds over constraints. In order to apply this theorem we observe that if M is connected, then the Lagrangian submanifold R_{21} is connected and maximal, inasmuch as it is the graph of a symplectomorphism, $\alpha_2 \circ \alpha_1^{-1}$.

8.8 Thermostatic potentials

For a simple and closed thermostatic system we have four fundamental control modes. They correspond to the four possible pairs of the fundamental observables (S, V, P, T) that are in involution: $Q_1 = (S, V)$, $Q_2 = (V, T)$, $Q_3 = (T, P)$, and $Q_4 = (P, S)$. Let us call *thermostatic potentials* the corresponding generating families of the Lagrangian set $\mathscr{E}$ of the equilibrium states.[6] The corresponding control one-forms θ_i, for which $d\theta_i = \omega$, are

$$\boxed{\begin{array}{ll} \theta_1 = T\,dS - P\,dV, & \theta_3 = V\,dP - S\,dT, \\ \theta_2 = \;-P\,dV - S\,dT, & \theta_4 = T\,dS + V\,dP \end{array}} \tag{8.60}$$

According to these definitions, the thermostatic potentials are generating families. They can be, in particular, Morse families or ordinary generating functions. This depends on the state equations of $\mathscr{E}$.

---

[5] For further comments on the Legendre transform see (Tulczyjew and Urbanski 1999).

[6] It is customary to call them *thermodynamic potentials*.

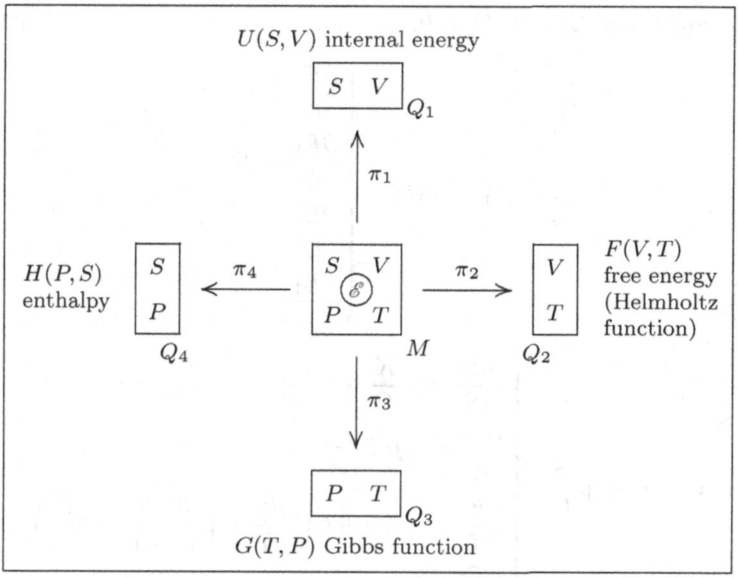

Fig. 8.12 The four control modes of thermostatics and the corresponding potentials

The internal energy is the first (and fundamental) thermostatic potential that we examined in Sect. 8.3. We have seen how to compute it for the ideal gas. That method of computation has been extended to the Van der Waals gas in Sect. 8.5. We found that for these gases, the internal energy exists as a global ordinary generating function. This existence is in fact equivalent to one of the following three items.

- The constitutive set $\mathscr{E}$ of the Van der Waals gas (in particular, of the ideal gas) is a section of π_1 Theorem 8.6.
- $\mathscr{E}$ has no singular points with respect to the projection π_1.
- Condition (8.38) holds at the points of $\mathscr{E}$, Remark 8.9.

What we have done for the internal energy can be repeated for the other three potentials, by applying *mutatis mutandis* Theorem 8.6 and Remark 8.6. This work can be synthesized as follows.

Theorem 8.12. *The set $\mathscr{E}$ of the equilibrium states of a closed simple thermostatic system, defined by two equations of state $E_1(S, V, P, T) = 0$ and $E_2(S, V, P, T) = 0$, admits a thermostatic potential TP_i as a global generating function, if and only if $\mathscr{E}$ has no singular points with respect to the projection π_i, that is, if and only if $\det \boldsymbol{M}_i \neq 0$ at all points of $\mathscr{E}$, according to Table 8.1.*

Let us apply this theorem to the ideal gas.

- State equations: $E_1 = 0$, $E_2 = 0$.

$$E_1 \doteq PV - nRT, \quad \begin{cases} \dfrac{\partial E_1}{\partial S} = 0, \\ \dfrac{\partial E_1}{\partial V} = P, \\ \dfrac{\partial E_1}{\partial T} = -nR, \\ \dfrac{\partial E_1}{\partial P} = V. \end{cases}$$

$$E_2 \doteq P\,V^{\gamma} - K\,\exp\frac{S}{C}, \quad \begin{cases} \dfrac{\partial E_2}{\partial S} = -\dfrac{K}{C}\,\exp\dfrac{S}{C}, \\ \dfrac{\partial E_2}{\partial V} = \gamma P\,V^{\gamma-1} \\ \dfrac{\partial E_2}{\partial T} = P\,V^{\gamma}\log V\,\dfrac{d\gamma}{dT} + \left(\dfrac{K}{C^2}\dfrac{dC}{dT} - \dfrac{dK}{dT}\right)\exp\dfrac{S}{C}, \\ \dfrac{\partial E_2}{\partial P} = V^{\gamma}. \end{cases}$$

- Free energy.

$$\boldsymbol{M}_2 = \begin{bmatrix} \dfrac{\partial E_1}{\partial S} & \dfrac{\partial E_1}{\partial P} \\ \dfrac{\partial E_2}{\partial S} & \dfrac{\partial E_2}{\partial P} \end{bmatrix} = \begin{bmatrix} 0 & V \\ -\dfrac{K}{C}\,\exp\dfrac{S}{C} & V^{\gamma} \end{bmatrix},$$

$$\det \boldsymbol{M}_2 = \frac{K}{C}\,V\,\exp\frac{S}{C} > 0.$$

- Gibbs function.

$$\boldsymbol{M}_3 = \begin{bmatrix} \dfrac{\partial E_1}{\partial S} & \dfrac{\partial E_1}{\partial V} \\ \dfrac{\partial E_2}{\partial S} & \dfrac{\partial E_2}{\partial V} \end{bmatrix} = \begin{bmatrix} 0 & P \\ -\dfrac{K}{C}\,\exp\dfrac{S}{C} & \gamma\,PV^{\gamma-1} \end{bmatrix},$$

$$\det \boldsymbol{M}_3 = \frac{K}{C}\,P\,\exp\frac{S}{C} > 0.$$

Table 8.1 Existence of thermostatic potentials

TP_i	π_i	$\det \boldsymbol{M}_i \neq 0$
$U(S,V)$ internal energy	$\pi_1 \colon (S,V,P,T) \to (S,V)$	$\boldsymbol{M}_1 = \begin{bmatrix} \dfrac{\partial E_1}{\partial T} & \dfrac{\partial E_1}{\partial P} \\ \dfrac{\partial E_2}{\partial T} & \dfrac{\partial E_2}{\partial P} \end{bmatrix}$
$F(V,T)$ free energy	$\pi_2 \colon (S,V,P,T) \to (V,T)$	$\boldsymbol{M}_2 = \begin{bmatrix} \dfrac{\partial E_1}{\partial S} & \dfrac{\partial E_1}{\partial P} \\ \dfrac{\partial E_2}{\partial S} & \dfrac{\partial E_2}{\partial P} \end{bmatrix}$
$G(T,P)$ Gibbs function	$\pi_3 \colon (S,V,P,T) \to (T,P)$	$\boldsymbol{M}_3 = \begin{bmatrix} \dfrac{\partial E_1}{\partial S} & \dfrac{\partial E_1}{\partial V} \\ \dfrac{\partial E_2}{\partial S} & \dfrac{\partial E_2}{\partial V} \end{bmatrix}$
$H(P,S)$ enthalpy	$\pi_4 \colon (S,V,P,T) \to (P,S)$	$\boldsymbol{M}_4 = \begin{bmatrix} \dfrac{\partial E_1}{\partial T} & \dfrac{\partial E_1}{\partial V} \\ \dfrac{\partial E_2}{\partial T} & \dfrac{\partial E_2}{\partial V} \end{bmatrix}$

- Enthalpy. It is convenient to replace the second state equation $E_2 = 0$ with

$$E_2^* \doteq A + (R + C)\,\log V + C\,\log P - S = 0;$$

see (8.41). According to Remark 8.10, $A =$ constant, thus we have

$$\frac{\partial E_2^*}{\partial T} = \frac{dC}{dT}\,\log(PV),$$

$$\boldsymbol{M}_4^* = \begin{bmatrix} \dfrac{\partial E_1}{\partial T} & \dfrac{\partial E_1}{\partial V} \\ \dfrac{\partial E_2^*}{\partial T} & \dfrac{\partial E_2^*}{\partial V} \end{bmatrix} = \begin{bmatrix} -nR & P \\ \dfrac{dC}{dT}\,\log(PV) & \dfrac{R+C}{V} \end{bmatrix},$$

$$\det \boldsymbol{M}_4^* = \frac{P}{T}\,\det \begin{bmatrix} -V & T \\ \dfrac{dC}{dT}\,\log(PV) & \dfrac{R+C}{V} \end{bmatrix} = -\frac{P}{T}\left(T\,\frac{dC}{dT}\,\log(PV) + R + C\right).$$

On $\mathscr{E}$ we have $PV = nRT$, thus the condition $\det \boldsymbol{M}_4^* = 0$ is equivalent to the differential equation in $C(T)$,

$$T\,\frac{dC}{dT}\,\log(nRT) + R + C = 0,$$

integrable by separation of variables. Its integral is

$$C(T) = n\,c_V(T) = \frac{\text{constant}}{|\log(nRT)|} - R.$$

However, this result is not in accordance with any experimental data of $c_V(T)$. Hence, $\det \boldsymbol{M}_4^* \neq 0$. The inconsistency of this solution is also due to the condition of existence $nRT \neq 1$, which depends on the mole number n.

All the above determinants are $\neq 0$, therefore we conclude that the Lagrangian submanifold $\mathscr{E}$ is not singular with respect to all projections π_i, hence, the following.

Theorem 8.13. *The ideal gas admits all the thermostatic potentials as ordinary generating functions.*

8.9 From internal energy to free energy

Let us look for the construction of a thermostatic potential by means of a Legendre transform, starting from the internal energy. For the sake of brevity, we consider only one case, namely the Legendre transform from α_1 to α_2, and we apply the results to the Van der Waals gas (the free energy for the ideal gas can be obtained by setting $a = b = 0$). Since

$$\theta_2 - \theta_1 = \,-\,S\,dT - T\,dS = \,-\,d(ST),$$

we have, according to the notation of Theorem 8.11,

$$\begin{cases} W_{21}(S,V,P,T) = \,-\,ST, \\ Q_1 = (S,V_1), \quad Q_2 = (V_2,T), \\ \pi_{21}(S,V,P,T) = ((V,T),(S,V))\,, \\ I_{21} = (\pi_2 \times \pi_1)(\Delta_M) = \{((V_2,T),(S,V_1)) \ \text{ such that } \ V_2 = V_1\}, \\ E_{21}\,((V_2,T),(S,V_1)) = \,-\,TS. \end{cases}$$

Then, all requirements of Theorem 8.11 are fulfilled. Because I_{21} is a submanifold defined by equation $V_2 - V_1 = 0$, it follows that the generating family of R_{21} is

$$L_{21}(V_2,T,S,V_1;\lambda) = \,-\,TS + \lambda(V_2 - V_1)$$

with supplementary variable $\lambda \in \mathbb{R}$. Thus, if $G_1 = U(S, V)$ is the generating function of $\mathscr{E}_1 = \alpha_1(\mathscr{E})$, then the generating family G_2 of $\mathscr{E}_2 = \alpha_2(\mathscr{E})$ is

$$G_2(V_2, T; S, V_1, \lambda) = L_{21} + U(S, V_1) = -TS + \lambda(V_2 - V_1) + U(S, V_1)$$

with supplementary variables (S, V_1, λ). However, because one of the associated equations is $V_2 - V_1 = 0$, this generating family is reducible to

$$\boxed{\mathscr{F}(V, T; S) = U(S, V) - TS} \tag{8.61}$$

Hence, we have proved the following.

Theorem 8.14. *If a simple, closed, thermostatic system admits a global internal energy $U(S, V)$, then the free energy is the generating family* (8.61) *with supplementary variable S.*

$\theta_2 = -P\,dV - S\,dT$, therefore it follows that $\mathscr{E}_2$ is described by the variational equation

$$-P\,\delta V - S\,\delta T = \delta(U(S, V) - TS)$$

equivalent to equations

$$P = -\frac{\partial U}{\partial V}, \quad 0 = \frac{\partial U}{\partial S} - T. \tag{8.62}$$

However, the generating family $\mathscr{F}(V, T; S)$ may be reducible to an ordinary global generating function $F(V, T)$.

Theorem 8.15. *If we can solve the second equation* (8.62) *with respect to S (i.e., if we can express S as a function of (V, T)) then the generating family* (8.61) *is reducible to a generating function $F(V, T)$ given by*

$$\boxed{F(V, T) = U\,(S(V, T), V) - T\,S(V, T)} \tag{8.63}$$

Proof. If the second equation (8.62) is solvable with respect to S, so that we can express S as a function $S = S(V, T)$, then we can eliminate the supplementary variable S in $\mathscr{F}(V, T; S)$. □

This means that the Lagrangian submanifold $\mathscr{E}$ of equilibrium states admits the free energy $F(V, T)$ as an ordinary generating function. definition (8.63) of $F(V, T)$ can be written in the form

$$\boxed{F(V, T) = \mathrm{stat}_S\,(U(S, V) - TS)} \tag{8.64}$$

where $\mathrm{stat}_S(\star)$ means *compute $\star$ at its stationary points with respect to S,* just in accordance with the second equation (8.62).

Example 8.10. The free energy of a Van der Waals gas. From the state equation (8.50)

$$(V - n\,b)\left(P + a\,\frac{n^2}{V^2}\right) - n\,RT = 0,$$

we get

$$\log(V - n\,b) + \log\left(P + a\,\frac{n^2}{V^2}\right) = \log(n\,RT),$$

and the expression of the entropy (8.51) becomes

$$\boxed{S(V,T) = A + nR\,\log(V - nb) + C\,\log(n\,RT)} \tag{8.65}$$

The assumption of Theorem 8.15 is fulfilled. We have

$$\exp\frac{S}{C} = \exp\frac{A}{C}\cdot(V - nb)^{\gamma-1}\,n\,RT),$$

because $nR/C = \gamma - 1$, and

$$K\,(V - nb)^{1-\gamma}\,\exp\frac{S}{C} = n\,RT,$$

because $K = \exp(-A/C)$. Hence, from the expression (8.56) of the internal energy

$$U(S,V) = \frac{K}{\gamma - 1}\,(V - nb)^{1-\gamma}\,\exp\frac{S}{C} - \frac{a\,n^2}{V}$$

we obtain

$$U(S(V,T),V) = \frac{nRT}{\gamma - 1} - \frac{a\,n^2}{V}.$$

Now we apply Eq. (8.63):

$$\begin{aligned}
F(V,T) &= \frac{nRT}{\gamma-1} - \frac{a\,n^2}{V} - T\,S(V,T) \\
&= \frac{nR}{\gamma-1}\,T + \frac{a\,n^2}{V} - T\,\left(A + \log(V - nb)^{nR} + C\,\log(n\,RT)\right) \\
&= C\,T - \frac{a\,n^2}{V} - T\,(A + nR\,\log(V - nb) + C\,\log(n\,R) + C\,\log T) \\
&= -\frac{a\,n^2}{V} - T\,(A + nR\,\log(V - nb) + C\,\log(n\,R) + C\,\log T - C)\,.
\end{aligned}$$

Hence, if we put

$$\boxed{D = C\,(1 - \log(nR)) - A} \tag{8.66}$$

we finally obtain

$$\boxed{F(V,T) = T\,[D - nR\,\log(V-nb) - C\,\log T] - \frac{a\,n^2}{V}} \tag{8.67}$$

Let us check the correctness of this result. Being $\theta_2 = -P\,dV - S\,dT$ we must have

$$P = -\frac{\partial F}{\partial V}, \quad S = -\frac{\partial F}{\partial T}.$$

Because

$$\frac{\partial F}{\partial V} = -\frac{n\,RT}{V-nb} + \frac{an^2}{V^2},$$

$$\frac{\partial F}{\partial T} = D - nR\,\log(V-nb) - C\,\log T - C,$$

we find

$$P = \frac{n\,RT}{V-nb} - \frac{an^2}{V^2},$$

which is just the first equation of state, and

$$\begin{aligned} S(V,T) &= nR\,\log(V-nb) + C\,(\log T + 1) - D \\ &= nR\,\log(V-nb) + C\,(\log T + 1) - C\,(1 - \log(nR)) + A \\ &= nR\,\log(V-nb) + C\,\log(nRT) + A, \end{aligned}$$

in accordance with Eq. (8.65) above. ◇

8.10 Simple open thermostatic systems

If we act on a thermostatic system also by adding or subtracting particles, then we say that the system is *open*. In this case the energy transferred to the system by the external device in a "quasi-static process" c is given by the integral

$$E_c = \int_c \theta,$$

of the one-form

$$\theta = T\,dS - P\,dV + \mu\,dn.$$

The quantity μ represents the *chemical potential*, and the *molar number* n is assumed to have continuous values. If we set

$$V = n\,v, \quad S = n\,s,$$

then v and s are the *molar volume* and the *molar entropy*, respectively. The states of the system are represented in the manifold

$$M = (s, v, P, T, n, \mu) = \mathbb{R}^6,$$

endowed with the symplectic form

$$\begin{aligned}\omega = d\theta &= dT \wedge dS + dV \wedge dP + d\mu \wedge dn \\ &= n\,(dT \wedge ds + dv \wedge dP) + (d\mu - vdP + sdT) \wedge dn.\end{aligned}$$

We say that the system is *simple* if the set of the equilibrium states $\mathscr{E} \subset M$ is an exact Lagrangian submanifold: there exists a function $W\colon \mathscr{E} \to \mathbb{R}$ such that $\theta|\mathscr{E} = dW$.

For an open system we have eight fundamental control modes, corresponding to the triples of the fundamental observables that are in involution (now we include the observables (n, μ)).

1 – Let us consider the control mode associated with the control manifold

$$Q_1 = (s, v, n)$$

and the control form $\theta_1 = \theta$. For a simple system we introduce the *molar internal energy* $u(s, v)$, so that the total internal energy is

$$U(S, V, n) = n\,u(s, v),$$

and the constitutive set $\mathscr{E}$ is described by the variational equation

$$T\,\delta S - P\,\delta V + \mu\,\delta n = \delta U,$$

equivalent to equations[7]

$$T = u_s(s, v), \quad P = -u_v(s, v), \quad \mu = u(s, v) + P - T\,s. \tag{8.68}$$

Note that the observables T and P do not depend on n, in accordance with their character of "intensive observables".

Example 8.11. For an ideal gas (see Sect. 8.4)

$$u(s, v) = \frac{K}{\gamma - 1}\, v^{1-\gamma} \exp\frac{s}{c}, \tag{8.69}$$

and for a Van der Waals gas

$$u(s, v) = \frac{K}{\gamma - 1}\,(v - b)^{1-\gamma} \exp\frac{s}{c} - \frac{a}{v}. \quad \diamondsuit \tag{8.70}$$

2 – Let us consider the control mode associated with the control manifold

[7] In the following u_s stands for $\partial u/\partial s$, and so on.

$$Q_2 = (v, T, n)$$

and the control form

$$\theta_2 = -P\,dV - S\,dT + \mu\,dn = (\mu - Pv)\,dn - nP\,dv - ns\,dT.$$

The constitutive set $\mathscr{E}$ is then described by the variational equation

$$(\mu - Pv)\,\delta n - nP\,\delta v - S\,\delta T = \delta F, \tag{8.71}$$

where F is the free energy,

$$F(V, T, n) = n\,f(v, T) \tag{8.72}$$

and $f(v, T)$ is the *molar free energy*. Equation (8.71) yields equations

$$P = -f_v(v, T), \quad s = -f_T(v, T), \quad \mu = f(v, T) + Pv. \tag{8.73}$$

The two descriptions (8.68) and (8.73) of $\mathscr{E}$ are equivalent if and only if the molar free energy is related to the molar internal energy by the Legendre transform.

In order to perform the Legendre transform according to Theorem 8.11 we list the following ingredients.

$$\theta_2 - \theta_1 = -P\,dV - S\,dT + \mu\,dn - (T\,dS - P\,dV + \mu\,dn) = -d(ST),$$

$$W_{21}(s, v, P, T, n, \mu) = -ST = -nsT, \quad Q_1 = (s, v_1, n_1), \quad Q_2 = (v_2, T, n_2),$$

$$\pi_{21}(s, v, P, T, n, \mu) = ((v, T, n), (s, v, n)),$$

$$I_{21} = \{((v_2, T, n_2), (s, v_1, n_1)) \text{ such that } v_2 = v_2,\ n_2 = n_1\},$$

$$E_{21}((v_2, T, n_2), (s, v_1, n_1)) = -n_1\,sT \text{ (or equivalently, } = -n_2\,sT).$$

Then,

$$L_{21}((v_2, T, n_2), (s, v_1, n_1); \lambda_1, \lambda_2) = -n_1 sT + \lambda_1(v_2 - v_1) + \lambda_2(n_2 - n_1),$$

and

$$G_2(v_2, T, n_2; s, v_1, n_1, \lambda_1, \lambda_2) = n_1\,u(s, v_1) - n_1 sT + \lambda_1(v_2 - v_1) + \lambda_2(n_2 - n_1).$$

This last generating family, with supplementary variables $s, v_1, n_1, \lambda_1, \lambda_2$, is reducible to

$$G_2(v, T, n; s) = n\left(u(s, v) - s\,T\right), \tag{8.74}$$

with the supplementary variable s only. This means that $\mathscr{E}_2$ is described by the variational equation

$$(\mu - Pv)\,\delta n - nP\,\delta v - ns\,\delta T = \delta G_2,$$

and the vanishing of the coefficient of δs yields equation

$$T = u_s(s, v).$$

If this equation is solvable with respect to s, we can remove s from G_2 defined in Eq. (8.74) and get an ordinary generating function of the kind (8.72).

8.11 Composite thermostatic systems

Let us consider the following mathematical model.

1. There are two symplectic manifolds, M and $\bar{M}$, representing the states of two physical systems.
2. There are two control modes $\alpha_c \colon M \to T^*Q_c$ and $\bar{\alpha}_c \colon \bar{M} \to T^*\bar{Q}_c$.
3. There is a control relation $R_c \subseteq Q_c \times \bar{Q}_c$, where Q_c plays the role of control manifold.
4. There is a Lagrangian submanifold $\bar{\mathscr{E}} \subset \bar{M}$ representing the equilibrium states of the system $\bar{M}$; this is assumed to be a Lagrangian submanifold generated by a function $\bar{V} \colon \bar{Q}_c \to \mathbb{R}$ with respect to the control mode $\bar{\alpha}_c$, so that $\bar{\alpha}_c(\bar{\mathscr{E}}) = d\bar{V}(\bar{Q}_c)$.

We consider the set of equilibrium states (the constitutive set) $\mathscr{E}_c \subset M$ of the system $\bar{M}$ under the control relation R_c. By considering the principle expressed by formula (8.8) and the notion of control mode described in Sect. 8.6, we assume that this set is defined by

$$\boxed{\mathscr{E}_c = \alpha_c^{-1}\left(\widehat{R}_c \circ \bar{\alpha}_c(\bar{\mathscr{E}})\right) \subset M} \tag{8.75}$$

This model is illustrated by the following diagram,

$$\begin{array}{ccccc}
\mathscr{E}_c \hookrightarrow M & & & & \bar{M} \hookleftarrow \bar{\mathscr{E}} \\
\Big\downarrow \alpha_c & & & & \bar{\alpha}_c \Big\downarrow \\
T^*Q_c & \xleftarrow{\;\widehat{R}_c\;} & & & T^*\bar{Q}_c \hookleftarrow d\bar{V}(\bar{Q}_c) \\
\Big\downarrow & & & & \Big\downarrow \\
Q_c & \xleftarrow{\;R_c\;} & & & \bar{Q}_c
\end{array}$$

Let us apply this mathematical model to a closed thermostatic system of n moles composed of N open subsystems in equilibrium.

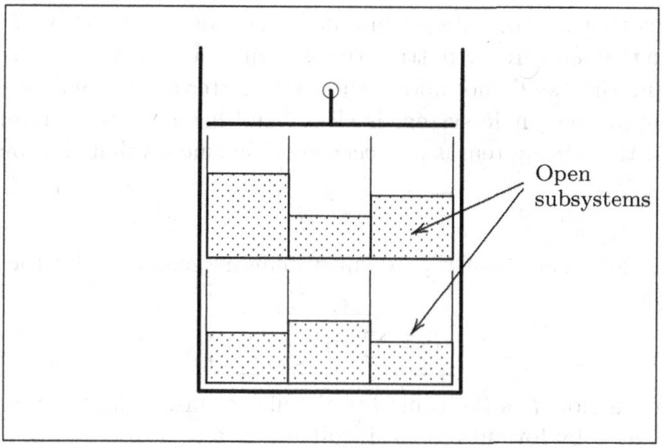

Fig. 8.13 Closed thermostatic system of n moles composed of N open subsystems

The state manifold of this *composite system* is

$$\bar{M} = \times_{i=1}^{N} M_i = \times_{i=1}^{N} (s_i, v_i, T_i, P_i, n_i, \mu_i).$$

We control this system by acting only on macroscopic observables of the space

$$M = (S, V, P, T).$$

The subsystems are assumed to be open; this means that transfer of particles between the subsystems is allowed.

8.11.1 Control volume and temperature

The control manifolds are

$$Q_2 = (V, T), \;\; \bar{Q}_2 = \times_{i=1}^{N} (v_i, T_i, n_i).$$

The control relation $R_2 \subseteq Q_2 \times \bar{Q}_2$ is defined by the fibration $\phi \colon \bar{Q}_2 \to Q_2$ described by equations

$$V = \sum_i n_i v_i, \;\; T = T_1,$$

and by the constraint $\Sigma_2 \subset \bar{Q}_2$ described by equations

$$\sum_i n_i = n, \;\; T_i = T_j.$$

This means that all the subsystems have the same temperature T (we can control the temperature by putting the system in a heat bath at the temperature T) and that we do not add or subtract matter to the whole system (the pot containing the whole system is closed and has a controlled volume V). We assume that the system is *homogeneous*: this means that the constitutive set

$$\bar{\mathscr{E}} \subset \bar{M}$$

of the complete control of the N subsystems is generated by the function (free energy)

$$\boxed{F = \textstyle\sum_i n_i\, f(v_i, T_i)} \tag{8.76}$$

where the function f is the same for all subsystems. From the virtual work principle stated by formula (8.75) it follows that

Theorem 8.16. *The constitutive set $\mathscr{E}_2$ is described by equations*

$$\begin{cases} \sum_i n_i = n, \\ S = -\sum_i n_i\, f_T(v_i, T), \\ V = \sum_i n_i v_i, \end{cases} \quad \begin{cases} -P = f_v(v_1, T) = \ldots = f_v(v_N, T), \\ P v_i + f(v_i, T) = P v_j + f(v_j, T). \end{cases} \tag{8.77}$$

This means that $(S, V, P, T) \in \mathscr{E}_2$ if and only if these equations are satisfied for some values of (v_i, n_i).

Proof. The generating family of the control relation is

$$G_{R_2} = \lambda_1(T - T_1) + \lambda_2(V - \sum_i n_i v_i) + \lambda_3 (\sum_i n_i - n) + \sum_{i \neq j} \lambda_{ij}\, (T_i - T_j).$$

The control form is

$$\theta_2 = \; - S\, dT - P\, dV.$$

From the principle (8.75), written for $c = 2$, and from the composition rule of generating families, it follows that $\mathscr{E}_2$ is described by the variational equation

$$\boxed{-\, S\, \delta T - P\, \delta V = \delta G_{R_2} + \delta F} \tag{8.78}$$

thus by equation

$$\begin{aligned} -S\, \delta T - P\, \delta V = \delta\big[&\lambda_1(T - T_1) + \lambda_2(V - \textstyle\sum_i n_i v_i) + \lambda_3 (\sum_i n_i - n) \\ &+ \textstyle\sum_{i \neq j} \lambda_{ij}(T_i - T_j)\big] + \delta\left(\sum_i n_i\, f(v_i, T_i)\right). \end{aligned} \tag{8.79}$$

The vanishing of the coefficients of $\delta\lambda_1$ and $\delta\lambda_{ij}$ of this last equation yields equations $T = T_1$ and $T_i = T_j$, respectively. Note that the temperatures T_i

play the role of supplementary variables in the generating family $G_R \oplus F$. Hence, Eq. (8.79) is reducible to

$$-S\,\delta T - P\,\delta V = \delta\left[\lambda_2\left(V - \sum_i n_i v_i\right) + \lambda_3\left(\sum_i n_i - n\right)\right] + \delta\left[\sum_i n_i\, f(v_i, T)\right].$$

The coefficients of $(\delta T, \delta V, \delta\lambda_2, \delta\lambda_3, \delta n_i, \delta v_i)$ yield, respectively, the following equations

$$\begin{cases} S = -\sum_i n_i\, f_T(v_i, T), \\ P = -\lambda_2, \\ 0 = V - \sum_i n_i v_i, \end{cases} \qquad \begin{cases} 0 = \sum_i n_i - n, \\ 0 = -\lambda_2\, v_i + \lambda_3 + f(v_i, T), \\ 0 = \lambda_2\, n_i + n_i\, f_v(v_i, T). \end{cases} \tag{8.80}$$

By eliminating the Lagrangian multipliers we obtain Eqs. (8.77). □

Remark 8.16. The control relation considered above fits with Case 4 of Sect. 8.1. Indeed, the fibration ϕ reduces to a fibration over the constraint Σ. This means that we can replace $\bar{Q}_2$ by $\Sigma_2 = (\mathsf{x}_i(v_i, n_i), T_\circ) \simeq \mathbb{R}_+^{2N} \times \mathbb{R}_+$, with all subsystems at the same temperature $T = T_\circ$. The control relation is now described by equations

$$R_2 \colon Q_2 \leftarrow \bar{Q}_2 \quad \begin{cases} V = \sum_i n_i v_i, \quad T = T_\circ & \text{(fibration)}, \\ n = \sum_i n_i & \text{(constraint)}, \end{cases}$$

and (8.79) is replaced by equation

$$\begin{aligned} -S\,\delta T - P\,\delta V = \delta &\Big[\lambda_1(T - T_\circ) + \lambda_2(V - \textstyle\sum_i n_i v_i) + \lambda_3\left(\textstyle\sum_i n_i - n\right)\Big], \\ &+ \delta\Big[\textstyle\sum_i n_i\, f(v_i, T_\circ)\Big], \end{aligned}$$

which reduces to

$$\begin{aligned} -S\,\delta T - P\,\delta V = \delta &\Big[\lambda_2\left(V - \textstyle\sum_i n_i v_i\right) + \lambda_3\left(\textstyle\sum_i n_i - n\right)\Big] \\ &+ \delta\Big[\textstyle\sum_i n_i\, f(v_i, T)\Big]. \end{aligned}$$

Then we find again Eqs. (8.77). ◇

Let us study Eqs. (8.77) of $\mathscr{E}_2$.

Theorem 8.17. *Let $(S, V, P, T) \in \mathscr{E}_2$. For each pair (a, b) of values of the molar volumes v_i satisfying equations* (8.77), *we have*[8]

$$\int_a^b \Big(P + f_v(v, T)\Big)\, dv = 0. \tag{8.81}$$

[8] (Janeczko 1983a, b).

Proof. For constant values of P and T, a primitive of the function $P+f_v(v,T)$ is $Pv+f(v,T)$. Because of the last equation (8.77), this primitive takes equal values at the endpoints of the interval of integration. □

Theorem 8.18. *For each equilibrium state described by equations* (8.77), *the molar volumes* (v_i) *are determined by the points* $(v,y) \in \mathbb{R}^2$ *of the graph of the function* $y = f(v,T)$ [9] *having a common tangent line.*

Proof. For $v_i \neq v_j$, from the last equation (8.77) it follows that

$$P(v_i - v_j) = f(v_j,T) - f(v_i,T),$$

thus,

$$P = -\frac{f(v_i,T) - f(v_j,T)}{v_i - v_j}.$$

Because of Eqs. $(8.77)_4$,

$$\frac{f(v_i,T) - f(v_j,T)}{v_i - v_j} = f_v(v_k,T),$$

for all v_k. □

Remark 8.17. The number of molar volumes (v_i) resulting from this theorem is the number of *phases* that may coexist in an equilibrium state. Note that the last equation (8.77) and the last equation (8.73) show that the subsystems have a common value of the chemical potential, $\mu_i = \mu_j$. ◇

The two theorems stated above give an explanation of the so-called *Maxwell convention*, or *Maxwell rule*, of the "equal areas" (see the discussion in (Poincaré 1892), (Fermi 1936) and (Huang 1987) and of the phenomenon of the *coexistence of phases*. A first "symplectic" approach to this matter can be found in (Janeczko 1983a, b). In the present approach, the Maxwell rule is a theorem following from the general variational principle expressed by formula 8.75)

Remark 8.18. If for all values of T the function $f(v,T)$ is a convex function of v, then any tangent line to its graph is tangent at a single point. This means that for each equilibrium state described by Eqs. (8.77) all v_i assume the same value depending on T: $v_i = v(T)$. It follows that $\mathscr{E}_2$ is a Lagrangian submanifold, generated by the function

$$F(V,T) = n\, f\left(\frac{V}{n}, T\right).$$

[9] Introduced in (8.76).

Indeed, according to the expression of the control form θ_2 in (8.50), Eqs. (8.77) reduce to equations

$$P = -F_V = -f_v\left(\frac{V}{n}, T\right), \quad S = -F_T = -n\, f_T\left(\frac{V}{n}, T\right).$$

In this case the thermostatic system (S, V, P, T) behaves as a simple closed system. In all other cases $\mathscr{E}_2$ may not be a submanifold, and we have coexistence of phases; that is, there are states corresponding to different values of the molar volumes v_i. Fig. 8.14 illustrates the case in which for a certain value of T the graph of $y = f(v, T)$ has two distinct points $v_1 \neq v_2$ with a common tangent.

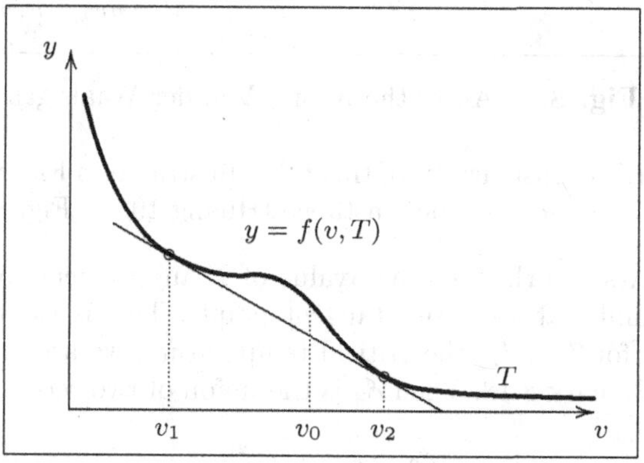

Fig. 8.14 The case of two points with a common tangent

Then the graph of the corresponding isotherm in the (V, P)-plane

$$P = -f_v\left(\frac{V}{n}, T\right)$$

is of the Van der Waals kind for $T < T_c$. If $V_1 = nv_1$ and $V_2 = nv_2$, then

$$\begin{aligned}\int_{V_1}^{V_2} (P - P_1)\, dV &= \int_{V_1}^{V_2} \left(f_v(\tfrac{V}{n}, T) - P_1\right) dV \\ &= n \int_{v_1}^{v_2} (f_v(v, T) - P_1)\, dv = 0.\end{aligned}$$

because of formula (8.81), Theorem 8.17. This is the *Maxwell rule*: all points on the horizontal segment defined by $P = P_1 = -f_v(v_1, T) = -f_v(v_2, T)$ correspond to further equilibrium states.

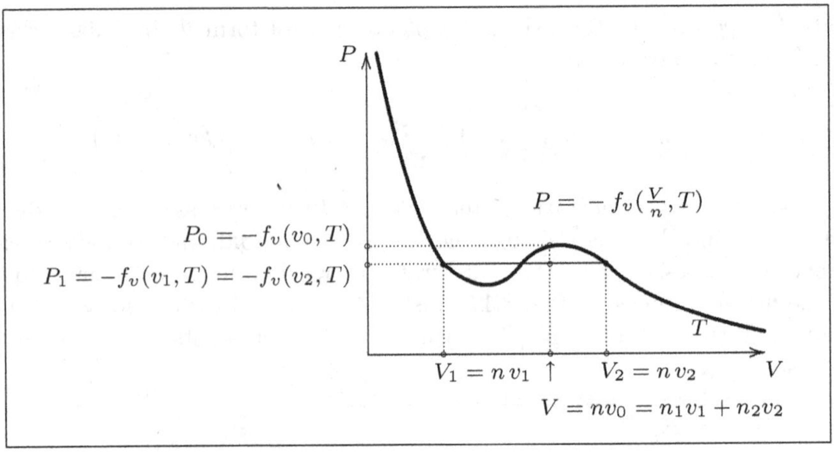

Fig. 8.15 An isotherm of a Van der Waals gas

The geometrical construction of this rule, illustrated in Fig. 8.15, is known as *Maxwell construction* ; see for instance (Huang 1987). Fig. 8.16. ◇

Remark 8.19. Assume that for each value of T any tangent to the graph of $y = f(v, T)$ admits at most two tangent points. This is the case of a Van der Waals gas: for $T < T_c$, the critical temperature, we are in the situation considered in Remark 8.18. Then $\mathscr{E}_2$ is the union of two sets,

$$\mathscr{E}_2 = \mathscr{E}_2^{(1)} \cup \mathscr{E}_2^{(2)}.$$

The first set $\mathscr{E}_2^{(1)}$ corresponds to the case of a single phase: $v_i = v = V/n$. It is the Lagrangian submanifold described by Eqs. (8.80). The second set $\mathscr{E}_2^{(2)}$ represents the equilibrium states with the coexistence of two phases $v_1 < v_2$. For these equilibrium states the values of the volume V belong to the open interval $V_1 < V < V_2$, with $V_i = n\,v_i$. Indeed, for $V \leq V_1$ or $V \geq V_2$ we necessarily have a single phase and the corresponding states belong to $\mathscr{E}_2^{(1)}$. According to Eqs. (8.77) the states of $\mathscr{E}_2^{(2)}$ are then described by equations

$$P = -f_v(v_1, T), \quad S = -n_1\, f_T(v_1, T) - n_2\, f_T(v_2, T).$$

The first equation shows that P has a unique value determined by T, when v_1 is expressed as a function of T itself. About the second equation we observe that the value of S depends on the mole numbers (n_1, n_2) of the two phases. However, these two numbers are determined by the value of V. Indeed, by solving the linear equations

$$n_1\, v_1 + n_2\, v_2 = V, \quad n_1 + n_2 = n,$$

we find

$$n_1 = \frac{n\,v_2 - V}{v_2 - v_1} = \frac{V_2 - V}{v_2 - v_1}, \quad n_2 = \frac{V - n\,v_1}{v_2 - v_1} = \frac{V - V_1}{v_2 - v_1}.$$

Thus,

$$S = \frac{1}{v_1 - v_2}\left[(V_2 - V)\,f_T(v_1, T) + (V - V_1)\,f_T(v_2, T)\right]. \quad \diamond$$

$$T_r = \frac{T}{T_c}, \quad T_c = \frac{8\,a}{27\,bR},$$

$$P_r = \frac{P}{P_c}, \quad P_c = \frac{a}{27\,b^2},$$

$$V_r = \frac{V}{V_c}, \quad V_c = 3\,b$$

$$P_r = \frac{8\,T_r}{3\,V_r - 1} - \frac{3}{V_r^2}$$

T_r

1.2
1.1
1.0
0.9

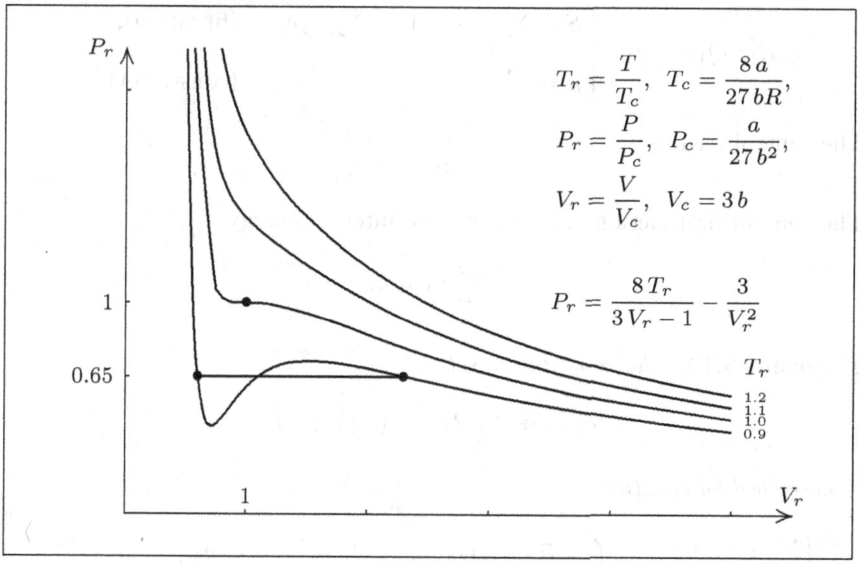

Fig. 8.16 Isotherms of Van der Waals in reduced coordinates

For a complete description of the set $\mathscr{E}_2^{(2)}$ it remains to express v_1 and v_2 as functions of T.

Remark 8.20. Starting from the expression of the molar internal energy, Eq. (8.70), and by performing the suitable Legendre transform, it can be shown that for a Van der Waals gas the Helmholtz molar function is

$$f(v, T) = c_V\,T \log\frac{e\,K}{R\,T} - R\,T\,\log(v - b) - \frac{a}{v}.$$

By studying the graph of $y = f(v, T)$, it can be shown that all v_i assume the same values for $T \geq T_c$ or two distinct values for $T < T_c$, where T_c is the critical temperature. $\quad \diamond$

8.11.2 Control entropy and volume

The control manifolds are

$$Q_1 = (S, V), \;\; \bar{Q}_1 = \times_{i=1}^{N}(s_i, v_i, n_i).$$

The control relation is defined by equations

$$R_1 \colon Q_1 \leftarrow \bar{Q}_1 \quad \begin{cases} S = \sum_i n_i s_i, \;\; V = \sum_i n_i v_i & \text{(fibration)}, \\ n = \sum_i n_i & \text{(constraint)}. \end{cases}$$

The control form is

$$\theta_1 = T\, dS - P\, dV.$$

The generating function of $\bar{\mathscr{E}}$ is the total internal energy

$$U = \sum_i n_i \, u(s_i, v_i).$$

Theorem 8.19. *The constitutive set*

$$\mathscr{E}_1 = \alpha_1^{-1} \left(\widehat{R}_1 \circ \bar{\alpha}_1(\bar{\mathscr{E}}) \right) \subset M$$

is described by equations

$$\begin{cases} \sum_i n_i = n, \\ S = \sum_i n_i \, s_i, \\ V = \sum_i n_i \, v_i, \end{cases} \quad \begin{cases} -P = u_v(s_1, v_1) = \ldots = u_v(s_N, v_N), \\ T = u_s(s_1, v_1) = \ldots = u_s(s_N, v_N), \\ u(s_i, v_i) + P v_i - T s_i = u(s_j, v_j) + P v_j - T s_j. \end{cases} \tag{8.82}$$

This means that $(S, V, P, T) \in \mathscr{E}_1$ if and only if these equations are satisfied for some values of (s_i, v_i, n_i). The proof of this theorem is similar to that of Theorem 8.16. Equation (8.78) is replaced by

$$T\, \delta S - P\, \delta V = \delta(G_{R_2} + U).$$

We can state a theorem similar to Theorem 8.18.

Theorem 8.20. *For each equilibrium state described by equations (8.82), the molar volumes (v_i) and the molar entropy are determined by the points $(s, v, y) \in \mathbb{R}^3$ of the graph of the function $y = u(s, v)$*[10] *having a common tangent plane.*

Proof. Let us consider two pairs (s_1, v_1) and (s_2, v_2) satisfying Eqs. (8.82). Due to the last equation,

---

[10] This graph is called *Gibbs surface.*

$$u(s_1, v_1) - u(s_2, v_2) = T\,(s_1 - s_2) - P\,(v_1 - v_2),$$

and by applying Eqs. $(8.82)_{4,5}$ we get

$$\begin{aligned} u(s_1, v_1) - u(s_2, v_2) &= (s_1 - s_2)\, u_s(s_1, v_1) + (v_1 - v_2)\, u_v(s_1, v_1) \\ &= (s_1 - s_2)\, u_s(s_2, v_2) + (v_1 - v_2)\, u_v(s_2, v_2). \end{aligned}$$

On the other hand, the equation of the tangent plane to the Gibbs surface at a point $P_1 = (s_1, v_1, u(s_1, v_1))$ is

$$(s - s_1)\, u_s(s_1, v_1) + (v - v_1)\, u_v(s_1, v_1) = y - u(s_1, v_1).$$

By setting $(s, v) = (s_2, v_2)$ we get the equation above. □

Remark 8.21. The comparison of (8.77) and (8.82) shows that $\mathscr{E}_1 = \mathscr{E}_2$ (i.e., that the equilibrium states under the two control relations coincide). Indeed, $(8.77)_2$ and $(8.82)_2$ are equivalent because of $(8.79)_3$,

$$s_i = -\, f_T(v_i, T).$$

The remaining equations are, respectively,

$$\begin{cases} -\,P = f_v(v_i, T) = f_v(v_j, T), \\ P v_i + f(v_i, T) = P v_j + f(v_j, T), \end{cases} \tag{8.83}$$

$$\begin{cases} -\,P = u_v(s_i, v_i) = u_v(u_j, v_j), \\ T = u_s(s_i, v_i) = u_s(s_j, v_j), \\ u(s_i, v_i) + P v_i - T s_i = u(s_j, v_j) + P v_j - T s_j. \end{cases} \tag{8.84}$$

The Legendre transform (8.64) can be written

$$f(v, T) = (u(s, v) - sT)\Big|_{s = s(v,T)}, \tag{8.85}$$

where the function $s = s(v, T)$ is the inverse of $T = u_s(s, v)$. It follows that for any fixed value of T,

$$f_v(v, T) = \left[u_v(s, v) + u_s(s, v)\, \frac{\partial s}{\partial v} - T\, \frac{\partial s}{\partial v} \right]_{s = s(v,T)} \cdot \Big[u_s(s, v) \Big]_{s = s(v,T)}.$$

This shows that $(8.83)_1$ is equivalent to $(8.84)_1$ and $(8.84)_2$. Finally, $(8.83)_2$ and $(8.84)_3$ are equivalent because of (8.85). ◇

Chapter 9
Supplementary Topics

Abstract This is a chapter of appendices, which develops some topics previously mentioned.

9.1 Regular distributions and Frobenius theorem

A *regular distribution* on a manifold Q is a subbundle Δ of the tangent bundle TQ, that is, a submanifold of TQ such that for each point $q \in Q$,

$$\Delta_q = \Delta \cap T_q Q,$$

is a subspace of constant dimension r, called the *rank* of the distribution. Hence, a distribution is a map that assigns at each point $q \in Q$ a subspace $\Delta_q \subset T_q Q$ of constant dimension r, in such a way that the union Δ of all Δ_q is a submanifold.

A vector field X on Q is said to be *compatible* with the distribution Δ if its image is contained in Δ:

$$X(q) \in \Delta_q, \quad \text{for all } q \in Q.$$

A one-form θ on Q is a *characteristic form* of Δ if it annihilates all vectors of Δ,

$$\langle \theta, v \rangle = 0, \quad \text{for all } v \in \Delta.$$

A regular distribution Δ of rank r can be locally described in three equivalent ways:

(i) By *equations*, that is, by $m = n - r$ independent homogeneous linear equations

$$\theta_i^a(q)\,\dot{q}^i = 0, \quad a = r+1, \ldots, n. \tag{9.1}$$

(ii) By a *basis of characteristic forms*, that is, by $n - r$ pointwise independent one-forms θ^a annihilating the vectors of Δ $(a = r+1, \ldots, n)$. Such a basis is formed by the one-forms

$$\theta^a = \theta_i^a\, dq^i$$

where the components θ_i^a are given by Eqs. (9.1).

(iii) By a *basis of generators*, that is, by r pointwise independent vector fields X_α $(\alpha = 1, \ldots, r)$ spanning at each point where they are defined, the subspaces Δ_q. We have

$$\langle \theta^a, X_\alpha \rangle = 0.$$

The set of all vector fields compatible with Δ forms a subspace $\mathscr{X}_\Delta$ of the space $\mathscr{X}(Q)$. We say that the distribution is *involutive* if $\mathscr{X}_\Delta$ is a Lie subalgebra that is, if it is closed in the Lie bracket,

$$X, Y \in \mathscr{X}_\Delta \implies [X, Y] \in \mathscr{X}_\Delta.$$

It follows that if (X_α) is a basis of generators, then the distribution Δ is involutive if and only if

$$[X_\alpha, X_\beta] = F_{\alpha\beta}^{\gamma} X_\gamma, \tag{9.2}$$

where $F_{\alpha\beta}^{\gamma}$ are functions on the domain of definition of the local generators.

If (θ^a) is a basis of local characteristic forms, then the distribution Δ is involutive if and only if

$$d\theta^a \wedge \theta^{r+1} \wedge \cdots \wedge \theta^n = 0. \tag{9.3}$$

Let us prove the equivalence of conditions (9.2) and (9.3) for a distribution of rank $r = n - 1$ (the proof for the general case is similar). In this case we have a single characteristic form θ and condition (9.3) becomes

$$d\theta \wedge \theta = 0.$$

If (X, Y, Z) are three vector fields, then the following identity can be proved by using the fundamental properties of the derivations i_X and d_X (Sect. 1.16).

$$i_X i_Y i_Z (d\theta \wedge \theta) = i_X\theta (d_Y i_Z \theta - d_Z i_Y \theta - i_{[Y,Z]}\theta) + \text{c.p.}$$

(c.p. = cyclic permutations of the vector fields). For each $X, Y, \in \mathscr{X}_\Delta$ and $Z \notin \mathscr{X}_\Delta$ we get

$$i_X i_Y i_Z (d\theta \wedge \theta) = -\, i_Z\theta\, i_{[X,Y]}\theta.$$

This shows that $d\theta \wedge \theta = 0$ if and only if $i_{[X,Y]}\theta = 0$ (i.e., $[X, Y] \in \mathscr{X}_\Delta$).

An *integral manifold* of a regular distribution Δ is a submanifold $S \subseteq Q$ such that $T_qS = \Delta_q$. A *maximal integral manifold* is a connected integral manifold that is not properly contained in another connected integral manifold. A maximal integral manifold may be an immersed submanifold. A distribution is *completely integrable* if for all $q \in Q$ there exists an integral manifold containing q.

Theorem 9.1. (i) *A regular distribution is completely integrable if and only if it is involutive.* (ii) *If a regular distribution is completely integrable, then for each point $q \in Q$ there is a unique maximal integral manifold containing q.*

This is known as *Frobenius theorem.*

An *integral function* of a distribution Δ is a smooth function $F\colon Q \to \mathbb{R}$ such that

$$\langle v, dF \rangle = 0, \quad \text{for all } v \in \Delta.$$

Locally, the integral functions are the solutions of the linear homogeneous partial differential equations

$$\langle X_\alpha, dF \rangle = X^i_\alpha\, \partial_i F = 0. \tag{9.4}$$

Theorem 9.2. *A regular distribution is completely integrable if and only if in a neighborhood of any point $q \in Q$ there exists a basis of $n - r$ integral functions (u^a).*

This means that the differentials (du^a) are pointwise independent and any other integral function is functionally dependent on (u^a). Note that a completely integrable distribution may have no global integral function.

Proof. If Δ is completely integrable, then the foliation of its integral manifolds can be locally parametrized by coordinates $(u^i) = (u^\alpha, u^a)$ such that the differentials du^a form a basis of local characteristic forms or, in other words, such that the integral manifolds are locally described by equations $u^a = \text{constant}$. As a consequence, the derivations $\partial/\partial u^\alpha$ form a basis of local generators. coordinates of this kind are said to be *adapted* to the distribution. In adapted coordinates Eqs. (9.4) read

$$\frac{\partial F}{\partial u^\alpha} = 0.$$

The most general solution of these equations is a function depending only on the coordinates (u^a). Conversely, if (u^a) is a basis of integral functions, then locally we can find other functions (u^α) in such a way that (u^a, u^α) is a coordinate system. Since $\partial u^a/\partial u^\alpha = 0$, the r derivations $(X_\alpha = \partial/\partial u^\alpha)$ form a basis of generators that are tangent to the r-dimensional submanifolds $u^a = \text{const}$. □

There is a remarkable symplectic interpretation of all these concepts, which leads to a simple proof of Frobenius theorem.

Let $\Delta^\circ \subset T^*Q$ be the subbundle of the covectors annihilating the vectors of Δ,

$$\Delta^\circ = \{p \in T^*Q \text{ such that } \langle v, p\rangle = 0, \text{ for all } v \in \Delta_q,\ q = \pi_Q(p)\}. \quad (9.5)$$

Lemma 9.1. *The distribution Δ is involutive if and only if Δ° is a coisotropic submanifold.*

Proof. It follows from (9.5) that a vector field X is compatible with Δ if and only if

$$P_X|\Delta^\circ = 0. \quad (9.6)$$

Hence, Δ° is described by equations $P_X = 0$ for X variable in the space $\mathscr{X}_\Delta$. It follows that Δ° is coisotropic if and only if

$$\{P_X, P_Y\}|\Delta^\circ = 0 \quad (9.7)$$

for all $X, Y \in \mathscr{X}_\Delta$. Due to the third equation (4.10), this equation is equivalent to

$$P_{[X,Y]}|\Delta^\circ = 0.$$

This shows that (9.7) holds if and only if $[X,Y] \in \mathscr{X}_\Delta$. □

Lemma 9.2. *If Δ° is coisotropic then the canonical lift $\widehat{X}$ of a vector field $X \in \mathscr{X}_\Delta$ is a characteristic vector field of Δ°.*

Proof. The Hamiltonian of $\widehat{X}$ is P_X. If $X \in \mathscr{X}_\Delta$ then (9.6) holds. This shows that the Hamiltonian of $\widehat{X}$ is constant on Δ°. Hence, $\widehat{X}$ is characteristic, Theorem 3.11. □

Lemma 9.3. *If Δ° is coisotropic then* (i) *the corresponding rays are r-dimensional submanifolds of Q tangent to Δ that is, integral manifolds of Δ, and* (ii) *they coincide with the characteristics lying on the zero-section of T^*Q.*

Proof. Let (X_α) be a local basis of generators of Δ. Due to Lemma 9.2, the canonical lifts $\widehat{X}_\alpha$ are pointwise independent and span the characteristic distribution of Δ°. Since they project onto the r independent vectors X_α, the characteristics projects onto r-dimensional submanifolds tangent to these vectors. Hence, the rays are the integral manifolds of Δ. This proves item (i). Item (ii) follows from the fact that on the zero-section, identified with Q, we have $\widehat{X}_\alpha|Q = X_\alpha$. □

Proof. Proof of Frobenius theorem. If we assume that Δ is involutive, then Δ° is coisotropic (Lemma 9.1) and the corresponding rays are r-dimensional integral manifolds of Δ (Lemma 9.3). Thus, Δ is completely integrable. Conversely, if Δ is completely integrable, then any basis of generators is tangent to the integral manifolds, so that also their Lie brackets are tangent, and (9.2) holds. This proves item (i) of Theorem 9.1. Item (ii) follows from item (ii) of Lemma 9.3, because two distinct maximal characteristics of a coisotropic submanifold have an empty intersection. □

Remark 9.1. An involutive (completely integrable) distribution provides an example of the Hamilton–Jacobi equation (i.e., of a coisotropic submanifold $C = \Delta^{\circ}$) of codimension $r \geq 1$. The complete solution $W(q, u)$ is a solution of a system of r independent linear homogeneous equations,

$$X_{\alpha}^{i}\, \partial_i W = 0,$$

depending on $n - r$ parameters $u = (u^a)$, which is in fact a parametrized family of integral functions of the distribution. Assume that the set U of the maximal integral manifolds of Δ has a differentiable structure such that the canonical projection $\pi\colon Q \to U$ is a submersion. Then the following can be proved. ◊

Theorem 9.3. *The reduced symplectic manifold $M = T^*Q/\Delta^{\circ}$ is symplectomorphic to the cotangent bundle T^*U and the symplectic reduction $R_{\Delta^{\circ}}$ is isomorphic to the canonical lift of the graph of π.*

9.2 Exact Lagrangian submanifolds

Let $\Lambda \subset T^*Q$ be a Lagrangian submanifold. Since Λ is isotropic, the pullback of the canonical symplectic form $d\theta_Q$ to Λ is the zero-two-form: $(d\theta_Q)|\Lambda = 0$. Since the differential operator commutes with the pullback, the pullback of the Liouville form θ_Q to Λ is a closed one-form,

$$d(\theta_Q|\Lambda) = 0.$$

Hence, for each $p \in \Lambda$ there is an open neighborhood $U_p \subseteq \Lambda$ and a function $W_p\colon U_p \to \mathbb{R}$ such that $\theta_Q|U_p = dW_p$. We call these functions the *local potentials* of Λ. We say that a Lagrangian submanifold $\Lambda \subset T^*Q$ is *exact* if it admits a *global potential*, that is, i.e., if there exists a function $W : \Lambda \to \mathbb{R}$ such that

$$\theta_Q|\Lambda = dW.$$

If $\iota\colon \Lambda \to T^*Q$ is the immersion of Λ, then this equation can be written

$$\iota^*\theta_Q = dW.$$

Let $\pi\colon \Lambda \to Q$ be the restriction of the cotangent fibration $\pi_Q\colon T^*Q \to Q$ to Λ. Then,

$$\pi = \pi_Q \circ \iota.$$

We observe that π is a differentiable function, because it is the composition of two differentiable functions, and that $T\pi_Q(v) = T\pi(v)$ for all $v \in T\Lambda$.

Theorem 9.4. *If Λ is generated by a function $G\colon Q \to \mathbb{R}$, then it is exact and $W = G \circ \pi = \pi^* G$ is a global potential.*

Proof. If $v \in T_p\Lambda$, then $\langle v, \theta_Q \rangle = \langle T\pi_Q(v), p \rangle = \langle T\pi(v), dG \rangle = \langle v, \pi^* dG \rangle = \langle v, dW \rangle$. □

Note that this theorem follows directly from formula (5.5), with $S = Q$: $\theta_Q|\Lambda = d\pi^* G$. Conversely, we have the following.

Theorem 9.5. *Let $\Lambda \subset T^*Q$ be an exact Lagrangian submanifold, with global potential W, such that:* (i) *$\pi\colon \Lambda \to Q$ is a surjective submersion;* (ii) *there exists a function $G\colon Q \to \mathbb{R}$ such that $W = \pi^* G = G \circ \pi$. Then Λ is the Lagrangian submanifold generated by the function G, $\Lambda = dG(Q)$.*

Proof. Let $p \in \Lambda$, $q = \pi(p)$, $v \in T_p\Lambda$, and $u = T\pi(v)$. Then,

$$\langle v, \theta_Q \rangle = \langle T\pi(v), p \rangle = \langle u, p \rangle$$

and

$$\langle v, dW \rangle = \langle v, d\pi^* G \rangle = \langle v, \pi^* dG \rangle = \langle T\pi(v), dG \rangle = \langle u, dG \rangle.$$

Bicause $\langle v, \theta_Q \rangle = \langle v, dW \rangle$ for all $v \in T\Lambda$, it follows that

$$\langle u, p \rangle = \langle u, dG \rangle \tag{9.8}$$

for all $u \in T\pi(T_p\Lambda)$. Inasmuch as π is a submersion, $T\pi(T_p\Lambda) = T_qQ$. Thus, $p = d_qG$. □

Remark 9.2. For simplicity we consider only a C^∞ Lagrangian submanifold, so that a global potential is a C^∞ function. However, there are cases in which this theorem holds with a generating function G that is not C^∞. An example is the Lagrangian submanifold $q = p^3$ of $T^*\mathbb{R}$, Example 4.1. Its parametric equations are $p = \lambda$, $q = \lambda^3$. The global potential is $W(\lambda) = \frac{3}{4}\lambda^4$, and the projection π is represented by equation $q = \lambda^3$. It is a one-to-one map, but it is not a diffeomorphism. The generating function is $G(q) = \frac{3}{4}q^{4/3}$, and this function does not admit the second derivative for $q = 0$. ◊

Remark 9.3. Assumption (i) in Theorem 9.5 does not imply (ii). An example is the curve $\Lambda \subset T^*\mathbb{S}_1 \sim \mathbb{S}_1 \times \mathbb{R}$ defined by parametric equations $\boldsymbol{u} = (\cos\lambda, \sin\lambda) \in \mathbb{S}_1$ and $\lambda \in \mathbb{R}$. ◊

Remark 9.4. Let us replace assumptions (i) and (ii) by: *π is a diffeomorphism.* Then the function $G = (\pi^{-1})^* W = W \circ \pi^{-1}$ is a C^∞ generating function of Λ. Indeed, $T_p\pi\colon T_p\Lambda \to T_qQ$ is an isomorphism for each $p \in \lambda$ and moreover,

$$\langle u, dG\rangle = \langle u, (\pi^{-1})^* dW\rangle = \langle T\pi^{-1}(u), dW\rangle = \langle v, dW\rangle = \langle v, \theta_Q\rangle = \langle u, p\rangle.$$

◇

Remark 9.5. We can replace assumption (i) by: *π is surjective and Λ is connected.* In this case condition (9.8) still holds for all $u \in T\pi(T_p\Lambda)$. This shows that if p is a regular point (i.e., $T\pi(T_p\Lambda) = T_qQ$) then $p = d_qG$. This means, in particular, that Λ cannot have two distinct regular points p over a same point $q \in Q$ (i.e., on a same fiber of T^*Q). Let us consider the caustic $\Gamma \subset Q$ of Λ and a point $q \in \Gamma$.

(i) Assume that q is an isolated point of Γ: there exists an open and connected neighborhood N_q of q not containing caustic points except q. Then Λ is generated by G over $N_q - \{q\}$. Λ is connected, therefore the exceptional point q is included by continuity.

(ii) Assume that in any open neighborhood N of q there are points which do not belong to the caustic. Then, in these points we have that Λ is the image of dG and, by an argument similar to that of (i), we conclude it is the image of dG in a neighborhood of q.

(iii) The case in which there exists an open neighborhood N of q all contained in Γ is not possible. Indeed, a caustic cannot contain open subsets.[1] To see this, let us consider a Morse family $F(q^i, u^\alpha)$, at least of class C^2, generating Λ in the neighborhood of a singular point. The caustic is the projection into Q of the intersection of the critical set Ξ, described by equations $\partial_\alpha F = 0$, with the set described by equation $\det[\partial_\alpha\partial_\beta F] = 0$. Hence, it is contained in the projection of Ξ. Since F is a Morse family, Ξ is (locally) a submanifold of dimension equal to the dimension of Q. It projects locally onto open subsets of Q if and only if it is locally a section of the trivial fibration $Q \times U \to Q$. But this is the case in which it is completely reducible to an ordinary generating function of class C^2, and this is against our assumption that it generates a neighborhood of Λ containing a singular point. Note that in the last part of this proof we need the existence of a Morse family of class C^2. This is certainly satisfied if Λ is of class C^2. ◇

Let us consider the case of a Lagrangian submanifold over a submanifold $S \subset Q$.

Theorem 9.6. *The Lagrangian submanifold $\Lambda = \widehat{(S, G)}$ generated by a function $G\colon \mathscr{S} \to \mathbb{R}$ on a submanifold $S \subset Q$ is exact with global potential $W = \pi^* G$, where $\pi\colon \Lambda \to S$ is the restriction of π_Q to Λ.*

Proof. As we have seen in Sect. 3.6, $\theta_Q|\Lambda = d\pi^* G$. □

Conversely, we have the following.

Theorem 9.7. *Let $\Lambda \subset T^*Q$ be an exact Lagrangian submanifold, with global potential W, which projects onto a submanifold $S = \pi_Q(\Lambda) \subseteq Q$. Assume*

[1] A caustic is a closed subset (Abraham and Robbins 1967).

that: (i) *the restriction* $\pi\colon \Lambda \to S$ *of* π_Q *to* Λ *is a submersion,* (ii) *there exists a function* $G\colon S \to \mathbb{R}$ *such that* $\pi^* G = W$*, and* (iii) Λ *is connected and maximal (i.e., it is not properly contained in a larger Lagrangian submanifold satisfying properties* (i) *and* (ii)). *Then* Λ *is generated by* G *on the constraint* S: $\Lambda = \widehat{(S, G)}$.

Proof. Since it projects onto S, Λ is made of covectors based on points of the submanifold S. Hence, it is contained in the coisotropic submanifold $C = T^*_S Q$. By the absorption principle it follows that it is made of characteristics of C. Because of (i), it is the union of maximal characteristics. Then its image by the reduction R_C, $\Lambda_0 = R_C \circ \Lambda \subset T^*S$, is a Lagrangian submanifold. Let $\rho\colon C \to T^*S$ be the surjective submersion underlying R_C. Since R_C is a canonical lift, $\rho^*\theta_S = \theta_Q|C$ and for each $v \in T\Lambda$ we have

$$\langle T\rho(v), \theta_S\rangle = \langle v, \rho^*\theta_S\rangle = \langle v, \theta_Q|C\rangle = \langle v, dW\rangle. \tag{9.9}$$

If v is tangent to a characteristic, then $T\rho(v) = 0$ and $\langle v, dW\rangle = 0$. This shows that the function W is constant on the characteristics (contained in Λ) so that it is reducible to a function W_0 on Λ_0. Let us consider the restriction of ρ to Λ, $\rho|\Lambda\colon \Lambda \to \Lambda_0$. It is a surjective submersion such that

$$W = (\rho|\Lambda)^* W_0. \tag{9.10}$$

It follows that

$$\langle v, dW\rangle = \langle v, (\rho|\Lambda)^* dW_0\rangle = \langle T(\rho|\Lambda)(v), dW_0\rangle.$$

Thus, because of (9.9), for each vector $u \in T\Lambda_0$ we have $\langle u, \theta_S\rangle = \langle u, dW_0\rangle$. This shows that $dW_0 = \theta_S|\Lambda_0$: Λ_0 is exact with potential function W_0. The projection $\pi = \pi_Q|\Lambda\colon \Lambda \to S$ is the composition of $\rho|\Lambda$ with $\pi_S|\Lambda_0\colon \Lambda_0 \to S$,

$$\pi = \pi_S|\Lambda_0 \circ \rho|\Lambda. \tag{9.11}$$

The map π is a surjective submersion (by assumption) as well as $\rho|\Lambda$, thus the map $\pi_S|\Lambda_0$ is also a surjective submersion. From $W = \pi^* G$ and (9.10), (9.11) it follows that

$$(\rho|\Lambda)^* W_0 = W = \pi^* G = (\rho|\Lambda)^*(\pi_S|\Lambda_0)^* G.$$

This shows that $W_0 = (\pi_S|\Lambda_0)^* G$. Hence, to the Lagrangian submanifold $\Lambda_0 \subset S$ we can apply Theorem 9.5, so that $\Lambda_0 = dG(S)$. Due to (5.25), we have $\Lambda = R_C^\top \circ \Lambda_0 = R_C^\top \circ dG(S) = \widehat{(S, G)}$. □

Remark 9.6. If in Theorem 9.7 the last assumption (iii) is not fulfilled, then we can conclude only that Λ is an open subset of $\widehat{(S, G)}$. ◇

9.3 Dual pairings

Let A and B be (real, finite-dimensional) vector spaces. A *dual pairing* between A and B is a bilinear map

$$\langle\,|\,\rangle\colon A \times B \to \mathbb{R}\colon (a,b) \mapsto \langle a|b\rangle,$$

satisfying the following regularity conditions

$$\begin{cases} \langle a|b\rangle = 0, \text{ for all } a \in A \text{ implies } b = 0, \\ \langle a|b\rangle = 0, \text{ for all } b \in B \text{ implies } a = 0. \end{cases}$$

With each subspace (or subset) $K \subseteq A$ we associate a subspace $K^{\P} \subseteq B$, which we call the *polar* of K in the dual pairing $\langle\,|\,\rangle$, defined by

$$K^{\P} = \{b \in B \mid \langle a|b\rangle = 0, \ \text{ for all } a \in K\}.$$

By the same symbol $H^{\P}$ we denote the polar of a subspace $H \subseteq B$.

A first example of dual pairing is the evaluation $\langle\,,\,\rangle$ between vectors of a space A and the covectors of the dual space $B = A^*$. We denote by $K^\circ \subset A^*$ the polar of $K \subset A$ in this canonical dual pairing

$$K^\circ = \{b \in A^* \mid \langle a, b\rangle = 0, \ \text{ for all } a \in K\}. \tag{9.12}$$

All dual pairings are *isomorphic* to this one, as shown by the following theorem.

Theorem 9.8. *Let $\langle\,|\,\rangle\colon A \times B \to \mathbb{R}$ be a dual pairing. The linear map $\psi\colon B \to A^*$ defined by*

$$\langle a, \psi(b)\rangle = \langle a, b\rangle \tag{9.13}$$

is an isomorphism, and for each subspace $K \subseteq A$,

$$\psi(K^{\P}) = K^\circ. \tag{9.14}$$

Proof. Assume that $\psi(b) = 0$. From (9.13) it follows that $\langle a|b\rangle = 0$, for all $a \in A$, and this implies $b = 0$, because of the regularity condition. Hence, the kernel of ψ is the zero vector only, and the map is injective. It follows in particular that $\dim B \leq \dim A^* = \dim A$. We can define in a similar way a linear map $\psi'\colon A \to B^*$, and by the regularity condition (which operates on both sides of the dual pairing) we conclude that it is injective, thus $\dim A \leq \dim B^* = \dim B$. It follows that $\dim A = \dim B$, and ψ is an isomorphism. Formula (9.14) is a direct consequence of (9.12) and (9.13). □

From this theorem and its proof we get the following.

Theorem 9.9. *In a dual pairing $\langle\,|\,\rangle\colon A \times B \to \mathbb{R}$ the spaces A and B have the same dimension.*

Due to Theorem 9.8 and formula (9.14), the polar operator ¶ has formal properties similar to those of ° (here 0_A and 0_B denote the zero-vectors of A and B, respectively),

$$\begin{cases} A^{\P} = 0_B, \ \ B^{\P} = 0_A, \ \ 0_B^{\P} = A, \ \ 0_A^{\P} = B, \\ \dim K + \dim K^{\P} = \dim A = \dim B, \\ K^{\P} \subseteq L^{\P} \iff L \subseteq K, \\ (K+L)^{\P} = K^{\P} \cap L^{\P}, \\ K^{\P} + L^{\P} = (K \cap L)^{\P}, \\ K^{\P\P} = K. \end{cases} \tag{9.15}$$

A second remarkable example of dual pairing is

$$\langle \, | \, \rangle \colon A \times A \to \mathbb{R} \colon (a, a') \mapsto \alpha(a', a),$$

where (A, α) is a symplectic vector space. We have denoted by $K^{\S}$ the polar of $K \subseteq A$ in this dual pairing (Sect. 3.1.1). The isomorphism $\flat \colon A \to A^*$ is the isomorphism ψ of Theorem 9.8.

The notion of dual pairing turns out to be useful in various applications, for instance, in the proof of the basic functorial rule (3.4). We use three lemmas.

Lemma 9.4. *Let $R \subseteq B \oplus A$ be a linear relation. A linear relation $R^{\bullet} \subseteq B^* \oplus A^*$ is defined by*

$$R^{\bullet} = \{(g, f) \in B^* \oplus A^* \text{ such that } \langle a, f \rangle = \langle b, g \rangle \text{ for all } (b, a) \in R\}. \tag{9.16}$$

The subspace $R^{\bullet}$ is the polar of R in the dual pairing

$$\langle \, | \, \rangle \colon (B \oplus A) \times (B^* \oplus A^*) \to \mathbb{R} \colon ((b, a), (g, f)) \mapsto \langle b, g \rangle - \langle a, f \rangle. \tag{9.17}$$

The proof is straightforward.

Lemma 9.5. *Let (A, α) and (B, β) be symplectic vector spaces and let $R \subseteq B \oplus A$ be a linear relation. If $\flat_A \colon A \to A^*$ and $\flat_B \colon B \to B^*$ are the natural isomorphisms defined by the symplectic forms α and β, respectively* (see 3.1), *then*

$$R^{\bullet} = (\flat_B \times \flat_A)(R^{\S}), \tag{9.18}$$

where $R^{\bullet} \subseteq B^* \oplus A^*$ is defined by (9.16) *and $R^{\S} \subseteq B \oplus A$ is defined in* (3.3).

Proof. Due to Eq. (3.1), we can rewrite Eq. (9.16) as follows (here, we denote by the same symbol $\sharp$ the inverse maps of $\flat_A$ and $\flat_B$),

$$R^\bullet = \{(g, f) \in B^* \oplus A^* \text{ such that } \alpha(f^\sharp, a') - \beta(g^\sharp, b') = 0 \\ \text{for all } (b', a') \in R\}.$$

This is equivalent to

$$R^\bullet = \{(b^\flat, a^\flat) \in B^* \oplus A^* \text{ such that } \alpha(a, a') - \beta(b, b') = 0 \\ \text{for all } (b', a') \in R\}.$$

Because of the definition (3.3) of $R^\S$, this equation is equivalent to (9.18). □

The reason why we consider $R^\bullet$ is explained by the following lemma.

Lemma 9.6. *If $R \subseteq B \oplus A$ and $S \subseteq C \oplus B$ are linear relations, then*

$$(S \circ R)^\bullet = S^\bullet \circ R^\bullet. \tag{9.19}$$

Proof. Because of (9.16), we have

$$\begin{cases} R^\bullet = \{(g, f) \in B^* \oplus A^* \text{ such that } \langle a, f\rangle = \langle b, g\rangle, \\ \qquad \text{for all } (b, a) \in R\}, \\ S^\bullet = \{(h, g) \in C^* \oplus B^* \text{ such that } \langle b, g\rangle = \langle c, h\rangle, \\ \qquad \text{for all } (c, b) \in S\}, \end{cases} \tag{9.20}$$

$$(S \circ R)^\bullet = \{(h, f) \in C^* \oplus A^* \text{ such that } \langle c, h\rangle = \langle a, f\rangle, \\ \text{for all } (c, a) \in S \circ R\}, \tag{9.21}$$

and

$$\begin{aligned} S^\bullet \circ R^\bullet = \{&(h, f) \in C^* \oplus A^* \text{ such that there exists } g \in B^* \\ &\text{with } (h, g) \in S^\bullet \text{ and } (g, f) \in R^\bullet \\ = \{&(h, f) \in C^* \oplus A^* \text{ such that there exists } g \in B^* \\ &\text{with } \langle c, h\rangle = \langle b, g\rangle \text{ and } \langle b', g\rangle = \langle a, f\rangle, \\ &\text{for all } (c, b) \in S \text{ and } (b', a) \in R\}. \end{aligned} \tag{9.22}$$

(i) Let $(h, f) \in S^\bullet \circ R^\bullet$. For any arbitrary element $(c, a) \in S \circ R$ there exists $b \in B$ such that $(c, b) \in S$ and $(b, a) \in R$. It follows from (9.22), with $b = b'$, that $\langle c, h\rangle = \langle a, f\rangle$. Because of (9.21), $(h, f) \in (S \circ R)^\bullet$. This proves the inclusion $S^\bullet \circ R^\bullet \subseteq (S \circ R)^\bullet$. (ii) To prove the inverse inclusion we consider the following dual pairing

$$(C \oplus B \oplus B \oplus A) \times (C^* \oplus B^* \oplus B^* \oplus A^*) \to \mathbb{R}: \\ ((c, b, b', a), (h, g, g', f)) \mapsto \langle c, h\rangle - \langle b, g\rangle + \langle b', g'\rangle - \langle a, f\rangle. \tag{9.23}$$

We denote by ¶ the corresponding dual operator. For this dual pairing we have

$$(S \oplus R)^{\P} = S^{\bullet} \oplus R^{\bullet}. \tag{9.24}$$

Indeed, due to (9.23),

$$\begin{aligned}(S \oplus R)^{\P} = \{(h, g, g', f) \text{ such that } &\langle c, h\rangle - \langle b, g\rangle + \langle b', g'\rangle - \langle a, f\rangle = 0, \\ &\text{for all } (c, b) \in S \text{ and } (b', a) \in R\}\end{aligned}$$

and, due to (9.20),

$$\begin{aligned}S^{\bullet} \oplus R^{\bullet} = \{(h, g, g', f) \text{ such that } &\langle c, h\rangle = \langle b, g\rangle,\ \langle b', g'\rangle = \langle a, f\rangle, \\ &\text{for all } (c, b) \in S \text{ and } (b', a) \in R\}.\end{aligned}$$

This second expression shows that $S^{\bullet} \oplus R^{\bullet} \subseteq (S \oplus R)^{\P}$. On the other hand, by the dimensional property of the dual polar operators, we have

$$\begin{aligned}\dim(S^{\bullet} \oplus R^{\bullet}) &= \dim S^{\bullet} + \dim R^{\bullet} = \mathrm{codim} S + \mathrm{codim} R \\ &= \dim C + 2 \dim B + \dim A - \dim S - \dim R,\end{aligned}$$

and

$$\begin{aligned}\dim(S \oplus R)^{\P} &= \mathrm{codim}\ (S \oplus R)^{\P} \\ &= \dim C + 2 \dim B + \dim A - \dim S - \dim R.\end{aligned}$$

Then $\dim(S^{\bullet} \oplus R^{\bullet}) = \dim(S \oplus R)^{\P}$ and (9.24) is proved. Let us consider the following two subspaces of $C \oplus B \oplus B \oplus A$,

$$L = \{(c, b, b, a)\},\ \ K = (S \oplus R) \cap L.$$

We remark that

$$K = \{(c, b, b, a) \text{ such that } (c, b) \in S,\ (b, a) \in R\} \tag{9.25}$$

so that

$$(c, b, b, a) \in K \ \Rightarrow\ (c, a) \in S \circ R. \tag{9.26}$$

The polar $L^{\P}$ is made of elements of the kind $(0, g, g, 0)$ with $g \in B^*$. Indeed,

$$\begin{aligned}L^{\P} &= \{(h, g, g', f) \text{ such that } \langle c, h\rangle - \langle b, g\rangle + \langle b', g'\rangle - \langle a, f\rangle = 0, \\ &\qquad \text{for all } (c, b, a) \in C \times B \times A\} \\ &= \{(h, g, g', f) \text{ such that } h = 0,\ f = 0,\ g = g'\}.\end{aligned}$$

Furthermore, from one of the rules (9.15) and from (9.24) we derive

$$K^{\P} = S^{\bullet} \oplus R^{\bullet} + L^{\P}. \tag{9.27}$$

If $(h, f) \in (S \circ R)^{\bullet}$ and $g \in B^*$, then $(h, g, g, f) \in K^{\P}$. Indeed, $(h, f) \in (S \circ R)^{\bullet}$ means

$$\langle c, h\rangle = \langle a, f\rangle, \quad \text{for all } (c, a) \in S \circ R,$$

and because of (9.25) and (9.26), for any $(c, b, b, a) \in K$ we have, in the dual pairing (9.23),

$$\langle (c, b, b, a) \mid (h, g, g, f)\rangle = \langle c, h\rangle - \langle b, g\rangle + \langle b, g\rangle - \langle a, f\rangle = 0.$$

It follows from (9.27) that there exist elements $\bar{g} \in B^*$ and $(h', g', g'', f') \in S^{\bullet} \oplus R^{\bullet}$ such that

$$(h, g, g, f) = (h', g', g'', f') + (0, \bar{g}, \bar{g}, 0).$$

From this equality we see that $g' = g''$, $h' = h$ and $f' = f$. Since $(h', g') \in S^{\bullet}$ and $(g'', f') \in R^{\bullet}$, we conclude that if $(h, f) \in (S \circ R)^{\bullet}$ then there exists a $g' \in B^*$ such that $(h, g') \in S^{\bullet}$ and $(g', f) \in R^{\bullet}$; that is, $(h, f) \in S^{\bullet} \circ R^{\bullet}$. This proves $(S \circ R)^{\bullet} \subseteq S^{\bullet} \circ R^{\bullet}$. □

Now we can prove Theorem 3.1.

Proof. Let (A, α), (B, β), and (C, γ) be symplectic vector spaces and let $R \subseteq B \oplus A$ and $S \subseteq C \oplus B$ be linear relations. Then we have

$$\begin{aligned}
(\flat_C \times \flat_B)(S) \circ & (\flat_B \times \flat_A)(R) \\
&= \{(f, g) \in C^* \oplus A^* \text{ such that there exists } h \in B^* \\
&\qquad \text{with } (f, h) \in (\flat_C \times \flat_B)(S) \text{ and } (h, g) \in (\flat_B \times \flat_A)(R)\} \\
&= \{(f, g) \in C^* \oplus A^* \text{ such that there exists } b \in B \\
&\qquad \text{with } (f^{\sharp}, b) \in S \text{ and } (b, g^{\sharp}) \in R\} \\
&= \{(f, g) \in C^* \oplus A^* \text{ such that } (f^{\sharp}, g^{\sharp}) \in S \circ R\} \\
&= (\flat_C \times \flat_A)(S \circ R).
\end{aligned}$$

This proves the identity

$$(\flat_C \times \flat_B)(S) \circ (\flat_B \times \flat_A)(R) = (\flat_C \times \flat_A)(S \circ R),$$

which holds for any two relations R and S. We can write it for $R^{\S}$ and $S^{\S}$,

$$(\flat_C \times \flat_B)(S^{\S}) \circ (\flat_B \times \flat_A)(R^{\S}) = (\flat_C \times \flat_A)(S^{\S} \circ R^{\S}).$$

Because of (9.18) and (9.19), it follows that

$$(\flat_C \times \flat_A)(S^{\S} \circ R^{\S}) = S^{\bullet} \circ R^{\bullet} = (S \circ R)^{\bullet} = (\flat_C \times \flat_A)(S \circ R)^{\S}.$$

The map $\flat_C \times \flat_A$ is an isomorphism, thus the functorial rule (3.4) is proved. □

9.4 Lagrangian splittings and canonical bases

A *Lagrangian splitting* of a symplectic vector space (A, α) is an ordered pair (L, M) of Lagrangian subspaces such that

$$L \cap M = 0.$$

This condition is equivalent to

$$L + M = A.$$

Indeed, from $A^{\S} = 0$, $L^{\S} = L$, and $M^{\S} = M$ it follows that

$$(L \cap M)^{\S} = L^{\S} + M^{\S} = L + M.$$

Hence, a Lagrangian splitting is a decomposition of A as a direct sum of two Lagrangian subspaces,

$$A = L \oplus M.$$

Theorem 9.10. *Let (L, M) be a Lagrangian splitting of (A, α). The map*

$$\langle\,|\,\rangle\colon L \times M \to \mathbb{R}\colon (l, m) \mapsto \alpha(m, l)$$

is a dual pairing.

Proof. Equation $\langle l|m\rangle = 0$ (i.e., $\alpha(l, m) = 0$) for each $l \in L$, means that $m \in L^{\S}$, that is $m \in L$. Since $L \cap M = 0$, it follows that $m = 0$. □

Let $\psi\colon M \to L^*$ be the isomorphism associated with this dual pairing. It is defined by

$$\langle l, \psi(m)\rangle = \alpha(m, l). \tag{9.28}$$

Let (e_i) be an ordered basis of the subspace L and let (ε^i) be its dual basis in the dual space L^*: $\langle e_i, \varepsilon^j\rangle = \delta_i^j$. It follows from (9.28) that the vectors $f^j = \psi^{-1}(\varepsilon^j)$ form a basis of M such that (e_i, f^j) is a *canonical basis* of (A, α); that is,

$$\alpha(e_i, e_j) = 0, \quad \alpha(f^i, f^j) = 0, \quad \alpha(e_i, f^j) = \delta_i^j. \tag{9.29}$$

Conversely, let (e_i, f^j) be a canonical basis and let I be a subset of the set $I_n = \{1, 2, \ldots, n\}$, $n = \frac{1}{2}\dim A$. Let us denote by L_I the subspace of L spanned by the vectors $(e_i, f^{\bar{i}})$ with $i \in I$ and $\bar{i} \in \bar{I}$, the complementary set of I in I_n. Then, we have the following theorem.

Theorem 9.11. *For each Lagrangian subspace L there exists a subset $I \subset I_n$ such that (L, L_I) is a Lagrangian splitting.*

For the proof we use the following lemma.

Lemma 9.7. *If E is a n-dimensional vector space with a basis (e_i) and $S \subset E$ is a subspace, then there exists a subset $I \subset I_n$ such that*

$$E = S \oplus S_I \quad \begin{cases} S_I \cap S = 0, \\ S + S_I = E, \end{cases}$$

where $S_I = \operatorname{span}(e_i,\ i \in I)$.

Proof. Assume that S is defined by the $m = n - r$ independent linear equations

$$S_{\alpha i} x^i = 0.$$

Up to an inessential reordering of the basis we assume that the submatrix $[S_{\alpha\beta}]$ $(\alpha, \beta = 1, \ldots, m)$ is regular with inverse matrix $[S^{\alpha\beta}]$, $S^{\alpha\beta} S_{\beta\gamma} = \delta^\alpha_\gamma$. From

$$S_{\alpha\beta}\, x^\beta + S_{\alpha\bar{\beta}}\, x^{\bar{\beta}} = 0 \ \ (\bar{\alpha}, \bar{\beta} = m + 1, \ldots, n)$$

it follows that the vectors $x \in S$ are characterized by equations

$$x^\beta = -\, S^{\beta\alpha}\, S_{\alpha\bar{\beta}}\, x^{\bar{\beta}}.$$

Let us consider the subspace $S' = \operatorname{span}(e_\alpha)$ made of vectors $y = y^\alpha e_\alpha$. Let $v = v^\alpha e_\alpha + v^{\bar{\alpha}} e_{\bar{\alpha}}$ be any vector of E. Let us set

$$\begin{cases} x^\beta = -\, S^{\beta\alpha}\, S_{\alpha\bar{\beta}}\, v^{\bar{\beta}}, \\ x^{\bar{\beta}} = v^{\bar{\beta}}, \\ y^\beta = v^\beta + S^{\beta\alpha}\, S_{\alpha\bar{\beta}}\, v^{\bar{\beta}}. \end{cases}$$

Then, $x = x^\beta e_\beta + x^{\bar{\beta}} e_{\bar{\beta}} \in S$ and $y = y^\beta e_\beta \in S'$ and moreover, $x + y = v$. This shows that $S + S' = E$. Let $y = y^\alpha e_\alpha = x \in S$. Then from $y^\alpha e_\alpha = x^\beta e_\beta + x^{\bar{\beta}} e_{\bar{\beta}}$ it follows that $x^{\bar{\beta}} = 0$ hence, $x^\beta = 0$. This shows that $S \cap S' = 0$. Note that $S' = S_I$ with $I = \{1, \ldots, m\}$. □

Proof. The subspace $E = L_{I_n} = \operatorname{span}(e_i)$ is Lagrangian (it is isotropic due to $(9.29)_1$ and of dimension n). The subspace $S = L \cap E$ is isotropic (it is the intersection of two isotropic subspaces). There exists at least a subset $I \subset I_n$ such that $S_I \cap S = 0$ and $S_I + S = E$, where $S_I = \operatorname{span}(e_i;\ i \in I)$ (Lemma 1). Note that $S_I \subseteq L_I$ is isotropic. As a consequence,

$$\begin{aligned}
&S \subseteq L \text{ and } S_I \subseteq L_I \Longrightarrow \quad L \subset S^{\S} \text{ and } L_I \subseteq S_I^{\S} \\
&\Longrightarrow \quad L \cap L_I \subseteq S^{\S} \cap S_I^{\S} = (S + S_I)^{\S} = E^{\S} = E \\
&\Longrightarrow \quad L \cap L_I = E \cap L \cap L_I = (E \cap L) \cap (E \cap L_I) = S \cap S_I = 0. \qquad \square
\end{aligned}$$

Now we observe that $(L_{\bar{I}}, L_I)$ is also a Lagrangian splitting, having a common element with (L, L_I). As a consequence, if we consider the two projections with respect to the complementary subspaces $(L_I, L_{\bar{I}})$, then L projects isomorphically onto $L_{\bar{I}}$.

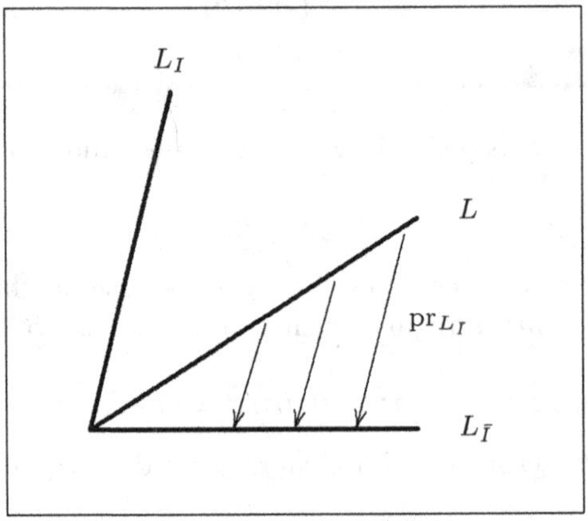

Fig. 9.1 Lagrangian splitting

Hence, we can prove the following theorem.[2]

Theorem 9.12. *Let*

$$\boldsymbol{L} = \begin{bmatrix} \boldsymbol{Q} \\ \boldsymbol{P} \end{bmatrix} = \begin{bmatrix} Q_k^i \\ P_{jk} \end{bmatrix} \tag{9.30}$$

be a $2n \times n$ *matrix with maximal rank* n *such that*

$$Q_k^i \, P_{ih} - P_{ik} \, Q_h^i = 0. \tag{9.31}$$

Then there exists a subset $I \subseteq I_n$ *such that the* $n \times n$ *submatrix*

$$\boldsymbol{S}_I = \begin{bmatrix} \boldsymbol{Q}^{\bar{I}} \\ \boldsymbol{P}_I \end{bmatrix} = \begin{bmatrix} Q_k^{\bar{I}} \\ P_{Ik} \end{bmatrix}, \quad I \subseteq \{1, 2, \ldots, n\}.$$

is regular.

Proof. Let (A, α) be a $2n$-dimensional symplectic vector space and let (e_i, f^j) be a canonical basis. For each $v \in A$ we have the representation $v = v^i e_i +$

[2] See also (Arnold 1967) and (Mishchenko et al. 1978).

$v_j f^j$. Let $L \subset A$ be the subspace described by parametric equations

$$v^i = Q^i_k \lambda^k, \quad v_j = P_{jk} \lambda^k, \quad (\lambda^k) \in \mathbb{R}^n, \tag{9.32}$$

with the matrix (9.30) of maximal rank. It follows that L is a subspace of dimension n. Condition (9.31) is equivalent to the isotropy of this subspace:

$$\begin{aligned} \alpha(u, v) &= \alpha(u^i e_i + u_j f^j, v^i e_i + v_j f^j) = u^i v_i - u_i v^i \\ &= (Q^i_k P_{jk} - P_{ik} Q^i_h) \lambda^k_u \lambda^h_v. \end{aligned}$$

Hence, L is a Lagrangian subspace and Eqs. (9.31) describe an isomorphism $\mathbb{R}^n \to L$. On the other hand, due to Theorem 9.11, there is an isomorphism $\mathbb{R}^n \to L_{\bar{I}}$. Because $v \in L_{\bar{I}}$ if and only if $v = \sum_{a \in \bar{I}} v^a e_a + \sum_{\alpha \in I} v_\alpha f^\alpha$, this last isomorphism is described by equations

$$v^a = Q^a_k \lambda^k, \quad v_\alpha = P_{\alpha k} \lambda^k,$$

it follows that

$$\det \begin{bmatrix} Q^a_k \\ P_{\alpha k} \end{bmatrix} \neq 0. \qquad \square$$

9.5 The Maslov–Hörmander theorem

Theorem 9.13. *Maslov–Hörmander theorem. If $\Lambda \subset T^*Q$ is a Lagrangian submanifold, then for each $p \in \Lambda$ there exists a Morse family generating Λ in a neighborhood of p.*

Proof.[3] Let us consider a Lagrangian immersion, Eqs. (4.14),

$$q^i = q^i(u^k), \quad p_i = p_i(u^k), \tag{9.33}$$

and the $2n \times n$ matrix with maximal rank

$$\begin{bmatrix} Q^i_k \\ P_{jk} \end{bmatrix} = \begin{bmatrix} \dfrac{\partial q^i}{\partial u^k} \\ \dfrac{\partial p_j}{\partial u^k} \end{bmatrix}. \tag{9.34}$$

The immersion is Lagrangian, thus Eqs. (4.16) hold (i.e., $Q^i_k P_{ij} - Q^i_j P_{ik} = 0$). Then we are in the condition of applying Theorem 9.12 and, up to a reordering of the coordinates (q^i), the matrix (9.34) admits a regular $n \times n$ submatrix of the kind

[3] See also (Libermann and Marle 1987) and (Weinstein 1977). The proof given here is taken from (Benenti 1988).

$$\begin{bmatrix} Q_k^a \\ \\ P_{\alpha k} \end{bmatrix} = \begin{bmatrix} \dfrac{\partial q^a}{\partial u^k} \\ \\ \dfrac{\partial p_\alpha}{\partial u^k} \end{bmatrix}, \quad a = 1, \ldots, m, \quad \alpha = m+1, \ldots, n,$$

so that the subsystem of (9.33),

$$q^a = q^a(u^k), \quad p_\alpha = p_\alpha(u^k),$$

can be solved (locally) with respect to (u^k): $u^k = u^k(q^a, p_\alpha)$. This means that we can take (q^a, p_α) as parameters of the immersion, so that a Lagrangian submanifold Λ can always be represented by local immersions of the kind[4]

$$q^\alpha = q^\alpha(q^b, p_\beta), \quad p_a = p_a(q^b, p_\beta). \tag{9.35}$$

The one-form

$$\theta = p_a\, dq^a - q^\alpha\, dp_\alpha$$

is such that $d\theta = \omega = dp_i \wedge dq^i$ (the canonical symplectic form). Its pullback to Λ is closed (because Λ is Lagrangian), thus locally exact. It follows that there exists a function $F(q^a, p_\alpha)$ such that (9.35) are equivalent to

$$p_a = \frac{\partial F}{\partial q^a}, \quad q^\alpha = -\frac{\partial F}{\partial p_\alpha}. \tag{9.36}$$

Let us consider the function

$$G(q^i; p_\alpha) = F(q^a, p_\alpha) + p_\alpha\, q^\alpha.$$

This is a Morse family on (q^i) with supplementary variables (p_α). Indeed, in the matrix

$$\left[\frac{\partial^2 G}{\partial p_\alpha \partial p_\beta} \;\middle|\; \frac{\partial^2 G}{\partial p_\alpha \partial q^i} \right]$$

we find the regular square matrix

$$\left[\frac{\partial^2 F}{\partial p_\alpha \partial q^\beta} \right] = [\delta_\beta^\alpha].$$

The equations of the Lagrangian set generated by this Morse family are

---

[4] The parameters (q^a, p_α) are local coordinates on the Lagrangian submanifold Λ. They are called *canonical coordinates* of Λ. For more information about the existence and the use of a *canonical atlas* of a Lagrangian submanifold see (Mishchenko et al. 1978).

$$\begin{cases} p_i = \dfrac{\partial G}{\partial q^i} & \rightarrow \begin{cases} p_a = \dfrac{\partial G}{\partial q^a} = \dfrac{\partial F}{\partial q^a}, \\ p_\alpha = \dfrac{\partial G}{\partial q^\alpha} = p_\alpha. \end{cases} \\ 0 = \dfrac{\partial G}{\partial p_\alpha} = q^\alpha + \dfrac{\partial F}{\partial p_\alpha}. \end{cases}$$

These equations coincide with Eqs. (9.36) of Λ. □

Chapter 10
Global Hamilton Principal Functions on $\mathbb{S}_2$ and $\mathbb{H}_2$

Abstract As a final argument of this book I propose a theme, simple in its formulation, but not so simple in its design: to see which are the principal Hamilton functions for the geodesics of two basic Riemannian manifolds of constant curvature. In drafting this chapter, I was pleasantly helped by Franco Cardin, University of Padua.

10.1 Vector calculus in the real three-space

In $\mathbb{R}^3 = (x, y, z)$ endowed with the natural Euclidean structure we consider the unit sphere $\mathbb{S}_2$, $x^2 + y^2 + z^2 - 1 = 0$. In $\mathbb{R}^3$ endowed with a Minkowski metric, with z a time-like coordinate, we consider the hyperboloid $\mathbb{H}_2$ of equation $z = \sqrt{1 + x^2 + y^2}$ made of all unit time-like vectors oriented to the future.

Both are two-dimensional Riemannian manifolds with constant curvature (positive and negative, respectively). We show that their eikonal equations admit global Hamilton principal functions, which are not Morse families. To this end, we need to recall some basic definitions and formulae of vector calculus in $\mathbb{R}^3$.

10.1.1 The metric tensor and the scalar product

In $\mathbb{R}^3$ we consider the ordered canonical basis $\boldsymbol{c}_i$,

$$\boldsymbol{c}_1 = \begin{bmatrix} 1 \\ 0 \\ 0 \end{bmatrix}, \quad \boldsymbol{c}_2 = \begin{bmatrix} 0 \\ 1 \\ 0 \end{bmatrix}, \quad \boldsymbol{c}_3 = \boldsymbol{t} = \begin{bmatrix} 0 \\ 0 \\ 1 \end{bmatrix}$$

and the metric tensors $\boldsymbol{g}_\varepsilon$, with $\varepsilon = \pm 1$, such that

$$\begin{cases} \boldsymbol{g}_\varepsilon(\boldsymbol{c}_i, \boldsymbol{c}_j) = 0, \quad i \neq j, \\ \boldsymbol{g}_\varepsilon(\boldsymbol{c}_1, \boldsymbol{c}_1) = \boldsymbol{g}_\varepsilon(\boldsymbol{c}_2, \boldsymbol{c}_2) = 1, \\ \boldsymbol{g}_\varepsilon(\boldsymbol{c}_3, \boldsymbol{c}_3) = \varepsilon = \pm 1. \end{cases}$$

We denote by

$$\boldsymbol{u} \cdot \boldsymbol{v} = \boldsymbol{g}_\varepsilon(\boldsymbol{u}, \boldsymbol{v})$$

the scalar product of two vectors, and use the notation

$$\boldsymbol{u}^2 = \boldsymbol{u} \cdot \boldsymbol{u}, \quad |\boldsymbol{u}| = \sqrt{|\boldsymbol{u}^2|}.$$

Two vectors are *orthogonal* if $\boldsymbol{u} \cdot \boldsymbol{v} = 0$. In this case we use the notation $\boldsymbol{u} \perp \boldsymbol{v}$. If

$$g_{ij} = \boldsymbol{c}_i \cdot \boldsymbol{c}_j,$$

then

$$\left[g_{ij}\right] = \begin{bmatrix} 1 & 0 & 0 \\ 0 & 1 & 0 \\ 0 & 0 & \varepsilon \end{bmatrix}.$$

For $\varepsilon = 1$ the metric is positive-definite (Euclidean). For $\varepsilon = -1$ the metric is hyperbolic (Minkowskian) and the vector $\boldsymbol{c}_3 = \boldsymbol{t}$ is time-like.

Let $(\boldsymbol{e}_a)$ be any basis. Its dual basis $(\boldsymbol{e}^a)$ is defined by

$$\boldsymbol{e}_a \cdot \boldsymbol{e}^b = \delta_a^b.$$

If

$$g_{ab} = \boldsymbol{e}_a \cdot \boldsymbol{e}_b, \quad g^{ab} = \boldsymbol{e}^a \cdot \boldsymbol{e}^b,$$

then the two symmetric matrices $\left[g_{ab}\right]$ and $\left[g^{ab}\right]$ are inverses of each other (i.e. $g^{ab} g_{bc} = \delta_c^a$) and we get the well-known rules of raising and lowering the indices: if $\boldsymbol{v} = v^a \boldsymbol{e}_a = v_a \boldsymbol{e}^a$, then

$$v_a = g_{ab} v^b, \quad v^a = g^{ab} v_b,$$

and

$$v_a = \boldsymbol{v} \cdot \boldsymbol{e}_a, \quad v^a = \boldsymbol{v} \cdot \boldsymbol{e}^a.$$

For the canonical basis,

$$\boldsymbol{c}^1 = \boldsymbol{c}_1, \quad \boldsymbol{c}^2 = \boldsymbol{c}_2, \quad \boldsymbol{c}^3 = \varepsilon\, \boldsymbol{c}_3,$$

$$v^1 = v_1, \quad v^2 = v_2, \quad v^3 = \varepsilon\, v_3.$$

10.1.2 The volume form

We define a volume three-form $V(\boldsymbol{u}, \boldsymbol{v}, \boldsymbol{w})$ by setting

$$V(\boldsymbol{c}_1, \boldsymbol{c}_2, \boldsymbol{c}_3) = \varepsilon.$$

As a consequence,

$$V(\boldsymbol{c}_h, \boldsymbol{c}_i, \boldsymbol{c}_j) = \varepsilon\, \varepsilon_{hij}$$

where ε_{hij} is the Levi-Civita symbol, and

$$V(\boldsymbol{u}, \boldsymbol{v}, \boldsymbol{w}) = \varepsilon\, \varepsilon_{hij}\, u^h v^i w^j = \varepsilon \begin{vmatrix} u^1 & u^2 & u^3 \\ v^1 & v^2 & v^3 \\ w^1 & w^2 & w^3 \end{vmatrix}.$$

For any arbitrary basis $(\boldsymbol{e}_a)$ we have

$$V(\boldsymbol{u}, \boldsymbol{v}, \boldsymbol{w}) = V_{abc}\, u^a v^b w^c = V^{abc}\, u_a v_b w_c$$

where

$$V_{abc} = V(\boldsymbol{e}_a, \boldsymbol{e}_b, \boldsymbol{e}_c), \quad V^{abc} = V(\boldsymbol{e}^a, \boldsymbol{e}^b, \boldsymbol{e}^c) = g^{ad} g^{be} g^{cf}\, V_{def}.$$

If the basis $(\boldsymbol{e}_a)$ is oriented as $(\boldsymbol{c}_i)$, that is, if

$$\boldsymbol{e}_a = A_a^i\, \boldsymbol{c}_i, \quad \det \boldsymbol{A} > 0, \quad \boldsymbol{A} = [A_a^i],$$

then

$$V_{abc} = \varepsilon \sqrt{|g|}\, \varepsilon_{abc}, \quad V^{abc} = \frac{1}{\sqrt{|g|}}\, \varepsilon^{abc}, \quad g = \det[g_{ab}], \tag{10.1}$$

where ε_{abc} and ε^{abc} are Levi-Civita symbols. It follows that

$$V^{abc}\, V_{abc} = 3!\, \varepsilon, \quad V^{abc}\, V_{dbc} = 2\, \varepsilon\, \delta_d^a, \quad V^{abc}\, V_{dec} = \varepsilon\, \delta_{de}^{ab} = \varepsilon\, (\delta_d^a \delta_e^b - \delta_e^a \delta_d^b).$$

To prove (10.1) we observe that from $g_{ab} = A_a^i A_b^j\, g_{ij}$ it follows that

$$g = \det[g_{ab}] = (\det \boldsymbol{A})^2 \det[g_{ij}] = (\det \boldsymbol{A})^2\, \varepsilon.$$

Thus, g has the same sign of ε and we can write $g = \varepsilon\, |g|$. Moreover, since $\boldsymbol{A}$ has a positive determinant,

$$\det \boldsymbol{A} = \sqrt{|g|}.$$

Hence,

$$V_{abc} = A_a^i A_b^j A_c^k V_{ijk} = \varepsilon\, \varepsilon_{ijk}\, A_a^i A_b^j A_c^k = \varepsilon\, \varepsilon_{abc} \det \boldsymbol{A} = \varepsilon \sqrt{|g|}\, \varepsilon_{abc},$$

$$V^{abc} = V_{def}\, g^{ad} g^{be} g^{cf} = \varepsilon \sqrt{|g|}\, \varepsilon_{def}\, g^{ad} g^{be} g^{cf}$$

$$= \varepsilon \sqrt{|g|}\, \varepsilon^{abc} \det[g^{ab}] = \varepsilon \sqrt{|g|}\, \varepsilon^{abc}\, \frac{1}{g} = \frac{1}{\sqrt{|g|}}\, \varepsilon^{abc}.$$

10.1.3 The cross-product

By means of the volume form we define the cross-product $\boldsymbol{u} \times \boldsymbol{v}$ of two vectors by setting

$$\boldsymbol{u} \times \boldsymbol{v} \cdot \boldsymbol{w} = V(\boldsymbol{u}, \boldsymbol{v}, \boldsymbol{w}).$$

With respect to any basis $(\boldsymbol{e}_a)$ we have

$$\boldsymbol{e}_a \times \boldsymbol{e}_b \cdot \boldsymbol{e}_c = V_{abc}, \quad (\boldsymbol{u} \times \boldsymbol{v})_c = \boldsymbol{u} \times \boldsymbol{v} \cdot \boldsymbol{e}_c = V_{abc}\, u^a\, v^b,$$

and

$$\boldsymbol{u} \times \boldsymbol{v} = V_{abc}\, u^a\, v^b\, \boldsymbol{e}^c = V^{abc}\, u_a\, v_b\, \boldsymbol{e}_c.$$

For the canonical basis, $\boldsymbol{c}_i \times \boldsymbol{c}_j = \varepsilon\, \varepsilon_{ijk}\, \boldsymbol{c}^k$; thus,

$$\begin{cases} \boldsymbol{c}_1 \times \boldsymbol{c}_2 = \boldsymbol{c}_3 = \boldsymbol{t}, \\ \boldsymbol{c}_2 \times \boldsymbol{c}_3 = \varepsilon \boldsymbol{c}_1, \\ \boldsymbol{c}_3 \times \boldsymbol{c}_1 = \varepsilon \boldsymbol{c}_2. \end{cases}$$

The cross product satisfies the following rules,

$$\begin{cases} \boldsymbol{u} \times \boldsymbol{v} = -\boldsymbol{v} \times \boldsymbol{u}, \\ \boldsymbol{u} \times \boldsymbol{v} \cdot \boldsymbol{w} = \boldsymbol{w} \times \boldsymbol{u} \cdot \boldsymbol{v} = \boldsymbol{v} \times \boldsymbol{w} \cdot \boldsymbol{u}, \\ \boldsymbol{u} \times \boldsymbol{v} \cdot \boldsymbol{w} = \boldsymbol{u} \cdot \boldsymbol{v} \times \boldsymbol{w}, \end{cases}$$

whatever ε. For the double cross product we have

$$(\boldsymbol{u} \times \boldsymbol{v}) \times \boldsymbol{w} = \varepsilon\, (\boldsymbol{u} \cdot \boldsymbol{w}\, \boldsymbol{v} - \boldsymbol{v} \cdot \boldsymbol{w}\, \boldsymbol{u}).$$

Indeed,

$$(\boldsymbol{u} \times \boldsymbol{v}) \times \boldsymbol{w} = V_{abc}\, u^a\, v^b\, \boldsymbol{e}^c \times \boldsymbol{w} = V_{abc}\, u^a\, v^b\, V^{cde}\, w_d\, \boldsymbol{e}_c$$

$$= V_{abc}\, V^{dec}\, u^a\, v^b\, w_d\, \boldsymbol{e}_e \varepsilon\, \delta_{ab}^{de}\, u^a\, v^b\, w_d\, \boldsymbol{e}_e$$

$$= \varepsilon\, (u^a\, w_a\, v^b\, \boldsymbol{e}_b - v^b\, w_b\, u^a\, \boldsymbol{e}_a.$$

As a consequence,

$$(\boldsymbol{n} \times \boldsymbol{a}) \cdot (\boldsymbol{n} \times \boldsymbol{b}) = (\boldsymbol{n} \times \boldsymbol{a}) \times \boldsymbol{n} \cdot \boldsymbol{b} = \varepsilon \left(\boldsymbol{n}^2\, \boldsymbol{a} \cdot \boldsymbol{b} - \boldsymbol{n} \cdot \boldsymbol{a}\, \boldsymbol{n} \cdot \boldsymbol{b}\right)$$

and

$$(\boldsymbol{n} \times \boldsymbol{a})^2 = \varepsilon \left(\boldsymbol{n}^2\, \boldsymbol{a}^2 - (\boldsymbol{n} \cdot \boldsymbol{a})^2\right).$$

10.1.4 Rotations

With a vector $\boldsymbol{n}$ such that

$$\gamma = \boldsymbol{n}^2 = \pm 1$$

we associate the linear operators $\mathscr{R}\colon \mathbb{R}^3 \to \mathbb{R}^3$ of the kind

$$\mathscr{R}(\boldsymbol{v}) = \boldsymbol{v} + \alpha\, \boldsymbol{n} \times \boldsymbol{v} + \beta(\boldsymbol{n} \cdot \boldsymbol{v}\, \boldsymbol{n} - \gamma \boldsymbol{v}), \quad \alpha, \beta \in \mathbb{R}.$$

It follows that $\mathscr{R}(\boldsymbol{n}) = \boldsymbol{n}$, and

$$\boldsymbol{v} \cdot \mathscr{R}(\boldsymbol{v}) = (1 - \beta\gamma)\, \boldsymbol{v}^2 + \beta(\boldsymbol{n} \cdot \boldsymbol{v})^2. \tag{10.2}$$

Hence,

$$\begin{aligned}
(\mathscr{R}(\boldsymbol{v}))^2 &= \boldsymbol{v}^2 + \alpha^2 (\boldsymbol{n} \times \boldsymbol{v})^2 + \beta^2 \left(\gamma(\boldsymbol{n} \cdot \boldsymbol{v})^2 + \gamma^2 (\boldsymbol{v})^2 - 2\gamma (\boldsymbol{n} \cdot \boldsymbol{v})^2\right) \\
&\quad + 2\beta \left((\boldsymbol{n} \cdot \boldsymbol{v})^2 - \gamma \boldsymbol{v}^2\right) \\
&= \boldsymbol{v}^2 + \alpha^2 \varepsilon \left(\gamma \boldsymbol{v}^2 - (\boldsymbol{n} \cdot \boldsymbol{v})^2\right) + \beta^2 \left(\gamma^2 \boldsymbol{v}^2 - \gamma(\boldsymbol{n} \cdot \boldsymbol{v})^2\right) \\
&\quad + 2\beta \left((\boldsymbol{n} \cdot \boldsymbol{v})^2 - \gamma \boldsymbol{v}^2\right) \\
&= \left(1 + \gamma(\varepsilon\alpha^2 + \gamma\beta^2 - 2\beta)\right) \boldsymbol{v}^2 + \left(2\beta - \varepsilon\alpha^2 - \gamma\beta^2\right) (\boldsymbol{n} \cdot \boldsymbol{v})^2,
\end{aligned}$$

and $(\mathscr{R}(\boldsymbol{v}))^2 = \boldsymbol{v}^2$ is equivalent to

$$\left(\gamma(\varepsilon\alpha^2 + \gamma\beta^2 - 2\beta)\right) \boldsymbol{v}^2 + \left(2\beta - \varepsilon\alpha^2 - \gamma\beta^2\right) (\boldsymbol{n} \cdot \boldsymbol{v})^2 = 0. \tag{10.3}$$

Let us consider the case $\gamma = \boldsymbol{n}^2 = 1$ and $(\mathscr{R}(\boldsymbol{v}))^2 = \boldsymbol{v}^2$. Then, for each vector $\boldsymbol{v} \perp \boldsymbol{n}$, $\boldsymbol{v}^2 \neq 0$, from (10.2) and (10.3) we obtain

$$\frac{\boldsymbol{v} \cdot \mathscr{R}(\boldsymbol{v})}{\boldsymbol{v}^2} = 1 - \beta \tag{10.4}$$

and

$$\varepsilon\alpha^2 + \beta^2 - 2\beta = 0. \tag{10.5}$$

In the Euclidean metric (i.e., for $\varepsilon = 1$) Eq. (10.5) implies $\alpha^2 \leq 1$, and $|1 - \beta| \leq 1$. Let us set $1 - \beta = \cos\theta$, $\beta = 1 - \cos\theta$. Then (10.5) implies $\alpha^2 = \sin^2\theta$. If we choose $\alpha = \sin\theta$, then we obtain the *Rodrigues formula* for the rotations in the Euclidean three-space,

$$\boxed{\mathscr{R}_{(\boldsymbol{n},\theta)}(\boldsymbol{v}) = \boldsymbol{v} + \sin\theta\, \boldsymbol{n} \times \boldsymbol{v} + (1 - \cos\theta)(\boldsymbol{n} \cdot \boldsymbol{v}\, \boldsymbol{n} - \boldsymbol{v})} \tag{10.6}$$

The choice $\alpha = \sin\theta$ (instead of $\alpha = -\sin\theta$) is in accordance with the conditions

$$\mathscr{R}_{(\boldsymbol{n},\pi/2)}(\boldsymbol{v}) = \boldsymbol{n} \times \boldsymbol{v}, \quad \boldsymbol{v} \perp \boldsymbol{n}.$$

It follows that for all $\boldsymbol{v} \neq 0$ orthogonal to $\boldsymbol{n}$,

$$\boxed{\sin\theta = \frac{\boldsymbol{v} \times \mathscr{R}(\boldsymbol{v})}{\boldsymbol{v}^2} \cdot \boldsymbol{n}, \quad \cos\theta = \frac{\boldsymbol{v} \cdot \mathscr{R}(\boldsymbol{v})}{\boldsymbol{v}^2}} \tag{10.7}$$

Note that θ is the *angle of rotation* (i.e., the angle between $\boldsymbol{v}$ and $\mathscr{R}(\boldsymbol{v})$) for all $\boldsymbol{v} \perp \boldsymbol{n}$. The unit vector $\boldsymbol{n}$ is the *axis of rotation.*

In the Minkowski metric (i.e., for $\varepsilon = -1$) Eq. (10.5) shows that $|1-\beta| \geq 1$ and, because of Eq. (10.4), we put $1 - \beta = \cosh\chi$; that is, $\beta = 1 - \cosh\chi$. This choice corresponds to the assumption that any time-like vector $\boldsymbol{v} \perp \boldsymbol{n}$ is time-equioriented with its image $\mathscr{R}(\boldsymbol{v})$; that is, $\boldsymbol{v} \cdot \mathscr{R}(\boldsymbol{v}) < 0$. In particular, $\chi = 0$ corresponds to $\mathscr{R}(\boldsymbol{v}) = \boldsymbol{v}$. Equation (10.5) implies $\alpha^2 = \sinh^2\chi$. If we choose $\alpha = \sinh\chi$ then we obtain the Rodrigues formula for rotations in the Minkowski three-space with a space-like axis $\boldsymbol{n}$,

$$\mathscr{R}_{(\boldsymbol{n},\chi)}(\boldsymbol{v}) = \boldsymbol{v} + \sinh\chi\, \boldsymbol{n} \times \boldsymbol{v} + (1 - \cosh\chi)(\boldsymbol{n} \cdot \boldsymbol{v}\, \boldsymbol{n} - \boldsymbol{v}).$$

It follows that for all nonlight-like $\boldsymbol{v}$ orthogonal to $\boldsymbol{n}$,

$$\sinh\chi = -\frac{\boldsymbol{v} \times \mathscr{R}(\boldsymbol{v})}{\boldsymbol{v}^2} \cdot \boldsymbol{n}, \quad \cosh\chi = \frac{\boldsymbol{v} \cdot \mathscr{R}(\boldsymbol{v})}{\boldsymbol{v}^2}.$$

The choice $\alpha = \sinh\chi$ (instead of $\alpha = -\sinh\chi$) is in accordance with the condition

$$\mathscr{R}_{(\boldsymbol{c}_1,\chi)}(\boldsymbol{c}_2) = \cosh\chi\, \boldsymbol{c}_2 + \sinh\chi\, \boldsymbol{c}_3.$$

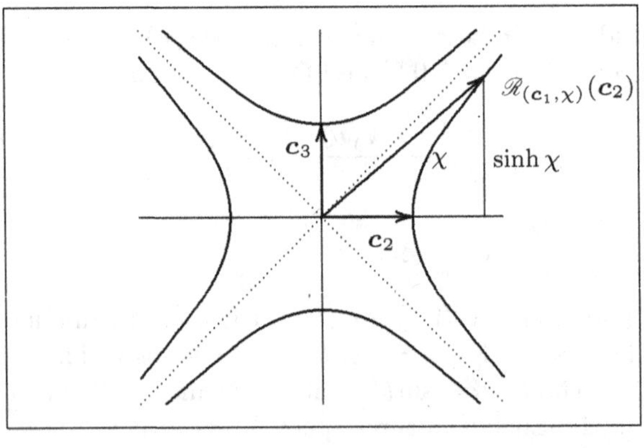

Fig. 10.1 Minkowski metric in the plane $(\boldsymbol{c}_2, \boldsymbol{c}_3)$

10.1.5 Symplectic structure of an orientable surface

Let us consider a surface $M \subset \mathbb{R}^3$ described by a parametric equation

$$\boldsymbol{x} = \boldsymbol{x}(u^1, u^2) = \boldsymbol{x}(u^\alpha).$$

The tangent vectors

$$\boldsymbol{e}_\alpha = \partial_\alpha \boldsymbol{x}$$

are assumed to be pointwise independent, so that they form a *tangent frame.* With this frame we associate the coefficients of the *first fundamental form*

$$A_{\alpha\beta} = \boldsymbol{e}_\alpha \cdot \boldsymbol{e}_\beta.$$

The *dual frame* is defined by

$$\boldsymbol{e}^\alpha = A^{\alpha\beta}\, \boldsymbol{e}_\beta, \quad \boldsymbol{e}^\alpha \cdot \boldsymbol{e}_\beta = \delta^\alpha_\beta.$$

The covariant components of a tangent vector $\boldsymbol{p}$ are

$$p_\alpha = \boldsymbol{p} \cdot \boldsymbol{e}_\alpha.$$

The *Christoffel symbols* and the coefficient of the *second fundamental vector-valued form* are defined by:

$$\partial_\alpha \boldsymbol{e}_\beta = \Gamma^\gamma_{\alpha\beta}\, \boldsymbol{e}_\gamma + \boldsymbol{B}_{\alpha\beta}, \quad \boldsymbol{B}_{\alpha\beta} \cdot \boldsymbol{e}_\gamma = 0.$$

A regular surface $M \subset \mathbb{R}^3$ is orientable if it admits a global orthogonal vector field $\boldsymbol{n} \neq 0$. We assume that $\boldsymbol{n}$ is a unit vector. Then $\boldsymbol{n} \cdot \boldsymbol{n} = \pm 1$, according to the signature of the metric in $\mathbb{R}^3$.

Let us consider the two-form σ on tangent vectors defined by

$$\sigma(\boldsymbol{u}, \boldsymbol{v}) = \boldsymbol{n} \cdot \boldsymbol{u} \times \boldsymbol{v}, \qquad \boldsymbol{u}, \boldsymbol{v} \in TM.$$

This is the *area two-form.* Its integral over a compact subset $U \subseteq M$ gives, by definition, the area of U. By setting

$$\sigma_{\alpha\beta} = \sigma(\boldsymbol{e}_\alpha, \boldsymbol{e}_\beta) = \boldsymbol{n} \cdot \boldsymbol{e}_\alpha \times \boldsymbol{e}_\beta, \tag{10.8}$$

we get

$$\sigma(\boldsymbol{u}, \boldsymbol{v}) = \boldsymbol{n} \cdot \boldsymbol{e}_\alpha \times \boldsymbol{e}_\beta\, u^\alpha\, v^\beta = \sigma_{\alpha\beta}\, u^\alpha v^\beta = \sigma_{12}(u^1 v^2 - u^2 v^1).$$

Since $v^\alpha = \langle \boldsymbol{v}, du^\alpha \rangle = \boldsymbol{v} \cdot \boldsymbol{e}^\alpha$, it follows that

$$\sigma = \tfrac{1}{2}\, \sigma_{\alpha\beta}\, du^\alpha \wedge du^\beta = \sigma_{12}\, du^1 \wedge du^2.$$

The area two-form is nondegenerate, thus it is a symplectic form (a two-form on a two-dimensional surface is obviously closed).

The Hamiltonian vector field $\boldsymbol{X}_f$ associated with a function $f(u^\alpha)$ on the surface is defined by equation

$$X_f^\alpha \, \sigma_{\alpha\beta} = -\partial_\beta f.$$

We observe that

$$\begin{aligned} X_f^\alpha \, \sigma_{\alpha\beta} = -\partial_\beta f \iff & X_f^\alpha \, \boldsymbol{n} \cdot \boldsymbol{e}_\alpha \times \boldsymbol{e}_\beta = -\partial_\beta f \\ \iff & \boldsymbol{n} \cdot \boldsymbol{X}_f \times \boldsymbol{e}_\beta = -\partial_\beta f \\ \iff & \boldsymbol{X}_f \cdot \boldsymbol{e}_\beta \times \boldsymbol{n} = -\partial_\beta f \\ \iff & \boldsymbol{X}_f \cdot \boldsymbol{n} \times \boldsymbol{e}_\beta = \partial_\beta f \\ \iff & \boldsymbol{X}_f \cdot \boldsymbol{n} \times \boldsymbol{e}_\beta \, A^{\beta\alpha} = A^{\beta\alpha} \, \partial_\beta f \\ \iff & \boldsymbol{X}_f \times \boldsymbol{n} \cdot \boldsymbol{e}^\alpha = (\nabla f)^\alpha \\ \iff & \boldsymbol{X}_f \times \boldsymbol{n} = \nabla f, \end{aligned}$$

where ∇ is the gradient operator on the surface. This shows that $\boldsymbol{X}_f$ is defined by the implicit equation $\boldsymbol{X}_f \times \boldsymbol{n} = \nabla f$. Since $\boldsymbol{X}_f$ is tangent to the surface, $\boldsymbol{X}_f \cdot \boldsymbol{n} = 0$. Then we have

$$\nabla f \times \boldsymbol{n} = (\boldsymbol{X}_f \times \boldsymbol{n}) \times \boldsymbol{n} = \varepsilon \, (\boldsymbol{X}_f \cdot \boldsymbol{n} \, \boldsymbol{n} - \boldsymbol{n} \cdot \boldsymbol{n} \, \boldsymbol{X}_f) = \; - \varepsilon \, \boldsymbol{n} \cdot \boldsymbol{n} \, \boldsymbol{X}_f .$$

It follows that

$$\boxed{\boldsymbol{X}_f = \varepsilon \, \nu \, \boldsymbol{n} \times \nabla f, \quad \nu \doteq \boldsymbol{n} \cdot \boldsymbol{n} = \pm 1} \tag{10.9}$$

This gives the explicit definition of $\boldsymbol{X}_f$. The Poisson bracket is defined by

$$\{f, g\}_\sigma = \sigma(\boldsymbol{X}_f, \boldsymbol{X}_g) = \boldsymbol{n} \cdot \boldsymbol{X}_f \times \boldsymbol{X}_g = \boldsymbol{n} \cdot (\boldsymbol{n} \times \nabla f) \times (\boldsymbol{n} \times \nabla g).$$

Thus,

$$\boxed{\{f, g\}_\sigma = \varepsilon \, \nu \, \boldsymbol{n} \cdot (\nabla f \times \nabla g)} \tag{10.10}$$

Let $f(\boldsymbol{x})$ be the restriction to $\boldsymbol{x} \in M$ of a function $\mathscr{F}(\boldsymbol{x})$ on $\mathbb{R}^3$. The gradient ∇f of a function f on a submanifold M of a Riemannian manifold $\mathbb{R}^3$ is simply the orthogonal projection to the tangent space of M of the gradient $\nabla \mathscr{F}$ of any (local) extension $\mathscr{F}$ of f; therefore we have

$$\nabla \mathscr{F}(\boldsymbol{x}) = \nabla f + h(\boldsymbol{x}) \, \boldsymbol{n}.$$

It follows that the equation

$$\boldsymbol{n} \cdot (\nabla f \times \nabla g) = \boldsymbol{n} \cdot (\nabla \mathscr{F} \times \nabla \mathscr{G})$$

holds on the surface. Then from (10.10) we get

$$\boxed{\{f, g\}_\sigma = \varepsilon \boldsymbol{n}^2\, \boldsymbol{n} \cdot (\nabla \mathscr{F} \times \nabla \mathscr{G})} \tag{10.11}$$

This formula gives the Poisson bracket of functions $f(\boldsymbol{x})$ on the surface in terms of local extensions $\mathscr{F}(\boldsymbol{x})$.

10.1.6 The Poisson bracket of functions of a special kind

With any (smooth) function $\mathscr{F}(\boldsymbol{x})$ on $\mathbb{R}^3$ we associate a function $F(\boldsymbol{q}, \boldsymbol{p})$ on $T^*\mathbb{R}^3$ defined by

$$F(\boldsymbol{q}, \boldsymbol{p}) = \mathscr{F}(\boldsymbol{x}), \quad \boldsymbol{x} = \boldsymbol{q} \times \boldsymbol{p}. \tag{10.12}$$

Let us compute the Poisson bracket of functions of this kind. In the canonical basis, we have $\boldsymbol{q} = q^i \boldsymbol{c}_i$, $\boldsymbol{p} = p^i \boldsymbol{c}_i$, and $x^i = V^{ijk}\, q_j\, p_k$. Hence,

$$\begin{cases} \dfrac{\partial x^i}{\partial q^l} = V^{ijk}\, g_{jl}\, p_k, \\ \dfrac{\partial x^i}{\partial p_l} = V^{ijl}\, q_j, \end{cases} \qquad \begin{cases} \dfrac{\partial G}{\partial q^l} = \dfrac{\partial \mathscr{G}}{\partial x^i}\dfrac{\partial x^i}{\partial q^l} = \dfrac{\partial \mathscr{G}}{\partial x^i} V^{ijk}\, g_{jl}\, p_k, \\ \dfrac{\partial F}{\partial p_l} = \dfrac{\partial \mathscr{F}}{\partial x^i}\dfrac{\partial x^i}{\partial p_l} = \dfrac{\partial \mathscr{F}}{\partial x^i} V^{ijl}\, q_j, \end{cases}$$

$$\begin{aligned} \frac{\partial F}{\partial p_l}\frac{\partial G}{\partial q^l} &= \frac{\partial \mathscr{F}}{\partial x^i} V^{ijl}\, q_j \frac{\partial \mathscr{G}}{\partial x^r} V^{rsk}\, g_{sl}\, p_k \frac{\partial \mathscr{F}}{\partial x^i}\frac{\partial \mathscr{G}}{\partial x^r} g^{ih}\, V_{hjs}\, V^{krs}\, q^j\, p_k \\ &= \varepsilon\, \frac{\partial \mathscr{F}}{\partial x^i}\frac{\partial \mathscr{G}}{\partial x^r} g^{ih}\, (\delta_h^k \delta_j^r - \delta_j^k \delta_h^r)\, q^j\, p_k \\ &= \varepsilon\, (\boldsymbol{p} \cdot \nabla \mathscr{F}\, \boldsymbol{q} \cdot \nabla \mathscr{G} - \boldsymbol{q} \cdot \boldsymbol{p}\, \nabla \mathscr{F} \cdot \nabla \mathscr{G}), \end{aligned}$$

and

$$\{F, G\} = \frac{\partial F}{\partial p_l}\frac{\partial G}{\partial q^l} - \frac{\partial G}{\partial p_l}\frac{\partial F}{\partial q^l} = \varepsilon\, (\boldsymbol{p} \cdot \nabla \mathscr{F}\, \boldsymbol{q} \cdot \nabla \mathscr{G} - \boldsymbol{q} \cdot \nabla \mathscr{F}\, \boldsymbol{p} \cdot \nabla \mathscr{G}).$$

Since

$$\begin{aligned} (\boldsymbol{q} \times \boldsymbol{p}) \cdot (\nabla \mathscr{F} \times \nabla \mathscr{G}) &= \boldsymbol{q} \cdot \boldsymbol{p} \times (\nabla \mathscr{F} \times \nabla \mathscr{G}) \\ &= \varepsilon\, \boldsymbol{q} \cdot (\nabla \mathscr{G} \cdot \boldsymbol{p}\, \nabla \mathscr{F} - \nabla \mathscr{F} \cdot \boldsymbol{p}\, \nabla \mathscr{G}) \\ &= \varepsilon\, (\nabla \mathscr{G} \cdot \boldsymbol{p}\, \nabla \mathscr{F} \cdot \boldsymbol{q} - \nabla \mathscr{F} \cdot \boldsymbol{p}\, \nabla \mathscr{G} \cdot \boldsymbol{q}), \end{aligned}$$

we find the formula

$$\boxed{\{F, G\} = -(\boldsymbol{q} \times \boldsymbol{p}) \cdot (\nabla \mathscr{F} \times \nabla \mathscr{G})} \tag{10.13}$$

that gives the Poisson bracket of functions $F(\boldsymbol{q}, \boldsymbol{p})$ on $T^*\mathbb{R}^3$ of the type (10.12).

10.2 The Hamilton principal function on $\mathbb{S}_2$

The basic objects are: (i) the space $\mathbb{R}^3$ endowed with the Euclidean metric; (ii) the configuration manifold $Q = \mathbb{S}_2 = (\boldsymbol{q}) \subset \mathbb{R}^3$, defined by $\boldsymbol{q}^2 = 1$; (iii) the cotangent bundle $T^*Q = T^*\mathbb{S}_2$: it consists of pairs $(\boldsymbol{q}, \boldsymbol{p})$, where $\boldsymbol{p}$ is interpreted as a vector tangent to $\mathbb{S}_2$ at the point $\boldsymbol{q}$, by setting $\langle \boldsymbol{v}, \boldsymbol{p} \rangle = \boldsymbol{v} \cdot \boldsymbol{p}$ for each vector $\boldsymbol{v}$ tangent to $\mathbb{S}_2$ at $\boldsymbol{q}$; and (iv) the eikonal equation (coisotropic submanifold) $C \subset T^*Q$ defined by equation $\boldsymbol{p}^2 = 1$.

The oriented geodesics on $\mathbb{S}_2$ are in one-to-one correspondence with the unit vectors $\boldsymbol{n}$. The geodesic corresponding to $\boldsymbol{n}$ is the intersection of $Q = \mathbb{S}_2$ with the plane $\Pi_{\boldsymbol{n}}$ orthogonal to $\boldsymbol{n}$ and passing through the origin. The orientation of this maximal circle is determined by the formula $\boldsymbol{p} = \boldsymbol{n} \times \boldsymbol{q}$, equivalent to $\boldsymbol{n} = \boldsymbol{q} \times \boldsymbol{p}$, where $\boldsymbol{q} \in Q$ and $\boldsymbol{p}$ is the unit vector tangent to the oriented circle. But the oriented geodesics are in one-to-one correspondence with the characteristics of C, thus we have the following.

Theorem 10.1. *The set M of the characteristics of C is a manifold diffeomorphic to the unit sphere $\mathbb{S}_2$. The one-to-one correspondence between characteristics $\gamma(\boldsymbol{n})$ of C and the unit vectors $\boldsymbol{n} \in \mathbb{S}_2$ is given by*

$$(\boldsymbol{q}, \boldsymbol{p}) \in \gamma(\boldsymbol{n}) \iff \begin{cases} \boldsymbol{q}^2 = 1 & (\boldsymbol{q} \in Q = \mathbb{S}_2), \\ \boldsymbol{p}^2 = 1 & (\boldsymbol{p} \in C), \\ \boldsymbol{q} \cdot \boldsymbol{p} = 0 & (\boldsymbol{p} \in T^*Q), \\ \boldsymbol{n} = \boldsymbol{q} \times \boldsymbol{p}. \end{cases} \tag{10.14}$$

It follows that two pairs $(\boldsymbol{q}_0, \boldsymbol{p}_0)$ and $(\boldsymbol{q}_1, \boldsymbol{p}_1)$ of T^*Q belong to the same characteristic $\gamma(\boldsymbol{n})$ if and only if the above equations are satisfied with $\boldsymbol{n} = \boldsymbol{q}_0 \times \boldsymbol{p}_0 = \boldsymbol{q}_1 \times \boldsymbol{p}_1$ or equivalently, if and only if $\boldsymbol{p}_0 = \boldsymbol{n} \times \boldsymbol{q}_0$ and $\boldsymbol{p}_1 = \boldsymbol{n} \times \boldsymbol{q}_1$. Since $\boldsymbol{q}_0$ and $\boldsymbol{q}_1$ are both orthogonal to $\boldsymbol{n}$, we can consider the rotation with axis $\boldsymbol{n}$ that maps $\boldsymbol{q}_0$ to $\boldsymbol{q}_1$. The axis $\boldsymbol{n}$ is determined (even in the case $\boldsymbol{q}_0 = \boldsymbol{q}_1$) by setting $\boldsymbol{n} = \boldsymbol{q}_0 \times \boldsymbol{p}_0$ (or $\boldsymbol{n} = \boldsymbol{q}_1 \times \boldsymbol{p}_1$). This proves the following.

Theorem 10.2. *A pair $((\boldsymbol{q}_0, \boldsymbol{p}_0), (\boldsymbol{q}_1, \boldsymbol{p}_1))$ belongs to the characteristic relation D_C if and only if $\boldsymbol{q}_0 \in \mathbb{S}_2$ and there exists a pair $(\boldsymbol{n}, \theta) \in \mathbb{S}_2 \times \mathbb{R}$ such that*

$$\begin{cases} \boldsymbol{q}_1 = \mathscr{R}_{(\boldsymbol{n},\theta)}(\boldsymbol{q}_0), \\ \boldsymbol{p}_1 = \mathscr{R}_{(\boldsymbol{n},\theta)}(\boldsymbol{p}_0), \\ \boldsymbol{q}_0 \cdot \boldsymbol{n} = \boldsymbol{q}_1 \cdot \boldsymbol{n} = 0, \\ \boldsymbol{p}_1 = \boldsymbol{n} \times \boldsymbol{q}_1, \end{cases} \tag{10.15}$$

where $\mathscr{R}_{(\boldsymbol{n},\theta)}$ is the rotation of axis $\boldsymbol{n}$ and angle θ,

$$\mathscr{R}_{(\boldsymbol{n},\theta)}(\boldsymbol{v}) = \boldsymbol{v} + \sin\theta\, \boldsymbol{n} \times \boldsymbol{v} + (1 - \cos\theta)(\boldsymbol{n} \cdot \boldsymbol{v}\, \boldsymbol{n} - \boldsymbol{v}).$$

Then we can prove the following.

Theorem 10.3. *The characteristic relation D_C is generated by the family*

$$S \colon (\mathbb{S}_2 \times \mathbb{S}_2; \mathbb{R} \times \mathbb{R} \times \mathbb{S}_2 \times \mathbb{R}^3) \longrightarrow \mathbb{R},$$

defined by

$$\begin{aligned} S(\boldsymbol{q}_1, \boldsymbol{q}_0; \lambda, \theta, \boldsymbol{n}, \boldsymbol{v}) = \theta + \lambda\ \left((\boldsymbol{q}_0 \cdot \boldsymbol{n})^2 + (\boldsymbol{q}_1 \cdot \boldsymbol{n})^2 + (\boldsymbol{v} - \boldsymbol{n} \times \boldsymbol{q}_1)^2\right) \\ + \boldsymbol{v} \cdot \left(\boldsymbol{q}_1 - \mathscr{R}_{(\boldsymbol{n},\theta)}(\boldsymbol{q}_0)\right). \end{aligned} \tag{10.16}$$

The critical set Ξ of S is described by equations

$$\boldsymbol{q}_0 \cdot \boldsymbol{n} = \boldsymbol{q}_1 \cdot \boldsymbol{n} = 0, \quad \boldsymbol{v} = \boldsymbol{n} \times \boldsymbol{q}_1, \quad \boldsymbol{q}_1 = \mathscr{R}(\boldsymbol{q}_0). \tag{10.17}$$

This generating family is not a Morse family.

Proof. The equations generated by S are

$$\begin{cases} 0 = \dfrac{\partial S}{\partial \lambda}, \\ 0 = \dfrac{\partial S}{\partial \boldsymbol{v}}, \end{cases} \quad \begin{cases} 0 = \dfrac{\partial S}{\partial \theta}, \\ 0 = \dfrac{\partial S}{\partial \boldsymbol{n}}, \end{cases} \quad \begin{cases} \boldsymbol{p}_0 = -\dfrac{\partial S}{\partial \boldsymbol{q}_0}, \\ \boldsymbol{p}_1 = \dfrac{\partial S}{\partial \boldsymbol{q}_1}. \end{cases} \tag{10.18}$$

The first four equations describe the critical set Ξ. For all $\boldsymbol{n} \in \mathbb{S}_2$ we have

$$\frac{\partial S}{\partial \boldsymbol{n}} = \nabla_{\boldsymbol{x}} S\big|_{\boldsymbol{x}=\boldsymbol{n}} (\boldsymbol{I} - \boldsymbol{n} \otimes \boldsymbol{n}) = P_{\boldsymbol{n}}(\nabla_{\boldsymbol{x}} S\big|_{\boldsymbol{x}=\boldsymbol{n}}),$$

where

$$P_{\boldsymbol{n}} = \boldsymbol{I} - \boldsymbol{n} \otimes \boldsymbol{n}.$$

Similar equations hold for $\boldsymbol{q}_0$ and $\boldsymbol{q}_1$. The first two equations (10.18) of the critical set read

$$0 = \frac{\partial S}{\partial \lambda} = (\boldsymbol{q}_0 \cdot \boldsymbol{n})^2 + (\boldsymbol{q}_1 \cdot \boldsymbol{n})^2 + (\boldsymbol{v} - \boldsymbol{n} \times \boldsymbol{q}_1)^2$$

and

$$0 = \frac{\partial S}{\partial \boldsymbol{v}} = 2\lambda(\boldsymbol{v} - \boldsymbol{n} \times \boldsymbol{q}_1) + \boldsymbol{q}_1 - \mathscr{R}(\boldsymbol{q}_0).$$

These are equivalent to Eqs. (10.17). We show below that the remaining Eqs. (10.18) of Ξ are identically satisfied. Due to Eqs. (10.17), on the critical set we have

$$\begin{aligned}\boldsymbol{p}_0 = -\frac{\partial S}{\partial \boldsymbol{q}_0} &= \frac{\partial}{\partial \boldsymbol{q}_0}(\boldsymbol{v} \cdot \mathscr{R}(\boldsymbol{q}_0)) = \frac{\partial}{\partial \boldsymbol{q}_0}(\boldsymbol{q}_0 \cdot \mathscr{R}^\top(\boldsymbol{v})) \\ &= \mathscr{R}^\top(\boldsymbol{v})(\boldsymbol{I} - \boldsymbol{q}_0 \otimes \boldsymbol{q}_0) = \mathscr{R}^\top(\boldsymbol{v}) - \mathscr{R}^\top(\boldsymbol{v}) \cdot \boldsymbol{q}_0\, \boldsymbol{q}_0 \\ &= \mathscr{R}^\top(\boldsymbol{v}) - \boldsymbol{v} \cdot \mathscr{R}(\boldsymbol{q}_0) = \mathscr{R}^\top(\boldsymbol{v}) - \boldsymbol{v} \cdot \boldsymbol{q}_1 \\ &= \mathscr{R}^\top(\boldsymbol{v}) - \boldsymbol{n} \times \boldsymbol{q}_1 \cdot \boldsymbol{q}_1 = \mathscr{R}^\top(\boldsymbol{v}),\end{aligned}$$

and

$$\boldsymbol{p}_1 = -\frac{\partial S}{\partial \boldsymbol{q}_1} = \boldsymbol{v}\,(\boldsymbol{I} - \boldsymbol{q}_1 \otimes \boldsymbol{q}_1) = \boldsymbol{v}.$$

Thus, because $\mathscr{R}^\top = \mathscr{R}^{-1}$,

$$\boldsymbol{p}_0 = \mathscr{R}^{-1}(\boldsymbol{p}_1).$$

All Eqs. (10.15) have been found. We show that the last two equations (10.18) of the critical set are identically satisfied. On the critical set the vector $\boldsymbol{q}_0 \times \boldsymbol{v}$ is parallel to $\boldsymbol{n}$,

$$\boldsymbol{q}_0 \times \boldsymbol{v} = \boldsymbol{q}_0 \times (\boldsymbol{n} \times \boldsymbol{q}_1) = \boldsymbol{q}_0 \cdot \boldsymbol{q}_1\, \boldsymbol{n} - \boldsymbol{q}_0 \cdot \boldsymbol{n}\, \boldsymbol{q}_1 = \boldsymbol{q}_0 \cdot \boldsymbol{q}_1\, \boldsymbol{n}. \tag{10.19}$$

so that

$$\begin{aligned}\frac{\partial S}{\partial \boldsymbol{n}} &= \frac{\partial}{\partial \boldsymbol{n}}(\boldsymbol{v} \cdot \mathscr{R}(\boldsymbol{q}_0)) \\ &= \sin\theta\, \frac{\partial}{\partial \boldsymbol{n}}(\boldsymbol{n} \cdot \boldsymbol{q}_0 \times \boldsymbol{v}) + (1 - \cos\theta)\, \frac{\partial}{\partial \boldsymbol{n}}(\boldsymbol{n} \cdot \boldsymbol{q}_0\, \boldsymbol{n} \cdot \boldsymbol{v}) \\ &= \sin\theta\, P_{\boldsymbol{n}}(\boldsymbol{q}_0 \times \boldsymbol{v}) = 0.\end{aligned}$$

Finally, because on the critical set $\boldsymbol{n} \cdot \boldsymbol{v} = 0$, we have

$$\begin{aligned}\frac{\partial S}{\partial \theta} &= 1 - \boldsymbol{v} \cdot (\cos\theta\, \boldsymbol{n} \times \boldsymbol{q}_0 + \sin\theta\, (\boldsymbol{n} \cdot \boldsymbol{q}_0\, \boldsymbol{n} - \boldsymbol{q}_0)) \\ &= 1 - \cos\theta\, \boldsymbol{v} \cdot \boldsymbol{n} \times \boldsymbol{q}_0 + \sin\theta\, \boldsymbol{v} \cdot \boldsymbol{q}_0,\end{aligned} \tag{10.20}$$

and, because of the second equation (10.7),

$$\begin{aligned} \boldsymbol{v} \cdot \boldsymbol{n} \times \boldsymbol{q}_0 &= (\boldsymbol{n} \times \mathscr{R}(\boldsymbol{q}_0)) \cdot (\boldsymbol{n} \times \boldsymbol{q}_0) \\ &= \mathscr{R}(\boldsymbol{q}_0) \cdot \boldsymbol{q}_0 = \boldsymbol{q}_0^2 \cos\theta \\ &= \cos\theta, \end{aligned} \tag{10.21}$$

Moreover, because of the first equation (10.7),

$$\boldsymbol{v} \cdot \boldsymbol{q}_0 = \boldsymbol{n} \times \mathscr{R}(\boldsymbol{q}_0) \cdot \boldsymbol{q}_0 = \boldsymbol{n} \cdot \mathscr{R}(\boldsymbol{q}_0) \times \boldsymbol{q}_0 = -\boldsymbol{q}_0^2 \sin\theta = -\sin\theta.$$

Thus,

$$\frac{\partial S}{\partial \theta} = 1 - \cos^2\theta - \sin^2\theta = 0.$$

Because

$$S_\lambda = \frac{\partial S}{\partial \lambda} = (\boldsymbol{q}_0 \cdot \boldsymbol{n})^2 + (\boldsymbol{q}_1 \cdot \boldsymbol{n})^2 + (\boldsymbol{v} - \boldsymbol{n} \times \boldsymbol{q}_1)^2,$$

on the critical set we have $dS_\lambda = 0$. This shows that S is not a Morse family. □

Theorem 10.4. *An equivalent reduced generating family of D_C is the function*

$$S' \colon (\mathbb{S}_2 \times \mathbb{S}_2; \mathbb{R} \times \mathbb{R} \times \mathbb{S}_2) \longrightarrow \mathbb{R}$$

defined by

$$S'(\boldsymbol{q}_1, \boldsymbol{q}_0; \lambda, \theta, \boldsymbol{n}) = \theta + \lambda \left((\boldsymbol{q}_0 \cdot \boldsymbol{n})^2 + (\boldsymbol{q}_1 \cdot \boldsymbol{n})^2 \right) - \boldsymbol{n} \times \boldsymbol{q}_1 \cdot \mathscr{R}_{(\boldsymbol{n},\theta)}(\boldsymbol{q}_0).$$

Proof. By means of equations $\boldsymbol{v} = \boldsymbol{n} \times \boldsymbol{q}_1$ of the critical set we can remove the extra variable $\boldsymbol{v}$ of S. Thus, we get the reduced generating family S'. □

Remark 10.1. On the critical set the generating family is reducible to $S = \theta$. This function is obviously symmetric in $(\boldsymbol{q}_0, \boldsymbol{q}_1)$, in accordance with the symmetry of the characteristic relation. Also the reduced generating family S' is not a Morse family. ◇

Remark 10.2. We consider the inclusion relation

$$R \subset \mathbb{S}_2 \times \mathbb{R}^3 = \{(\boldsymbol{q}, \boldsymbol{x}) \text{ such that } \boldsymbol{q} = \boldsymbol{x}\}.$$

The canonical lift $\widehat{R} \subset T^*\mathbb{S}_2 \times T^*\mathbb{R}^3$ is a symplectic reduction whose inverse image $\widehat{R}^\top \circ (T^*\mathbb{S}_2)$ is the coisotropic submanifold $T^*_{\mathbb{S}_2}\mathbb{R}^3$ of the covectors $\boldsymbol{p} \in \mathbb{R}^3$ based at points of $\mathbb{S}_2$. The fibers of this reduction are the equivalence classes of the equivalence relation

$$\boldsymbol{p} \sim \boldsymbol{p}' \iff \boldsymbol{p}, \boldsymbol{p}' \text{ based at the same point } \boldsymbol{q} \in \mathbb{S}_2, \ \ \boldsymbol{p} - \boldsymbol{p}' \perp \mathbb{S}_2,$$

that is,

$$
\boldsymbol{p} \sim \boldsymbol{p}' \iff \begin{cases} \boldsymbol{p}, \boldsymbol{p}' \text{ based at the same point } \boldsymbol{q} \in \mathbb{S}_2 \\ \text{and} \\ (\boldsymbol{p} - \boldsymbol{p}') \times \boldsymbol{q} = 0. \end{cases} \tag{10.22}
$$

A second symplectic reduction we have to consider is the characteristic reduction associated with the coisotropic submanifold C,

$$
R_C \colon M \leftarrow T^*Q.
$$

This is the graph of the surjective submersion which maps a point of C to the characteristic that contains this point. This reduction defines a reduced symplectic form ω on the space M of the characteristics. ♢

Theorem 10.5. *The reduced symplectic form ω is the opposite of the standard symplectic form σ on $\mathbb{S}_2$ defined by*

$$
\sigma(\boldsymbol{u}, \boldsymbol{v}) = \boldsymbol{n} \cdot \boldsymbol{u} \times \boldsymbol{v}, \tag{10.23}
$$

where $\boldsymbol{n} \in \mathbb{S}_2$ and $(\boldsymbol{u}, \boldsymbol{v})$ are vectors tangent to $\mathbb{S}_2$ at the point $\boldsymbol{n}$.

Proof. (I) With any arbitrary (smooth) function $\mathscr{F}(\boldsymbol{x})$ on $\mathbb{R}^3$ we associate a function $F(\boldsymbol{q}, \boldsymbol{p})$ on $T^*\mathbb{R}^3$,

$$
F(\boldsymbol{q}, \boldsymbol{p}) = \mathscr{F}(\boldsymbol{q} \times \boldsymbol{p}), \tag{10.24}
$$

and a function $f(\boldsymbol{n})$ on $M = \mathbb{S}_2$,

$$
f(\boldsymbol{n}) = \mathscr{F}(\boldsymbol{n}) \tag{10.25}
$$

(this is simply the restriction of $\mathscr{F}$ to the sphere).

(II) The functions F on $T^*\mathbb{R}^3$ are constant on the fibers of the first reduction $\widehat{R}$, because on a fiber we have $\boldsymbol{q} \times \boldsymbol{p} = \boldsymbol{q} \times \boldsymbol{p}'$ and $\boldsymbol{q} \in \mathbb{S}_2$; see (10.22). Hence, these functions are reducible to functions on $T^*\mathbb{S}_2$ by taking $\boldsymbol{q}^2 = 1$ and $\boldsymbol{p} \cdot \boldsymbol{q} = 0$.

(III) When restricted to the submanifold C, by taking $\boldsymbol{p}^2 = 1$, a function $F(\boldsymbol{q}, \boldsymbol{p})$ of the kind (10.24) is constant on each characteristic $\gamma(\boldsymbol{n})$ because of (10.14), so that it is reducible to a function $f(\boldsymbol{n}) = \mathscr{F}(\boldsymbol{n})$, with $\boldsymbol{n} = \boldsymbol{q} \times \boldsymbol{p}$ (note that $\boldsymbol{n}^2 = 1$, because $\boldsymbol{q}$ and $\boldsymbol{p}$ are orthogonal unit vectors).

(IV) Let us consider the standard symplectic form (10.23) on the sphere $M = \mathbb{S}_2$. By formula (10.10) we get the Poisson bracket

$$
\{f, g\}_\sigma = \boldsymbol{n} \cdot (\nabla f \times \nabla g).
$$

(V) We recall that, according to the general theory of the symplectic reductions, the Poisson bracket of two functions $f(\boldsymbol{n})$ on the reduced symplectic manifold (M, ω) is defined by

$$\{f, g\}_\omega(\boldsymbol{n}) = \{F, G\}(\boldsymbol{q}, \boldsymbol{p}), \tag{10.26}$$

where $(\boldsymbol{q}, \boldsymbol{p}) \in \gamma(\boldsymbol{n}) \subset C \subset T^*\mathbb{S}_2$ and where F and G are any two functions on $T^*\mathbb{R}^3$ constant on the characteristics γ of C. Because of (III) the functions F and f defined by (10.24) and (10.25) by means of any function $\mathscr{F}$ fit with this scheme. Hence, $(\boldsymbol{q}, \boldsymbol{p}) \in \gamma(\boldsymbol{n})$ means in particular that $\boldsymbol{n} = \boldsymbol{q} \times \boldsymbol{p}$; see (10.14). Then, by applying formulae (10.13) and (10.11) for $\varepsilon = 1$, which now reads $\{f, g\}_\sigma = \boldsymbol{n} \cdot \nabla\mathscr{F} \times \nabla\mathscr{G}$, we get

$$\begin{aligned}\{f, g\}_\omega(\boldsymbol{n}) = \{F, G\}(\boldsymbol{q}, \boldsymbol{p}) &= -\boldsymbol{q} \times \boldsymbol{p} \cdot (\nabla\mathscr{F} \times \nabla\mathscr{G}) \\ &= -\boldsymbol{n} \cdot \nabla\mathscr{F} \times \nabla\mathscr{G} = -\{f, g\}_\sigma(\boldsymbol{n}).\end{aligned} \tag{10.27}$$

This holds for all functions $\mathscr{F}(\boldsymbol{x})$. We remark that any function $f(\boldsymbol{n})$ on the sphere M admits an extension $\mathscr{F}(\boldsymbol{x})$ to $\mathbb{R}^3$, such that $\mathscr{F}(\boldsymbol{n}) = f(\boldsymbol{n})$. We can, for instance, extend the function f by constant values along the half-lines issued from the origin of $\mathbb{R}^3$ (the origin must be excluded, but this exclusion is irrelevant). Thus, the equality

$$\{f, g\}_\sigma(\boldsymbol{n}) = -\{f, g\}_\omega(\boldsymbol{n})$$

holds for all f and g on M. This shows that $\omega = -\sigma$. □

Remark 10.3. In Eq. (10.23) defining σ the normal vector $\boldsymbol{n}$ is oriented outside the sphere. If we choose it pointing to the center, then we get $\sigma = \omega$. ◇

10.3 The Hamilton principal function on $\mathbb{H}_2$

The basic objects are: (i) the space $\mathbb{R}^3$ endowed with the Minkowskian metric, with z time-like; (ii) the configuration manifold $Q = \mathbb{H}_2 = (\boldsymbol{q}) \subset \mathbb{R}^3$, defined by

$$\begin{cases} \boldsymbol{q}^2 = -1, \\ \boldsymbol{q} \cdot \boldsymbol{c}_3 < 0, \end{cases} \iff \begin{cases} x^2 + y^2 - z^2 + 1 = 0, \\ z > 0. \end{cases}$$

The Minkowskian metric induces on $\mathbb{H}_2$ a positive-definite metric and $\boldsymbol{q} \in \mathbb{H}_2$ implies $\boldsymbol{q} \perp \mathbb{H}_2$. Indeed, for any curve $\boldsymbol{q}(t) \in \mathbb{H}_2$ we have $\dot{\boldsymbol{q}} \cdot \boldsymbol{q} = 0$ thus, $\dot{\boldsymbol{q}}$ (tangent to $\mathbb{H}_2$) is space-like (every nonzero-vector orthogonal to a time-like vector is space-like). (iii) The cotangent bundle $T^*Q = T^*\mathbb{H}_2 = (\boldsymbol{q}, \boldsymbol{p})$, where $\boldsymbol{p}$ is a vector tangent to $\mathbb{H}_2$ at $\boldsymbol{q}$. (iv) The eikonal equation (coisotropic submanifold) $C \subset T^*Q$ defined by equation $\boldsymbol{p}^2 = 1$. Note that the covectors $(\boldsymbol{q}, \boldsymbol{p}) \in T^*\mathbb{H}_2$ can be interpreted as vectors $\boldsymbol{p}$ tangent to $\mathbb{H}_2$ by setting $\langle \boldsymbol{v}, \boldsymbol{p} \rangle = \boldsymbol{v} \cdot \boldsymbol{p}$ for each vector $\boldsymbol{v}$ tangent to $\mathbb{H}_2$ at the point $\boldsymbol{q}$. (v) The hyperboloid $\mathbb{K}_2$ of the unit space-like vectors $\boldsymbol{n}$, $\boldsymbol{n}^2 = 1$. The metric induced on $\mathbb{K}_2$ is Lorentzian.

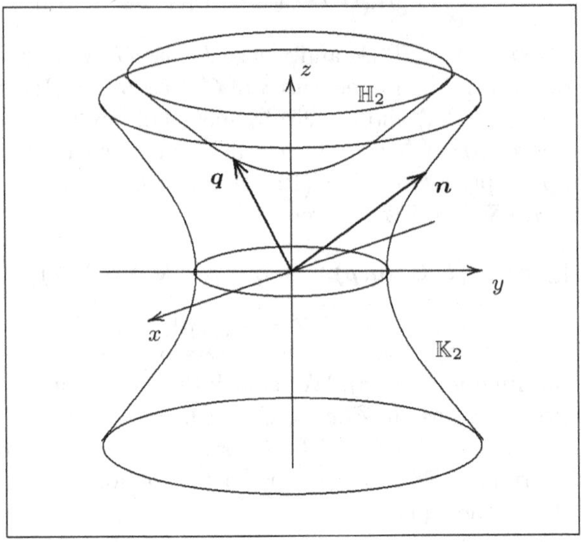

Fig. 10.2 The hyperboloids $\mathbb{H}_2$ and $\mathbb{K}_2$ in the Minkowski three-space

Theorem 10.6. *The set M of the characteristics of C is a manifold diffeomorphic to $\mathbb{K}_2$. The one-to-one correspondence between characteristics $\gamma(\boldsymbol{n})$ of C and the unit vectors $\boldsymbol{n} \in M$ is given by*

$$(\boldsymbol{q}, \boldsymbol{p}) \in \gamma(\boldsymbol{n}) \iff \begin{cases} \boldsymbol{q}^2 = 1 \;\; (\boldsymbol{q} \in Q = \mathbb{H}_2), \\ \boldsymbol{p}^2 = 1 \;\; (\boldsymbol{p} \in C), \\ \boldsymbol{q} \cdot \boldsymbol{p} = 0 \;\; (\boldsymbol{p} \in T^*Q), \\ \boldsymbol{n} = \boldsymbol{q} \times \boldsymbol{p}. \end{cases} \tag{10.28}$$

Proof. The geodesics of Q are the orbits of spontaneous motions (no active force). These motions admit the first integral

$$\boldsymbol{n} = \boldsymbol{q} \times \dot{\boldsymbol{q}}. \tag{10.29}$$

Indeed,

$$\dot{\boldsymbol{n}} = \boldsymbol{q} \times \dot{\boldsymbol{q}} = \boldsymbol{q} \times \boldsymbol{R},$$

where $\boldsymbol{R}$ is the reaction force orthogonal to $\mathbb{H}_2$. Since also $\boldsymbol{q}(t)$ is orthogonal to $\mathbb{H}_2$, it follows that $\dot{\boldsymbol{n}} = 0$. From (10.29) it follows that: (i) $\dot{\boldsymbol{q}} \perp \boldsymbol{n}$, so for any fixed $\boldsymbol{n}$, the corresponding geodesic has velocity $\dot{\boldsymbol{q}}$ orthogonal to $\boldsymbol{n}$; (ii) $\boldsymbol{n}$ is space-like because it is orthogonal to the time-like vector $\boldsymbol{q}$ ($\boldsymbol{n} \neq 0$, inasmuch as $\dot{\boldsymbol{q}}$ cannot be parallel to $\boldsymbol{q}$ unless $\dot{\boldsymbol{q}} = 0$. We can consider only geodesic motions with unit velocity, $\dot{\boldsymbol{q}}^2 = 1$. Due to (10.29), this is equivalent to assuming $\boldsymbol{n}^2 = 1$; that is, $\boldsymbol{n} \in \mathbb{K}_2$. The characteristics are in one-to-one correspondence with the oriented geodesics, thus the set M is identified with $\mathbb{K}_2$ and equations (10.28) follow by replacing $\dot{\boldsymbol{q}}$ with $\boldsymbol{p}$. □

As a corollary of Theorem 10.6 we have the following.

Theorem 10.7. *A pair $((\boldsymbol{q}_0, \boldsymbol{p}_0), (\boldsymbol{q}_1, \boldsymbol{p}_1))$ belongs to the characteristic relation D_C if and only if $\boldsymbol{q}_0 \in \mathbb{H}_2$ and there exists a pair $(\boldsymbol{n}, \chi) \in \mathbb{K}_2 \times \mathbb{R}$ such that*

$$\begin{cases} \boldsymbol{q}_1 = \mathscr{R}_{(\boldsymbol{n},\chi)}(\boldsymbol{q}_0), \\ \boldsymbol{q}_0 \cdot \boldsymbol{n} = \boldsymbol{q}_1 \cdot \boldsymbol{n} = 0, \end{cases} \qquad \begin{cases} \boldsymbol{p}_1 = \mathscr{R}_{(\boldsymbol{n},\chi)}(\boldsymbol{p}_0), \\ \boldsymbol{p}_1 = \boldsymbol{n} \times \boldsymbol{q}_1, \end{cases}$$

where $\mathscr{R}_{(\boldsymbol{n},\chi)}$ is the rotation of axis $\boldsymbol{n}$ and pseudo-angle χ,

$$\mathscr{R}_{(\boldsymbol{n},\chi)}(\boldsymbol{v}) = \boldsymbol{v} + \sinh\chi\, \boldsymbol{n} \times \boldsymbol{v} + (1 - \cosh\chi)(\boldsymbol{n} \cdot \boldsymbol{v}\, \boldsymbol{n} - \boldsymbol{v}).$$

Then we can prove the following.

Theorem 10.8. *The characteristic relation D_C is generated by the family*

$$S\colon (\mathbb{H}_2 \times \mathbb{H}_2\, ; \mathbb{R} \times \mathbb{R} \times \mathbb{K}_2 \times \mathbb{R}^3) \longrightarrow \mathbb{R},$$

defined by

$$\begin{aligned} S(\boldsymbol{q}_0, \boldsymbol{q}_1; \lambda, \chi, \boldsymbol{n}, \boldsymbol{v}) = \chi + \lambda\ &\left((\boldsymbol{q}_0 \cdot \boldsymbol{n})^2 + (\boldsymbol{q}_1 \cdot \boldsymbol{n})^2 + (\boldsymbol{v} - \boldsymbol{n} \times \boldsymbol{q}_1)^2\right) \\ &+ \boldsymbol{v} \cdot \left(\boldsymbol{q}_1 - \mathscr{R}_{(\boldsymbol{n},\chi)}(\boldsymbol{q}_0)\right) \end{aligned} \tag{10.30}$$

The critical set is described by equations

$$\boldsymbol{q}_0 \cdot \boldsymbol{n} = \boldsymbol{q}_1 \cdot \boldsymbol{n} = 0, \quad \boldsymbol{v} = \boldsymbol{n} \times \boldsymbol{q}_1, \quad \boldsymbol{q}_1 = \mathscr{R}(\boldsymbol{q}_0).$$

This generating family is not a Morse family.

Note that in (10.30) the by the symbol $(\boldsymbol{u})^2_+$ we mean the scalar product $\boldsymbol{u} \cdot \boldsymbol{u}$ in the Euclidean metric.

Proof. The proof follows the same pattern of that concerning $\mathbb{S}_2$, with the following variants. (i) In (10.19) we used the double cross-product formula, thus the second and the last terms should be multiplied by ε; in fact, this has no consequence and we again get $\partial S/\partial \boldsymbol{n} = 0$; (ii) formula (10.20) is replaced by a similar formula with $\cosh\chi$ and $-\sinh\chi$ instead of $\cos\theta$ and $\sin\theta$,

$$\frac{\partial S}{\partial \chi} = 1 - \cosh\chi\, \boldsymbol{v} \cdot \boldsymbol{n} \times \boldsymbol{q}_0 - \sinh\chi\, \boldsymbol{v} \cdot \boldsymbol{q}_0;$$

(iii) formula (10.21) involves the scalar product of two cross products, so that the second equality is multiplied by $\varepsilon = -1$; we get

$$\boldsymbol{v} \cdot \boldsymbol{n} \times \boldsymbol{q}_0 = -\mathscr{R}(\boldsymbol{q}_0) \cdot \boldsymbol{q}_0 = -\boldsymbol{q}_0^2 \cosh\chi = \cosh\chi,$$

with $\boldsymbol{v} = \boldsymbol{n} \times \boldsymbol{q}_1 = \boldsymbol{n} \times \mathscr{R}(\boldsymbol{q}_0)$. In the present case

$$\boldsymbol{v} \cdot \boldsymbol{q}_0 = \boldsymbol{n} \cdot \mathscr{R}(\boldsymbol{q}_0) \times \boldsymbol{q}_0 = \boldsymbol{q}_0^2 \sinh \chi = -\sinh \chi.$$

As a consequence,

$$\frac{\partial S}{\partial \chi} = 1 - \cosh^2 \chi + \sinh^2 \chi = 0.$$

□

We have theorems similar to Theorems 10.4 and 10.5 for the sphere.

Theorem 10.9. *An equivalent reduced generating family is*

$$S' : (\mathbb{H}_2 \times \mathbb{H}_2 \,;\, \mathbb{R} \times \mathbb{R} \times \mathbb{K}_2) \longrightarrow \mathbb{R},$$

defined by

$$\begin{aligned} &S'(\boldsymbol{q}_0, \boldsymbol{q}_1; \lambda, \chi, \boldsymbol{n}) = \\ &= \chi + \lambda \left[(\boldsymbol{q}_0 \cdot \boldsymbol{n})^2 + (\boldsymbol{q}_1 \cdot \boldsymbol{n})^2 \right] - \boldsymbol{n} \times \boldsymbol{q}_1 \cdot \mathscr{R}_{(\boldsymbol{n},\chi)}(\boldsymbol{q}_0). \end{aligned} \tag{10.31}$$

The proof is similar to that of $\mathbb{S}_2$.

Theorem 10.10. *The reduced symplectic manifold is* $(\mathbb{K}_2, \omega)$, *where the reduced symplectic form* ω *coincides with the standard symplectic form* σ *of* $\mathbb{K}_2$ *defined by*

$$\sigma(\boldsymbol{u}, \boldsymbol{v}) = \boldsymbol{n} \cdot \boldsymbol{u} \times \boldsymbol{v}.$$

Proof. The proof is similar, *mutatis mutandis*, to that of $\mathbb{S}_2$ till Eq. (10.27), which now gives

$$\begin{aligned} \{f, g\}_\omega(\boldsymbol{n}) = \{F, G\}(\boldsymbol{q}, \boldsymbol{p}) &= -\, \boldsymbol{q} \times \boldsymbol{p} \cdot (\nabla \mathscr{F} \times \nabla \mathscr{G}) \\ &= -\, \boldsymbol{n} \cdot \nabla \mathscr{F} \times \nabla \mathscr{G} \{f, g\}_\sigma(\boldsymbol{n}), \end{aligned}$$

being, for $\varepsilon = -1$, $\{f, g\}_\sigma = -\, \boldsymbol{n} \cdot \nabla \mathscr{F} \times \nabla \mathscr{G}$. □

Note that in the case of the eikonal equation of $\mathbb{H}_2$ the reduced symplectic form coincides with the standard area two-form on $\mathbb{K}_2$ associated with the normal vector $\boldsymbol{n}$. Moreover, the symplectic manifold $(\mathbb{K}_2, \sigma)$ is now symplectomorphic to a cotangent bundle, as shown by the following.

Theorem 10.11. *Let* $\mathbb{D}_2$ *be the cylinder in the Minkowski space* $\mathbb{R}^3$, *with axis* z *and intersecting the* (x, y)*-plane in the unit circle* $\mathbb{S}_1$, $x^2 + y^2 = 1$. *Then the map* $\phi \colon \mathbb{K}_2 \to \mathbb{D}_2$ *defined by*

$$\phi(\boldsymbol{n}) = \frac{\boldsymbol{n} + \boldsymbol{n} \cdot \boldsymbol{c}_3\, \boldsymbol{c}_3}{\sqrt{1 + (\boldsymbol{n} \cdot \boldsymbol{c}_3)^2}}$$

is a symplectomorphism from $(\mathbb{K}_2, \sigma)$ *to the cotangent bundle* $T^*\mathbb{S}_1 \sim \mathbb{S}_1 \times \mathbb{R}$.

Remark 10.4. The map ϕ is the radial orthogonal projection, with respect to the z-axis, from $\mathbb{K}_2$ to $\mathbb{D}_2$. Note that the vector $\boldsymbol{n} + \boldsymbol{n} \cdot \boldsymbol{c}_3\, \boldsymbol{c}_3$ is the (x, y)-component of $\boldsymbol{n}$, being orthogonal to $\boldsymbol{c}_3$, and that its square is

$$(\boldsymbol{n} + \boldsymbol{n} \cdot \boldsymbol{c}_3\, \boldsymbol{c}_3)^2 = \boldsymbol{n}^2 - (\boldsymbol{n} \cdot \boldsymbol{c}_3)^2 + 2\,(\boldsymbol{n} \cdot \boldsymbol{c}_3)^2 = 1 + (\boldsymbol{n} \cdot \boldsymbol{c}_3)^2. \qquad \diamondsuit$$

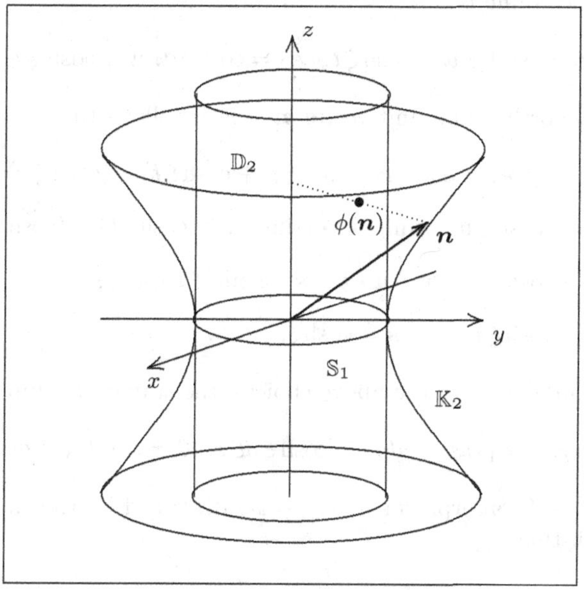

Fig. 10.3 The hyperboloid $\mathbb{K}_2$ and the cylinder $\mathbb{D}_2$ in the Minkowski three-space

Proof. Let $\mathbb{D}_2 \sim \mathbb{S}_1 \times \mathbb{R}$ be described by the parametric equation

$$\boldsymbol{r} = \boldsymbol{u}(\theta) + z\,\boldsymbol{t} = \cos\theta\; \boldsymbol{c}_1 + \sin\theta\; \boldsymbol{c}_2 + z\; \boldsymbol{c}_3, \quad (u^1, u^2) = (z, \theta).$$

The associated tangent frame is

$$\boldsymbol{e}_1 = \boldsymbol{t}, \quad \boldsymbol{e}_2 = \partial_\theta \boldsymbol{u} = \boldsymbol{s}.$$

We choose the orthogonal unit vector $\boldsymbol{n} = \boldsymbol{u}$. Then, according to (10.8),

$$\sigma_{12} = \boldsymbol{n} \cdot \boldsymbol{e}_1 \times \boldsymbol{e}_2 = \boldsymbol{u} \cdot \boldsymbol{t} \times \boldsymbol{s}$$

In the Minkowskian metric (i.e., for $\varepsilon = -1$)

$$\boldsymbol{u} \cdot \boldsymbol{t} \times \boldsymbol{s} = \boldsymbol{c}_1 \cdot \boldsymbol{c}_3 \times \boldsymbol{c}_2 = 1.$$

Thus, $\sigma_{12} = 1$ and the standard symplectic form on $\mathbb{D}_2$ is

$$\sigma_{\mathbb{D}_2} = \sigma_{12}\, du^1 \wedge du^2 = dz \wedge d\theta.$$

If we consider $T^*\mathbb{S}_1 \sim \mathbb{S}_1 \times \mathbb{R}$ with canonical coordinates (θ, z), then the Liouville form is $\theta_{\mathbb{S}_1} = z\, d\theta$ and the symplectic form $d\theta_{\mathbb{S}_1} = dz \wedge d\theta$ coincides with the area two-form. Let $\mathbb{K}_2 \sim \mathbb{S}_1 \times \mathbb{R}$ be described by the parametric equation

$$\boldsymbol{r} = \cosh\xi\, \boldsymbol{u}(\theta) + \sinh\xi\, \boldsymbol{t}, \quad (u^1, u^2) = (\xi, \theta).$$

The associated frame is

$$\boldsymbol{e}_1 = \sinh\xi\, \boldsymbol{u} + \cosh\xi\, \boldsymbol{t}, \quad \boldsymbol{e}_2 = \cosh\xi \partial_\theta\, \boldsymbol{u} = \cosh\xi\, \boldsymbol{s}.$$

We choose the orthogonal unit vector $\boldsymbol{n} = \boldsymbol{r}$. It follows that

$$\begin{aligned}
\sigma_{12} &= \boldsymbol{n} \cdot \boldsymbol{e}_1 \times \boldsymbol{e}_2 = \boldsymbol{r} \cdot (\sinh\xi\, \boldsymbol{u} + \cosh\xi\, \boldsymbol{t}) \times (\cosh\xi\, \boldsymbol{s}) \\
&= (\cosh\xi\, \boldsymbol{u} + \sinh\xi\, \boldsymbol{t}) \cdot (\sinh\xi\, \boldsymbol{u} + \cosh\xi\, \boldsymbol{t}) \times (\cosh\xi\, \boldsymbol{s}) \\
&= \cosh\xi\, (\cosh^2\xi\, \boldsymbol{u} \cdot \boldsymbol{t} \times \boldsymbol{s} + \sinh^2\xi\, \boldsymbol{t} \cdot \boldsymbol{u} \times \boldsymbol{s}) \\
&= \cosh\xi\, \boldsymbol{u} \cdot \boldsymbol{t} \times \boldsymbol{s} = \cosh\xi.
\end{aligned}$$

Then, in accordance with the above choices, the standard symplectic form is

$$\sigma_{\mathbb{K}_2} = \sigma_{12}\, du^1 \wedge du^2 = \cosh\xi\, d\xi \wedge d\theta = d\sinh\chi \wedge d\theta.$$

We look for a diffeomorphism $\phi\colon \mathbb{K}_2 \to \mathbb{D}_2$ described by the single equation $z = z(\xi)$ such that

$$\phi^* \sigma_{\mathbb{K}_2} = \sigma_{\mathbb{D}_2}.$$

This last condition is equivalent to equation $dz(\xi) \wedge d\theta = d\sinh\xi \wedge d\theta$. Then we can choose $z(\xi) = \sinh\xi$. The diffeomorphism ϕ so defined is the radial orthogonal projection with respect to the z-axis restricted to $\mathbb{K}_2$. □

References

Abraham R., Marsden J. E.: Foundations of Mechanics. Benjamin Cummings (1978).

Abraham R., Robbins J.: Transversal Mapping and Flows. Benjamin, New York (1967).

Arnold V. I.: Characteristic class entering in quantization conditions. Funct. Anal. Appl. **1**, 1–13 (1967).

Arnold V. I.: Mathematical Methods of Classical Mechanics. Springer GTM **60** (1978).

Arnold V. I.: Chapitres Supplémentaires de la Théorie des Équations Différentielles Ordinaires. Edition Mir, Moscou (1980).

Arnold V. I., Novikov S.: Dynamical Systems IV. E.M.S. Springer (1990) (Russian original edition, 1985).

Arnold V. I., Varchenko A., Goussein–Zadé S.: Singularités des Applications Différentiables, 1.re partie, Classification des Points Critiques, des Caustiques et des Fronts d'Onde. Editions Mir, Moscou (1986).

Benenti S.: (a) Symplectic relations in Analytical Mechanics. Atti Accad. Sci. Torino Cl. Sc. Fis. Mat. Nat. **117**, Suppl. I, 39–91 (1983). (b) The category of symplectic reductions. Proceedings of the International Meeting on Geometry and Physics. Florence, October 1982, Modugno M. Ed., Pitagora Editrice, Bologna, 11–41 (1983). (c) Linear symplectic relations, in Symplectic Geometry. Crumeyrolle A., Grifone J. Eds., Research Notes in Math. **80**, Pitman Advanced Publishing Program (1983).

Benenti S.: Relazioni Simplettiche, la Trasformazione di Legendre e la Teoria di Hamilton–Jacobi. Quaderni U.M.I. **33**, Pitagora Editrice, Bologna (1988).

Benenti S., Tulczyjew W. M.: (a) Relazioni lineari binarie tra spazi vettoriali di dimensione finita. Memorie Accad. Sc. Torino **3**, 67–113 (1979). (b) Relazioni lineari simplettiche, Memorie Accad. Sc. Torino **5**, 71–140 (1979).

Benenti S., Tulczyjew W. M.: The geometrical meaning and the globalization of the Hamilton–Jacobi method. Lecture Notes in Math. **836**, 484–497 (1980).

Benenti S., Tulczyjew W. M.: (a) Remarques sur les reductions symplectiques, C.R.A.S. **294**, 561–564 (1982). (b) Sur le théorème de Jacobi en mécanique analytique. C.R.A.S. Paris **294**, 677–680 (1982).

Bott R.: Nondegenerate critical manifolds. Ann. Mat. **60**, 248–261 (1954).

Buchdahl H. A.: An Introduction to Hamiltonian Optics. Cambridge University Press (1970).

Callen H. B.: Thermodynamics and an Introduction to Thermostatistics. John Wiley (1985).

Carathéodory C., Untersuchungen über die Grundlagen der Thermodynamik. Math. Ann. **67**, 335–386 (1909).

Cardin F.: On the geometrical Cauchy problem for the Hamilton–Jacobi equation. Nuovo Cimento **104** (5), 525–544 (1989).

Cardin F.: The global finite structure of generic envelope loci for Hamilton–Jacobi equations. J. Math. Phys. **43**, 417–430 (2002).

Chaperon M.: On generating families. The Floer Memorial Volume, Hofer H., Taubes C.H., Weinstein A., Zehnder E. Eds., Progress in Mathematics **133**, 283–296, Birkäuser (1995).

Cordani B.: Conformal regularization of the Kepler problem. Comm. Math. Phys. **103**, 403–413 (1986).

Cordani B.: The Kepler Problem. Group theoretical aspects, regularization and quantization, with application to the study of perturbations. Progress in Mathematical Physics **29**, Birkäuser (2003).

Dubois J.-G., Dufour J.-P.: La théorie des catastrophes. I. La machine à catastrophe. Ann. Inst. H. Poincaré **20**, 113–134 (1974).

Dubois J.-G., Dufour J.-P.: La théorie des catastrophes. III. Caustiques de l'optique géométrique. Ann. Inst. H. Poincaré **24**, 243–260 (1976).

Dubois J.-G., Dufour J.-P.: La théorie des catastrophes. V. Transformées de Legendre et thermodynamique. Ann. Inst. H. Poincaré **29**, 1–50 (1978).

Fermi E.: Thermodynamics. Italian edition: Termodinamica, Boringhieri (1972).

Frölicher A., Nijenhuis A.: Theory of vector-valued differential forms. Nederl. Akad. Wetensch. Proc. **59**, 338–359 (1956).

Guillemin V., Sternberg S.: Geometric Asymptotics. A.M.S. Math. Surveys **14** (1977).

Guillemin V., Sternberg S., Symplectic Techniques in Physics. Cambridge Univ. Press (1984).

Hamilton W. R.: Theory of systems of rays. Trans. R. Irish Acad. **15**, 69–174 (1828).

Hermann R.: Geometry, Physics and Systems. Dekker, New York (1973).

Hörmander L.: Fourier integral operators. Acta Math. **127**, 79–183 (1971).

Huang K.: Statistical Mechanics. Wiley and Sons (1987).

Janeczko S.: (a) Geometric approach to coexistence of phases and singularities of Lagrangian submanifolds. Geometrical methods in physics, Proceedings, J.E. Purkině University, Brno (1983). (b) Geometrical approach to phase transitions and singularities of Lagrangian submanifolds. Demonstratio Mathematica **16**, 487–502 (1983). (c) Geometric approach to equilibrium thermodynamics, thermostatics of phase transitions. Notes of a lecture delivered at the Istituto di Fisica Matematica J. L. Lagrange, University of Torino (1983).

Kalnins E. G.: Separation of Variables for Riemannian Spaces of Constant Curvature. Pitman Monographs and Surveys in Pure and Applied Mathematics **28**. Longman Scientific & Technical, John Wiley & Sons, Inc., New York (1986).

Kalnins E. G., Miller W. Jr., Pogosyan G. S.: Superintegrability on the two-dimensional hyperboloid. J. Math. Phys., **38** (10), 5416–5433 (1997).

Kalnins E. G., Miller W. Jr., Hakobyan Ye. M., Pogosyan G. S.: Superintegrability on the two-dimensional hyperboloid. II. J. Math. Phys. **40** (5), 2291–2306 (1999).

Kijowski J., Tulczyjew W. M.: A Symplectic Framework for Fields Theories. Lecture Notes in Physics **107** (1979).

Koslov V. V.: Dynamical Systems X – General Theory of Vortices. Springer (2003).

Lang S.: Differential Manifolds. Addison Wesley, Reading (1972).

Ławruk B., Sniatycki J., Tulczyjew W. M.: Special symplectic spaces. J. Diff. Equations **17**, 477–497 (1975).

Leray J.: The meaning of Maslov's asymptotic method: the need of Plank's constant in mathematics. Bull. Am. Math. Society **5** (1), 15–27 (1981).

Levi–Civita T., Amaldi U.: Lezioni di Meccanica Razionale, Vol. II, parte II. Zanichelli (1989).

Libermann P., Marle C.-M.: Symplectic Geometry and Analytical Mechanics. Reidel (1987).

Lichnerowicz A., C.R.A.S. Paris **280**, 523–527 (1975).

Luneburg R. K.: Mathematical Theory of Optics. University of California Press, Berkeley and Los Angeles (1964).

Marmo G., Morandi G., Mukunda N.: A Geometrical Approach to the Hamilton–Jacobi Form of Dynamics and Its Generalizations. Rivista del Nuovo Cimento **13** (1990).

Maslov V. P.: Théorie des Perturbations et Mèthodes Asymptotiques. Gauthier-Villar, Paris (1971).

Mishchenko A. S., Shatalov V. E., Sternin B. Y.: Lagrangian Manifolds and the Maslov Operator. Springer (1990) (Russian original edition, 1978).

Moser J.: Regularization of Kepler's problem and averaging method on a manifold. Comm. Pure Appl. Math. **23**, 609–636 (1970).

Mrugała R.: Lie, Jacobi, Poisson and quasi-Poisson structures in thermodynamics. Tensor **56**, 37–45 (1995).

Petersen P.: Riemannian Geometry. Springer (1998).

Pham Mau Quan: Géométrie du problème de Kepler: orbites et variétés des orbites. C.R.A.S. Paris **291**, 299–301 (1980).

Pham Mau Quan: (a) Régularization riemannienne des singularités d'équations differentielles. C.R.A.S. Paris **296**, 241–244 (1983). (b) Riemannian regularization of singularities. Application to the Kepler problem. Proceedings of the IUTAM–ISIMM Symposium on Modern developments in Analytical Mechanics, Torino, June 7–11, 1982. Atti dell'Accademia delle Scienze di Torino **117**, 341–348 (1983).

Poincaré H.: Thermodynamique. Cours de la Faculté des Sciences de Paris. G. Carré, Paris (1892).

Poston T., Stewart I.: Catastrophe Theory and its Applications. Pitman (1978).

Sniatycki J., Tulczyjew W. M.: Generating forms of Lagrangian submanifolds. Indiana Univ. Math. J. **22** (3) (1972).

Souriau J. M.: Structures des Systèmes Dynamiques. Dunod, Paris (1970).

Souriau J. M.: Géométrie globale du problème á deux corps. Proceedings of the IUTAM-ISIMM Symposium on Modern developments in Analytical Mechanics, Torino, June 7–11, 1982. Atti dell'Accademia delle Scienze di Torino **117**, 369–418 (1983).

Straumann N., Jetzer P., Kaplan J.: Topics on Gravitational Lensing, Napoli Series on Physics and Astrophysics, Vol. 1, Bibliopolis, Naples (1998).

Synge J.L.: Relativity: the General Theory. North-Holland (1960).

Synge J. L.: Geometrical Optics. An Introduction to Hamilton's Method. Cambridge University Press (1962).

Théret D.: A complete proof of Viterbo's uniqueness theorem on generating functions. Topology and Its Applications **96**, 249–266 (1999).

Tulczyjew W. M.:Hamiltonian systems, Lagrangian systems and the Legendre transformation, Istituto Nazionale di Alta Matematica, Symposia Mathematica **14**, 247–258 (1974).

Tulczyjew W. M.: Relations symplectiques et les équations d'Hamilton–Jacobi relativistes, C.R.A.S. Paris **281**, 545–547 (1975).

Tulczyjew W. M.: (a) The Legendre transformation. Ann. Inst. H. Poincaré **27**, 101–114 (1977). (b) A symplectic formulation of relativistic particle dynamics, Acta Phys. Polonica **B 8**, 431–447 (1977).

Tulczyjew W. M.: Geometric Formulation of Physical Theories, Statics and Dymamics of Mechanical Systems. Bibliopolis, Napoli (1989).

Tulczyjew W. M., Urbanski P.: A slow and careful Legendre transformation for singular Lagrangians, Acta Phys. Polonica **B 30**, 2909–2978 (1999).

Vaisman I.: Lectures on the Geometry of Poisson Manifolds. Progress in Mathematics **118**, Birkäuser (1994).

Vinogradov A. M., Kuperschmidt B. A.: The structures of Hamiltonian mechanics. Russian Math. Surveys **32**, 177–243 (1977).

Viterbo C.: Symplectic topology as the geometry of generating functions. Math. Ann. **292**, 685–710 (1992).

Weinstein A.: Symplectic manifolds and their Lagrangian submanifolds, Advances in Math. **6**, 329–346 (1971).

Weinstein A.: Lagrangian submanifolds and Hamiltonian systems. Ann. of Math. **98**, 377–410 (1973).

Weinstein A.: Lectures on symplectic manifolds. C.B.M.S. Conf. Series in Math., A.M.S. **29** (1977).

Weinstein A.: Symplectic geometry. Bull. Amer. Math. Soc. **5** (1), 1–13 (1981).

Weinstein A.: Poisson geometry, Diff. Geometry and its Applications **9**, 213–238 (1998).

Whittaker E. T.: A Treatise on the Analytical Dynamics of Particles and Rigid Bodies. Cambridge University Press (1927).

Wolf J. A.: Spaces of Constant Curvature. Publish or Perish (1984).

Index